CAMBRIDGE STUDIES
IN MATHEMATICAL BIOLOGY: 13
Editors
C. CANNINGS
Department of Probability and Statistics, University of Sheffield, UK
F. C. HOPPENSTEADT
College of Natural Sciences, Michigan State University, East Lansing, USA
L. A. SEGEL
Weizmann Institute of Science, Rehovot, Israel

THE THEORY OF THE CHEMOSTAT

T0276007

CAMBRIDGE STUDIES
IN MATHEMATICAL BIOLOGY

1 Brian Charlesworth *Evolution in age-structured populations*
2 Stephen Childress *Mechanics of swimming and flying*
3 C. Cannings and E. A. Thompson *Genealogical and genetic structure*
4 Frank C. Hoppensteadt *Mathematical methods of population biology*
5 G. Dunn and B. S. Everitt *An introduction to mathematical taxonomy*
6 Frank C. Hoppensteadt *An introduction to the mathematics of neurons*
7 Jane Cronin *Mathematical aspects of Hodgkin–Huxley neural theory*
8 Henry C. Tuckwell *Introduction to theoretical neurobiology*
  Volume 1 *Linear cable theory and dendritic structures*
  Volume 2 *Non-linear and stochastic theories*
9 N. MacDonald *Biological delay systems*
10 A. G. Pakes and R. A. Maller *Mathematical ecology of plant species competition*
11 Eric Renshaw *Modelling biological populations in space and time*
12 Lee A. Segel *Biological kinetics*

HAL L. SMITH
Arizona State University

PAUL WALTMAN
Emory University

The theory of the chemostat
Dynamics of microbial competition

CAMBRIDGE
UNIVERSITY PRESS

CAMBRIDGE UNIVERSITY PRESS
Cambridge, New York, Melbourne, Madrid, Cape Town, Singapore, São Paulo

Cambridge University Press
The Edinburgh Building, Cambridge CB2 8RU, UK

Published in the United States of America by Cambridge University Press, New York

www.cambridge.org
Information on this title: www.cambridge.org/9780521470278

First published 1995
This digitally printed version 2008

A catalogue record for this publication is available from the British Library

Library of Congress Cataloguing in Publication data
Smith, Hal L.
The theory of the chemostat : dynamics of microbial competition /
Hal L. Smith, Paul Waltman.
p. cm. – (Cambridge studies in mathematical biology)
Includes bibliographical references and index.
ISBN 0-521-47027-7
1. Microbial growth – Mathematical models. 2. Chemostat.
3. Microbial ecology – Mathematical models. I. Waltman, Paul E.
II. Title. III. Series.
QR84.5.S65   1994
576′.15 – dc20   94-19083
                                                      CIP

ISBN 978-0-521-47027-8 hardback
ISBN 978-0-521-06734-8 paperback

To
Kathi and Ruth

Contents

Preface   *page* xi

1 The Simple Chemostat   **1**
   1. Introduction   1
   2. Derivation of the Basic Equations of Growth   3
   3. Dynamical Systems   7
   4. Analysis of the Growth Equations   13
   5. Competition   14
   6. The Experiments   19
   7. Discussion   20

2 The General Chemostat   **28**
   1. Introduction   28
   2. Liapunov Theory   28
   3. General Monotone Response and Many Competitors   30
   4. Different Removal Rates   34
   5. Nonmonotone Uptake Functions   37
   6. Discussion   42

3 Competition on Three Trophic Levels   **43**
   1. Introduction   43
   2. The Model   44
   3. A Simple Food Chain   47
   4. Elementary Floquet Theory   51
   5. The Food Chain Continued   53
   6. Bifurcation from a Simple Eigenvalue   59
   7. Competing Predators   65
   8. Numerical Example   68
   9. A Long Food Chain   69
  10. Discussion   71

4 The Chemostat with an Inhibitor — **78**
1. Introduction — 78
2. The Model — 79
3. The Conservation Principle — 81
4. Rest Points and Stability — 86
5. Competition without an Interior Equilibrium — 91
6. Three-Dimensional Competitive Systems — 93
7. Competition with an Interior Equilibrium — 96
8. Discussion — 99

5 The Simple Gradostat — **101**
1. Introduction — 101
2. The Model — 103
3. The Set of Rest Points — 106
4. Growth without Competition — 111
5. Local Stability — 114
6. Order Properties — 119
7. Global Behavior of Solutions — 121
8. Numerical Example — 125
9. Discussion — 126

6 The General Gradostat — **129**
1. Introduction — 129
2. The Conservation Principle — 133
3. Growth without Competition — 137
4. Competition — 140
5. The Standard Gradostat — 151
6. Discussion — 157

7 The Chemostat with Periodic Washout Rate — **159**
1. Introduction — 159
2. Periodic Differential Equations — 162
3. The Conservation Principle — 164
4. Periodic Competitive Planar Systems — 169
5. Coexistence — 171
6. Discussion — 180

8 Variable-Yield Models — **182**
1. Introduction — 182
2. The Single-Population Growth Model — 183
3. The Competition Model — 188
4. The Conservation Principle — 192
5. Global Behavior of the Reduced System — 195
6. Competitive Exclusion — 203
7. Discussion — 206

9 A Size-Structured Competition Model   **208**
1. Introduction   208
2. The Single-Population Model   209
3. Reduction to Ordinary Differential Equations   214
4. Competition   219
5. Average Cell Size   222
6. The Conservation Principle   225
7. The Steady-State Size Distribution   226
8. Discussion   228

10 New Directions   **231**
1. Introduction   231
2. The Unstirred Chemostat   231
3. Delays in the Chemostat   238
4. A Model of Plasmid-Bearing, Plasmid-Free
Competition   243

11 Open Questions   **248**

Appendices   **253**
A   Matrices and Their Eigenvalues   255
B   Differential Inequalities   261
C   Monotone Systems   268
D   Persistence   277
E   Some Techniques in Nonlinear Analysis   282
F   A Convergence Theorem   294

References   299
Author index   309
Subject index   311

Preface

The chemostat is a basic piece of laboratory apparatus, yet it occupies a central place in mathematical ecology. Its importance stems from the many roles it plays. It is a model of a simple lake, the ideal place to study competition in its most primitive form – exploitative competition. It is also used as a model of the wastewater treatment process. In its commercial form the chemostat plays a central role in certain fermentation processes, particularly in the commercial production of products by genetically altered organisms (e.g., in the production of insulin). The theoretical literature is scattered; papers appear in mathematical, biological, and chemical engineering journals. In addition to being known as a chemostat, other common names are "continuous culture" and, in the engineering literature, CSTR (continuously stirred tank reactor). This monograph is devoted to the theoretical description of ecological models based on the chemostat and to problems that can be described in that context, drawing from literature in all fields and presented within a common framework and with consistent notation.

In order to understand the mathematical importance of the chemostat, one must look at the broader picture of the subject of nonlinear differential equations. Linear differential equations have been studied for more than two hundred years; their solutions have a rich structure that has been well worked out and exploited in physics, chemistry, and biology. A vast and challenging new world opens up when one turns to nonlinear differential equations. There is an almost incomprehensible variety of nonlinearities to be studied, and there is little common structure among them. Models of the physical and biological world provide classes of nonlinearities that are worthy of study. Some of the classic and most studied nonlinear differential equations are those associated with the simple pendulum. Other famous equations include those associated with the names of

Liénard and van der Pol, whose work led to significant advances in the understanding of nonlinear differential equations. The equations of celestial mechanics motivated the original work of Poincaré and the foundation of the "geometric" theory of nonlinear differential equations. Although models in physics are often accepted in spite of their restrictive (and perhaps unrealistic) hypotheses, this has not been the case with equations that arise in biology. For example, the equations of motion for the simple pendulum assume a perfect (frictionless) bearing, movement in a vacuum, and the constancy of gravity. None of these conditions is exactly satisfied, yet physicists and engineers study the pendulum and related equations for what they can learn about oscillations. In biology there are relatively few accepted mathematical models. The chemostat is one – and, in microbial ecology, perhaps the only – such model that does seem to have wide acceptance. This is true, at least in part, because the parameters are measurable and the experiments confirm that the mathematics and the biology are in agreement. The equations of population biology, in general, and those of the chemostat-like models, in particular, provide an interesting class of nonlinear systems of differential equations that are worthy of study for their mathematical properties. The study of these equations has already lead to the discovery of new mathematics. The theory of cooperative and competitive dynamical systems has been motivated, at least in part, by problems in population biology. The theory of persistence in dynamical systems grew directly from such considerations.

The chemostat models an open system, and although (strictly speaking) the exact assumptions of the model may be limited to laboratory environments, it can serve as a paradigm for more complicated naturally occurring open systems. In almost any competition problem, one can ask "What happens in the chemostat?" as a first step. The textbook Lotka–Volterra equations have an inherent shortcoming: they incorporate parameters modeling the competitive effect of one population upon the other, which can only be estimated by growing the organisms together. This negates any predictive value, although solution curves can sometimes be successfully fit to real data. With the chemostat, the relevant parameters can be measured by growing the organisms separately in "batch" culture, and predictions can be made as to the outcome of the competition in continuous culture. It would seem, and we hope to show, that the chemostat equations provide a better understanding of competition at its simplest level and are thus more appropriate for textbooks.

In describing the theoretical aspects of the chemostat, we were faced with a decision regarding how much generality to present. The basic nutrient uptake term that is most accepted comes from a model, usually credited to Monod, that appears in enzyme kinetics as the Michaelis-Menten model. Yet the mathematics frequently allows for more general functions than this particular nonlinearity. We have resolved this issue using the following principle: We retain the specifics of the Monod model as long as significantly stronger results (in the sense of describing behavior in the complete parameter space) can be obtained with its use. When the problems reach a level of complexity where stronger results cannot be obtained using the Monod model, we make use of more general classes of nonlinearities that subsume the Monod nonlinearity. The problems in Chapters 3, 4, and 5 retain the Monod formulation because the results are complete as a result of this assumption. The problems considered in Chapters 6, 7, 8, and 9 are treated more generally, since little is gained by the Monod assumption.

There is always the problem of how much mathematics to assume. We have tried to write the material so that the reader who is uninterested in the mathematical fine points may read the discussions and the statements of the theorems while skipping the mathematical proofs. However, one of our objectives is to show that much of the theory of chemostat-like models is based on a rigorous mathematical foundation.

The reader whose primary interests are not mathematical will gain the most by reading Chapters 1, 3, 4, and 5 – omitting all the proofs, but paying particular attention to the theorems, the figures, and the discussion sections. Chapter 8 is also accessible. Chapters 6 and 7 make strong use of the mathematical tools contained in the appendices. While the mathematical level is not high in Chapter 2, the point of this chapter is to show that the results of Chapter 1 hold in greater (mathematical) generality. We have tried to warn the reader when particularly technical proofs appear. After Chapter 1, the other chapters may be read independently, although reading Chapter 5 before Chapter 6 is recommended.

Two types of mathematics have been moved to the appendices. First of all, there is commonly known material which may not be easily found in a single source; an example is Appendix A, which is devoted to matrices. On the other hand, some very abstract mathematics (e.g. Appendix E, "Some Techniques in Nonlinear Analysis") is needed briefly in proofs, but it is not reasonable to assume this material as a prerequisite; the appendix gives the relevant definitions and theorems and refers the reader to a source of further detail. Some theorems that are used frequently – for example, comparison theorems and a result of Selgrade – are proved in an appendix.

A conscious attempt has been made to make this monograph as mathematically self-contained as possible. To achieve this, we have often chosen to avoid using more abstract mathematical results, which might have allowed a briefer (although less satisfying) mathematical treatment, in favor of a more elementary approach. Thus we have strived to use results that either are proved or can be given adequate motivation. For example, many of the results in Chapters 5 through 9 are treated by using comparison theorems and a result of Selgrade and without using the full strength of monotone dynamical systems theory.

The mathematical results in this monograph share a common feature: the attempt to describe the global behavior of solutions of the relevant differential equations. By this we mean the attempt to describe the behavior of all (or almost all) solutions, not just those in some small neighborhood of the equilibrium solutions. Such global results are essential in order to draw broad biological conclusions. For example, the principle of competitive exclusion is based on experimental results (those of Gause [Ga]) and on a global analysis of the simple Lotka–Volterra model of competition. A local stability analysis is inadequate for such a conclusion. In Chapters 1 and 2 a global analysis of the model of competition in the chemostat lends further support for the competitive exclusion principle. Most of the results of the remaining chapters feature variations in the model that will lead to a conclusion other than competitive exclusion; this also requires global mathematical results.

Chapter 1 describes the basic apparatus, proves the competitive exclusion property of the chemostat, and discusses the definitive experiments that show the mathematics and the biology to be in agreement. The idea of dynamical systems is introduced here. We make no use of the deep properties of the theory of dynamical systems, although we do feel strongly that this is the proper language for problems from population biology. Most of the deterministic models in population biology make use of ordinary differential equations, functional differential equations, or reaction-diffusion equations; the language of dynamical systems is appropriate for all three. This language is used throughout the book. Chemostats have a certain conservation property – indeed, the presence of this property may be fundamental to the definition of a chemostat – which requires some mathematics to treat properly. A simpler proof of the result for a three-dimensional system of chemostat equations is given in Chapter 1, and Appendix F treats a special case of a very general theorem of Thieme. When the limit sets are equilibria, this theorem is sufficient to make rigorous the reduction in complexity that follows from the conservation property. In Chapter 2 it is shown that the mathematical results for the simple chemostat hold in much greater generality. Specifically, they do not depend on the particular Monod formulation. Chapters 3 through 9 deal with specific theoretical problems that can be modeled in a chemostat setting.

Chapter 3 contains a chemostat version of the predator–prey problem, which is obtained by adding a third trophic level to the model. Both a food chain (which is the basic predator–prey model) and competing predators at the top level are considered. Chapter 4 studies competition in the chemostat when an inhibitor is present. This problem is of importance in commercial production with biological reactors, since the genetically altered organism may be an inferior competitor to the unaltered or "wild" type. It is common to introduce antibiotic resistance in the altered organism and an antibiotic into the feed bottle of the chemostat. This chapter is also relevant to detoxification problems. Chapter 5 deals with a variation of the chemostat known as the *gradostat*. The purpose of the gradostat is to create an environment with a nutrient gradient and to study growth and competition along it. The presence of nutrient gradients is well known in ecological problems in nature. Chapter 5 studies a special case where there are only two vessels – that is, two levels to the gradient. Chapter 6 considers the same problem in much greater generality and with the potential for more interesting gradients. The gradostat may become a standard model, like the chemostat, for growth along a gradient. Chapter 7 leaves the domain of autonomous differential equations and introduces a time-varying environment into the problem. This is an attempt to model the cycles that appear in nature, such as day–night cycles or seasonal ones. The (not unexpected) result is oscillations; however, they are treated by very elegant methods from monotone dynamical systems (originally applied by deMottoni and Schiaffino [DS]). Chapter 8 treats the internal-stores or variable-yield model, introduced to remove the basic hypothesis of constant yield made in the chemostat. Chapter 9 presents a structured model that takes into account the size of individuals in the population. Structured models have received increased attention recently in population modeling, and here the general approach of such modeling is indicated in the context of the chemostat.

Chapters 10 and 11 are intended to steer the interested reader into open problems in the subject. Chapter 10 deals with three specific types of problems where preliminary analysis of the model has been carried out but many open questions remain. It is felt that these new directions are worthy of serious study. Some of the problems are modeling problems and some are mathematical ones. In two cases the models are not ordinary differential equations, but rather functional differential equations and partial differential equations. In Chapter 11 specific technical open questions are

Preface

mentioned. Many of these appear in the text proper, but are highlighted here to call attention to mathematical questions whose answers would help to complete the theory.

The choice of material reflects the authors' personal mathematical interests, so it is hoped that we have used both the engineering and biological literature appropriately. Although some of the results contained in this monograph are new, we have not aimed for the most general possible results, especially when achieving them would complicate and obscure the basic ideas. However, when sharper results than those presented here are known, we have tried to point out the appropriate references.

The material was used as a basis for a topics course at the Georgia Institute of Technology and one at Emory University. Both authors wish to thank their students for many comments and discussions. Several colleagues read the entire manuscript, and we especially wish to express our sincere thanks for this effort: S. B. Hsu, Y. Kuang, W. Rivera, and J. Wu. Several others read portions of the manuscript as it was developed; we wish to acknowledge their comments with greatest appreciation and thanks: J. Cushing, K. Crowe, S. Ellermeyer, R. Pennington, S. S. Pilyugin, J. Suresh, and G. Wolkowicz. The figures were prepared by Bruce Long of Arizona State University, whose very fine work is much appreciated.

One of the authors (H.L.S.) would like to acknowledge the support of the faculty and staff at the Center for Dynamical Systems and Nonlinear Analysis at the Georgia Institute of Technology, and especially Professor Jack Hale, for providing a pleasant and stimulating environment for the sabbatical leave during which this collaboration occurred. Special thanks should also go to Lorrain Ruff and Annette Rohrs for answering this author's innumerable questions about $\TeX$.

1

The Simple Chemostat

1. Introduction

Competition modeling is one of the more challenging aspects of mathematical biology. Competition is clearly important in nature, yet there are so many ways for populations to compete that the modeling is difficult to carry out in any generality. On the other hand, the mathematical idea seems quite simple: when one population increases, the growth rate of the others should diminish (or at least not increase), a concept that is quite easily expressed by partial derivatives of the specific growth rates. If an ecosystem is modeled by a system of ordinary differential equations – for example, by

$$y_i' = y_i f_i(y),$$

where $i = 1, 2, ..., n$, f_i is a continuously differentiable function defined on $\mathbb{R}^n$, and $y = (y_1, y_2, ..., y_n)$ – then competition is expressed by the condition

$$\frac{\partial f_i}{\partial y_j} \leq 0$$

when $i \neq j$. Dynamical systems with such properties have been studied extensively; see Hirsch [Hi1; Hi2] and Smith [S3].

Such models easily reflect the direct impact of one population upon the other – for example, the production by one population of a metabolic product that inhibits the growth of the other. The simplest form of competition, however, occurs when two or more populations compete for the same resource, such as a common food supply or a growth-limiting nutrient. This is called *exploitative competition*. A simple example of this type of competition occurs in a laboratory device, called a *chemostat* or

1

a *continuous culture*, that models competition in a very simple lake. This device is important in ecological studies because the mathematics is tractable and the relevant experiments are possible (although by no means easy). Its place in theoretical ecology is well documented in the surveys of Bungay and Bungay [BB], Cunningham and Nisbet [CN2], Fredrickson and Stephanopoulos [FSt], Jannash and Mateles [JM], Taylor and Williams [TW], Veldcamp [V], Waltman [W1; W2], and Waltman, Hubbell, and Hsu [WHH]. The chemostat model also plays a role in wastewater treatment problems – two examples are D'Ans, Kokotović, and Gottlieb [DKG] and La Riviere [La] – and in the study of recombinant problems in genetically altered organisms, for example, in Stephanopoulis and Lapidus [SLa] and Stewart and Levin [SL2]. Moreover, the chemostat model is the starting point for many variations that yield more realistic biological models and interesting mathematical problems. The following quotations reflect the importance of the chemostat.

The chemostat is the best laboratory idealization of nature for population studies. It is a dynamic system with continuous material inputs and outputs, thus modeling the open system character and temporal continuity of nature. The input and removal of nutrient analogs the continuous turnover of nutrients in nature. The washout of organisms is equivalent to non-age specific death, predation or emigration which always occurs in nature. [Wi]

An ecosystem is so complex, so difficult to comprehend, that any attempt to understand the interactions of the component parts *in situ* is frequently doomed to failure because of a lack of rigorous controls. Under such circumstances the behavior displayed by one component may be ascribed to any number of phenomena. Consequently, if we wish to understand the mechanisms by which populations interact we must study them under simplified, controllable laboratory conditions. These should be modeled for theoretical insight, and under ideal circumstances the behavior displayed should be predictable under a variety of conditions imposed by the experimentalist.

   From such a perspective, mixed microbial cultures inhabiting simple continuous culture devices are ideal model systems for the study of many ecological phenomena. Unfortunately, population biology has neglected this whole field of research for far too long and without good reason; for micro-organisms are not only economically and ecologically important, their world is every bit as fascinating as that of higher forms of life that are the stable diet of our researchers. Indeed, they can provide unique insights unavailable from almost any other experimental approach. [De]

The name "chemostat" seems to have originated with Novick and Szilard [NS].

   In this monograph the basic literature on competition in the chemostat is collected and explained from a common viewpoint. The subject is by no means complete, but sufficient progress has been made to warrant exposition in a single place. There are also many biological situations that can be modeled by similar techniques. It is also hoped that successful analysis of the models presented here will help to convince biologists of the importance and utility of modern mathematics in ecological studies.

2. Derivation of the Basic Equations of Growth

The apparatus consists of three connected vessels as shown in the schematic in Figure 2.1. The leftmost vessel is called the *feed bottle* and contains all of the nutrients needed for growth of a microorganism – all in excess except for one, which is referred to as the *limiting* nutrient. The second vessel is called the *culture vessel*, and it is here that the "action" takes place. The third vessel is the overflow or *collection* vessel; it is here that the products of the culture vessel are collected. It will contain nutrient, organisms, and perhaps products produced by those organisms. Note that measurements can be made on the contents of the collection vessel without disturbing the action in the culture vessel. Since some nutrient is always in shortest supply, we focus on that limiting nutrient, hereafter simply called the nutrient, and ignore the others that are present in surplus quantities. We emphasize that Figure 2.1 is a schematic; the actual realization of the device can take many forms.

The contents of the feed bottle are pumped at a constant rate into the culture vessel; the contents of the culture vessel are pumped at the same constant rate into the collection vessel. Let V denote the volume of culture vessel (V has units of l^3, where l stands for length), and let F denote the volumetric flow rate (F has units of l^3/t, where t is time). The concentration of the input nutrient, denoted by $S^{(0)}$, is kept constant. Concentration has units of mass$/l^3$.

The culture vessel is charged with a variety of microorganisms, so it contains a mixture of nutrient and organisms. The culture vessel is well

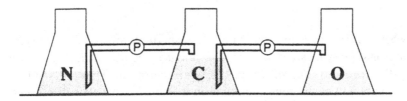

Figure 2.1. A schematic of the simple chemostat. (From [W2], Copyright 1990, Rocky Mountain Mathematics Consortium. Reproduced by permission.)

stirred, and all other significant parameters (e.g. temperature) affecting growth are kept constant. Since the output is continuous, the chemostat is often referred to as a "continuous culture" in contrast with the more common "batch culture."

We seek to write differential equations for this model, and begin by considering just one organism growing in the chemostat. (A more complete derivation can be found elsewhere; see e.g. Herbert, Elsworth, and Telling [HET].) The rate of change of the nutrient can be expressed as

rate of change = input − washout − consumption,

while that of the organism can be expressed as

rate of change = growth − washout.

Let $S(t)$ denote the concentration of nutrient in the culture vessel at time t. Thus $VS(t)$ denotes the amount of nutrient in the vessel at that time. The rate of change of nutrient is the difference between the amount of nutrient being pumped into the vessel per unit time and the amount of nutrient being pumped out of the vessel per unit time. If there were no organisms, and hence no consumption, then the equation for the nutrient would be

$$(VS)'(t) = S^{(0)}F - S(t)F,$$

where the prime denotes the derivative with respect to time. Note that the units on each side are mass/time. Since V is constant, the quantity on the left can be written as $VS'(t)$ and both sides divided by V. The quantity F/V, called the *dilution* (or *washout*) *rate*, is denoted by D and has units of $1/t$. The equation then becomes

$$S'(t) = S^{(0)}D - S(t)D.$$

The formulation of the consumption term, based on experimental evidence, goes back at least to Monod [Mo1; Mo2]. The term takes the form

$$\frac{mSx}{a+S},$$

where x is the concentration of the organism (units are mass/l^3), m is the maximal growth rate (units are $1/t$), and a is the Michaelis–Menten (or half-saturation) constant with units of concentration. The form (and the terminology) of the consumption term is that of enzyme kinetics, where S would be a substrate. Both a and m can be measured experimentally. Since it is generally accepted by microbial ecologists, and since it contains parameters that can be measured, the Michaelis–Menten (or Monod) formulation is most often used as the uptake function, but the mathematical results are valid for much more general functions. Simple monotonicity in S, with a limit as S tends to infinity, is usually sufficient. Trying to squeeze the greatest mathematical generality from the theorems could interfere with our presentation, so the emphasis here is on the Monod formulation. (A partial justification is given in Chapter 2, where it is shown that – for the simple chemostat – a more general response function does not introduce any new types of behavior.)

As noted, the form of the consumption term depends on experimental evidence and does not rest on any physiological basis. The uptake of nutrient is a very complex phenomenon from the standpoint of molecular biology. Indeed, the transport of the nutrient through the cell wall is itself a very complex phenomenon. Dawes and Sutherland [DSu] give a descriptive (i.e. nonmathematical) introduction to microbial physiology and its complexities. Koch [Ko] considers the uptake and the factors affecting growth in considerable detail. The Monod and other similar formulations give an aggregate description of the nutrient uptake; to do otherwise would make the modeling problem very difficult. One can, however, take into account that the uptake by "larger" cells is more than that of "smaller" ones [Cu2].

The differential equation for S takes the form

$$S' = (S^{(0)} - S)D - \frac{mS}{a+S}\frac{x}{\gamma}, \tag{2.1}$$

while that of the corresponding equation for the microorganism, assuming growth is proportional to consumption, is

$$x' = x\left(\frac{mS}{a+S} - D\right), \tag{2.2}$$

where γ is a "yield" constant reflecting the conversion of nutrient to organism. The constant γ can be determined (in batch culture) by measuring

$$\frac{\text{mass of the organism formed}}{\text{mass of the substrate used}}$$

and hence is dimensionless. (We will scale it out in the simple chemostat, but it is important for multiple-nutrient problems.) That γ is a constant is a hypothesis; this hypothesis will be reconsidered in Chapter 8. The assumption that reproduction is proportional to nutrient uptake is a vast simplification. The cell cycle is a very complex phenomenon, and entire books have been devoted to its description (see e.g. Murray and Hunt [MH]). Incorporating the essentials of the cell cycle into the chemostat model would be an interesting problem. From the mathematical point of view, this introduces a delay between nutrient uptake and cell division. Comments on the delay models can be found in Chapter 10, and their proper incorporation in microbial models is very much an open problem.

The appropriate initial conditions are $S(0) \geq 0$ and $x(0) > 0$. The number of parameters in the system is excessive, so some scaling is in order. First of all, note that $S^{(0)}$ and D (the input concentration and the washout rate) are under the control of the experimenter. The $S^{(0)}$ term has units of concentration and D has units of reciprocal time. Equations (2.1) and (2.2) may be rewritten as

$$\frac{S'}{S^{(0)}} = \left(1 - \frac{S}{S^{(0)}}\right)D - \frac{mS/S^{(0)}}{a/S^{(0)} + S/S^{(0)}}\frac{x}{S^{(0)}\gamma};$$

$$\frac{x'}{S^{(0)}\gamma} = \left(\frac{x}{S^{(0)}\gamma}\right)\left(\frac{mS/S^{(0)}}{a/S^{(0)} + S/S^{(0)}} - D\right).$$

By measuring S, a, and x/γ in units of $S^{(0)}$ and time in units of D^{-1}, one obtains the following nondimensional differential equations (note that m and a have changed their meanings):

$$S' = 1 - S - \frac{mSx}{a+S},$$

$$x' = x\left(\frac{mS}{a+S} - 1\right), \tag{2.3}$$

$$S(0) \geq 0, \quad x(0) > 0.$$

This sort of scaling will occur frequently in the problems that follow. The constants m and a can be regarded as the "natural" parameters of the organism in this particular environment. We have standardized the environment, scaling out the factors that can be changed by the experimenter; hence the use of natural parameters expressed in (2.3). This unclutters the mathematics from the "real" world and focuses attention on the selection of the parameters a and m. This, of course, is in marked contrast to the point of view of a person who wishes to perform an experiment. There the parameters a and m are given; they come from the organism selected. An experimenter wishes to tune the chemostat to make the organisms grow. Thus, particularly in the engineering literature, one finds an emphasis on presenting results in the form of "operating diagrams," graphs that show where to operate the chemostat. Since the emphasis here is theoretical, the scaling just described will be used whenever possible; results in terms of the original parameters can easily be obtained by reinterpreting the parameters.

3. Dynamical Systems

Although the system of equations (2.3) is simple enough to handle directly, we pause here to introduce some mathematical material that will be important in the remainder of the book. The reader who is not interested in mathematical tidiness may just note the definitions and go on to the next section. [CL] and [H2] are standard references for the material presented here. The focus throughout the book will be on the "dynamical systems" point of view. Dynamical systems are used primarily as a language, not because we need many deep results from that subject. The language, however, does seem natural for the problems considered. The dynamical system will be defined in terms of $\mathbb{R}^n$, but the natural (and most efficient) formulation is that of a metric space. In a later chapter we will use the space $C[0,1]$, the space of continuous functions on the interval $[0,1]$ with the usual sup norm, and the definition will be expanded at that time.

The most basic concept is that of a dynamical (or a semidynamical) system. Let $\pi: M \times \mathbb{R} \to M$ be a function of two variables, where M is $\mathbb{R}^n$ and $\mathbb{R}$ denotes the real numbers. (We use M for the first variable or state space to suggest that the results are true in greater generality.) The function π is said to be a *continuous dynamical system* if π is continuous and has the following properties:

(i) $\pi(x,0) = x$;
(ii) $\pi(x,t+s) = \pi(\pi(x,t),s)$.

An ordinary differential equation of the form

$$y' = f(y), \tag{3.1}$$

with $y \in \mathbb{R}^n$ and $f: \mathbb{R}^n \to \mathbb{R}^n$ and where f is continuously differentiable, generates such a system by defining $\pi(x,t)$ to be the value $y(t)$, where $y(t)$ is the solution of (3.1) satisfying the initial condition $y(0) = x$. (We are tacitly assuming, of course, that all initial value problems for (3.1) exist for all time.) When (ii) holds only for positive s and t, π is said to be a *semidynamical system*.

Given a point x, the set $\{\pi(x,t), t \geq 0\}$ is called the *positive orbit* or *positive trajectory* through the point and is denoted by $\gamma^+(x)$. If only nonpositive t are considered, the set is called the *negative orbit* or *negative trajectory* through the point and is denoted by $\gamma^-(x)$. The union of the positive and negative orbits is simply called the *orbit* or *trajectory* through the point, denoted by $\gamma(x)$. For emphasis, the latter is sometimes called the *full* orbit. For biological systems one wants to determine the eventual behavior – the asymptotic properties – of trajectories. Biological models require that trajectories remain positive (concentrations or populations are positive numbers) and that trajectories do not tend to infinity with increasing time. If a set S is such that all trajectories that begin (have their initial condition) in S remain in S for all positive time, then S is said to be *positively invariant*. (If trajectories remain in S for both positive and negative time, S is said to be *invariant*.) Hence the basic condition for positivity (of the dependent variables) can be stated as "the positive cone is positively invariant for the dynamical system generated by (3.1)." The dynamical system is said to be *dissipative* if all positive trajectories eventually lie in a bounded set. This is sufficient to ensure that all solutions of (3.1) exist for all positive time.

Let $\{t_n\}$ be a sequence of real numbers which tends to infinity as n tends to infinity. (Such a sequence is sometimes called an *extensive* sequence.) If $P_n = \pi(x, t_n)$ converges to a point P, then P is said to be an *omega limit point* of x. (More correctly, P is an omega limit point of the positive trajectory $\gamma^+(x)$; both references will be used, but since there is a unique trajectory through each point x, the abuse of terminology will cause no confusion when dealing with systems of the form (3.1).) The set of all such omega limit points is called the *omega limit set* of x, denoted $\omega(x)$. If the system is dissipative, the omega limit set is a non-empty, compact, connected, invariant set. Moreover, the orbit $\gamma^+(x)$ is asymptotic to the omega limit set of x in the sense that the distance from $\pi(x, t)$ and $\omega(x)$ tends to zero as t tends to infinity.

Now let $\{t_n\}$ be a sequence of real numbers which tends to negative infinity as n tends to infinity. If $P_n = \pi(x, t_n)$ converges to a point P, then P is said to be an *alpha limit point* of x. The set of all such alpha limit points is called the *alpha limit set* of x, denoted $\alpha(x)$. It enjoys similar properties if the trajectory lies in a compact set for $t < 0$.

A particularly important class of solutions are the constant ones, which are called steady states, *rest points*, or equilibrium points. In terms of (3.1), such a solution is a zero of $f(y)$, that is, a vector $y^* \in \mathbb{R}^n$ such that $f(y^*) = 0$. In the terminology of dynamical systems, a rest point is an element $p \in M$ such that $\pi(p, t) = p$ for all $t \in \mathbb{R}$. Similarly, a periodic orbit is one that satisfies $\pi(p, t + T) = \pi(p, t)$ for all t and for some fixed number T. The corresponding solution of (3.1) will be a periodic function.

If the omega limit set is particularly simple – a rest point or a periodic orbit – this gives information about the asymptotic behavior of the trajectory. An invariant set which is the omega limit set of a neighborhood of itself is called a (local) *attractor*. If (3.1) is two-dimensional then the following theorem is very useful, since it severely restricts the structure of possible attractors.

THEOREM (Poincaré–Bendixson). *If (3.1) is two-dimensional and if $\gamma^+(x)$ remains in a closed and bounded region of the plane without rest points, then either $\gamma^+(x)$ is a periodic orbit (and $\gamma^+(x) = \omega(x)$) or $\omega(x)$ is a periodic orbit.*

Although this is the classical statement of this theorem, a simple consequence is often useful. This is sometimes called the Poincaré–Bendixson trichotomy.

THEOREM. *Let $\gamma^+(y_0)$ be a positive semi-orbit of (3.1) which remains in a closed and bounded subset K of $\mathbb{R}^2$, and suppose that K contains only a finite number of rest points. Then one of the following holds:*

(i) *$\omega(y_0)$ is a rest point;*
(ii) *$\omega(y_0)$ is a periodic orbit;*
(iii) *$\omega(y_0)$ contains a finite number of rest points and a set of trajectories γ_i whose alpha and omega limit sets consist of one of these rest points for each trajectory γ_i.*

Figure 3.1 illustrates the possibilities. Additionally, if a two-dimensional system has a periodic orbit then it must have a rest point "inside" that orbit. These simple facts (and their generalizations) play an important role in the analysis presented here.

While the Poincaré–Bendixson theorem yields the existence of limit cycles, it is often important to know when limit cycles do *not* exist. For two-dimensional systems, a result in this direction which complements the Poincaré–Bendixson theorem is called the Dulac criterion. Its proof is a direct application of the classical Green's theorem in the plane (after an assumption that the theorem is false) and will not be given here; a good reference is [ALGM].

THEOREM (Dulac criterion). *Suppose that (3.1) is two-dimensional. Let Γ be a simply connected region in $\mathbb{R}^2$ and let $\beta(x)$ be a continuously differentiable scalar function defined on Γ. If $\nabla(f(x)\beta(x))$ is of one sign (excluding zero) in the region Γ then there are no periodic orbits in Γ.*

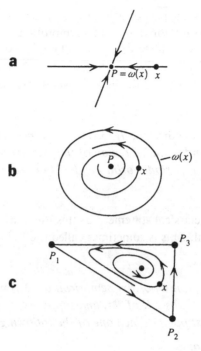

Figure 3.1. Examples of limit sets for planar systems: **a** a rest point; **b** a periodic
orbit; **c** multiple rest points with connecting orbits.

By ∇ is meant the gradient of the resulting two-dimensional vector function.

Local stability considerations also play a role in the analysis. For simplicity denote the solution of the autonomous system (3.1) through the
point y_0 at time $t = 0$ by $\phi(t, y_0)$. Let $\|\cdot\|$ denote the Euclidean norm
in $\mathbb{R}^n$. The solution $\phi(t, y_0)$ is said to be *stable* if, for any $\epsilon > 0$, there
exists a $\delta > 0$ such that if $\|y_0 - x_0\| < \delta$ then $\|\phi(t, y_0) - \phi(t, x_0)\| < \epsilon$ for
all $t > 0$. The solution $\phi(t, y_0)$ is said to be *asymptotically stable* if it is
stable and if there is a neighborhood N of y_0 such that if $x_0 \in N$ then
$\lim_{t \to \infty} \|\phi(t, x_0) - \phi(t, y_0)\| = 0$. We shall be concerned with the case where
$\phi(t, y_0)$ is a constant solution or rest point, that is, where $\phi(t, y_0) = y_0$
for all t. We usually use y^* to denote a rest point. Note that a rest point
y^* is asymptotically stable if it is stable and an attractor.

The system

$$x' = f_y(y^*)x \qquad\qquad (3.2)$$

is said to be the *linearization* of (3.1) around the rest point y^*, where
$f_y(y^*)$ is the Jacobian matrix

$$\left[\frac{\partial f_i}{\partial x_j}\right]\Big|_{y=y^*}.$$

This matrix is called the *variational matrix* at y^*.

If all of the eigenvalues of the variational matrix have negative real parts, then y^* is an asymptotically stable rest point of (3.1). When this happens it is possible to find an arbitrarily small neighborhood around the rest point such that, on the boundary of the neighborhood, all trajectories cross the boundary from outside to inside.

If an omega limit set contains an asymptotically stable rest point P, then that point is the entire omega limit set. If all of the eigenvalues of the variational matrix have positive real part then the rest point is said to be a *repeller*; such a rest point cannot be in the omega limit set of any trajectory other than itself. If k eigenvalues have positive real part and $n-k$ eigenvalues have negative real part then there exist two sets: $M^+(P)$, called the *stable* manifold and defined by

$$M^+(P) = \{x \mid \lim_{t\to\infty} \pi(x,t) = P\};$$

and $M^-(P)$, called the *unstable* manifold and defined by

$$M^-(P) = \{x \mid \lim_{t\to-\infty} \pi(x,t) = P\}.$$

The sets $M^+(P)$ and $M^-(P)$ are locally manifolds of dimension $n-k$ and k, respectively, and all trajectories with initial conditions on these sets tend to the rest point as t tends to infinity (stable) or as t tends to negative infinity (unstable). One should think of these manifolds as surfaces in the appropriate space. On these surfaces, trajectories tend to the rest point as t tends either to positive or to negative infinity. (To assist with the notation, the reader should associate the plus sign on M^+ with positive time and the minus sign on M^- with negative time.) In particular, a single eigenvalue with positive real part makes the rest point unstable. The corresponding eigenvectors generate the tangent space to the respective manifolds. When no eigenvalue of the variational matrix has zero real part, the rest point is said to be *hyperbolic*.

Let P, Q be hyperbolic rest points (not necessarily distinct). P is said to be *chained* to Q, written $P \to Q$, if there exists an element x, $x \notin P \cup Q$, such that $x \in M^-(P) \cap M^+(Q)$. A finite sequence $P_1, P_2, ..., P_k$ of hyperbolic rest points will be called a *chain* if $P_1 \to P_2 \to \cdots \to P_k$ ($P_1 \to P_1$ if $k=1$). The chain will be called a *cycle* if $P_k = P_1$. A chain reflects the connections between equilibrium states. A cycle will turn out to be an "undesirable" type of connection.

The following theorem is often useful.

THEOREM (Butler–McGehee). *Suppose that P is a hyperbolic rest point of (3.1) which is in $\omega(x)$, the omega limit set of $\gamma^+(x)$, but is not the entire omega limit set. Then $\omega(x)$ has nontrivial (i.e., different from P) intersection with the stable and the unstable manifolds of P.*

Figure 3.1c, where P is any of the three equilibria, illustrates the theorem. A short proof (due to McGehee) can be found in the appendix of [FW2], but it requires some advanced concepts from dynamical systems. The result is very general and theorems in the same spirit can be found in [BW], [BFW], [DRS], [T1], or [HaW] in very abstract settings. Note that the linearization around such a point P cannot have all of its eigenvalues with negative real part else $P = \omega(x)$; it also cannot have all eigenvalues with positive real part, for a repeller cannot be in the omega limit set of a point other than itself. Hence the stable and unstable manifolds are not empty. The intuition behind the result is that an orbit cannot "sneak" into and out of a neighborhood of P infinitely often without having accumulation points on the stable and unstable manifolds. The proof simply makes this idea precise. (The proof may be skipped on first reading.)

Proof of the Butler–McGehee Theorem. Since P is a hyperbolic equilibrium, there exists a bounded open set U containing P, but not x, with the property that if $\pi(y, t) \in U$ for all $t > 0$ $(t < 0)$, then y belongs to the local stable (unstable) manifold $M^+(P)$ $(M^-(P))$; see [H2]. (P is the largest invariant set in U, or U isolates P from any other invariant sets.) By taking a smaller open set V, $P \in V \subset \bar{V} \subset U$, we have that $\pi(y, t) \in \bar{V}$ for all $t > 0$ $(t < 0)$ implies $y \in M^+(P)$ $(M^-(P))$.

Since $P \in \omega(x)$, there exists a sequence $\{t_n\}$, $\lim_{n \to \infty} t_n = \infty$, such that $\lim_{n \to \infty} x_n = \lim_{n \to \infty} \pi(x, t_n) = P$. It follows that $x_n \in V$ for all large n. Since $x \notin M^+(P)$, else $\omega(x) = P$, from the property of the neighborhood V one may conclude that there exist positive numbers r_n, s_n such that $r_n < t_n$, $\pi(x_n, t) \in V$ for $-r_n < t < s_n$, and $\pi(x_n, -r_n)$, $\pi(x_n, s_n) \in \partial V$. By the continuity of $\pi(x, t)$, solutions that start near P must remain near P; hence it follows that r_n and s_n tend to infinity as n tends to infinity. However, $\bar{V}$ is compact, so (passing to a subsequence if necessary) one may conclude that $\lim_{n \to \infty} \pi(x_n, -r_n) = q \in \bar{V}$ and $\lim_{n \to \infty} \pi(x_n, s_n) = \hat{q} \in \bar{V}$. We continue the proof for q; the other case is similar.

It is claimed that $\pi(q, t) \in \bar{V}$ for all $t > 0$. Recall that $\lim_{n \to \infty} q_n = q$ where $q_n = \pi(x_n, -r_n)$. Fix $t > 0$. By the continuity of π, $\lim_{n \to \infty} \pi(q_n, t) = \pi(q, t)$. Since $-r_n < t - r_n < 0$ for all large n, $\pi(q_n, t) = \pi(x_n, t - r_n) \in V$

for all large n. It follows that $\pi(q, t) \in \bar{V}$. Since $t > 0$ was arbitrary, the claim is established.

Since $\pi(q, t) \in \bar{V}$ for all $t > 0$, we have $q \in M^+(P)$ by the isolating property of V cited in the first sentence of the proof. However, $q \in \overline{\gamma^+(x)} = \gamma^+(x) \cup \omega(x)$. Since $q \in M^+(P)$, $q \notin \gamma^+(x)$ and hence $q \in \omega(x)$, which establishes one case of the theorem. $\square$

For many of the systems of interest here, the dynamics restricted to the various boundaries of the positive cone in $\mathbb{R}^n$ will be dynamical systems in their own right – the boundaries will be invariant sets. It may happen that a rest point P will be asymptotically stable when regarded as a rest point of the lower-dimensional dynamical system and yet have unstable components when the full system is considered. If the entire stable manifold is contained in the boundary, then the Butler–McGehee theorem can be used to conclude that no trajectory from the interior of the positive cone can have P as an omega limit point. Indeed, the omega limit set cannot equal P because the initial point does not belong to the stable manifold of P. If the limit set contains P then it would also contain a point of the stable manifold distinct from P, by the Butler–McGehee theorem, and would therefore contain the closure of the entire orbit through this point since the omega limit set is closed and invariant. However, this typically leads to a contradiction, since orbits in the stable manifold of P are either unbounded or their limit sets contain equilibria that can be readily excluded from the original limit set (e.g. are repellors). Section 5 will use the theorem in this way.

4. Analysis of the Growth Equations

For system (2.3) the positive cone is positively invariant (see Appendix B, Proposition B.7). In simpler terms, if the system is given positive initial conditions then the two components of the solution remain positive for all finite time. Moreover, if one adds the two equations and defines $\Sigma = 1 - S - x$, then one obtains a single equation

$$\Sigma' = -\Sigma$$

with $\Sigma(0) > 0$. It follows at once that $\lim_{t \to \infty} \Sigma(t) = 0$ and that the convergence is exponential. This not only gives the required dissipativeness but also leads to the simplification of the system by the elimination of one variable. From $\lim_{t \to \infty} [S(t) + x(t)] = 1$, one can conclude that the omega limit set of the system (2.3) must lie in this set, and trajectories on the omega limit set must satisfy

$$x' = x\left[\frac{m(1-x)}{1+a-x} - 1\right], \quad 0 \le x \le 1. \tag{4.1}$$

It might seem at first that it was extremely fortuitous that the aforementioned limit should exist. However, there is a simple, intuitive explanation. If there were no organisms in the model – that is, if only nutrient were present in the equation – then the nutrient would satisfy

$$S' = 1-S \quad \text{and} \quad \lim_{t \to \infty} S(t) = 1.$$

The scaling in the system has expressed the concentration of organism in terms of its "nutrient equivalent." Since nothing is created or destroyed in the system, the sum should satisfy the same equation. The quantity Σ (more accurately, $1-\Sigma$) reflects this. Indeed, if all of the variables in the model are properly accounted for, this will always be true for the sum of the variables in a chemostat.

Since all trajectories of the original system are asymptotic to their omega limit set, in analyzing this equation it is sufficient to determine the asymptotic behavior of (2.3). From a more intuitive viewpoint this is merely starting on the manifold $S+x = 1$, to which all solutions must tend; the mathematical support for this is rigorously established later (see the proof of Theorem 5.1 or Appendix F). Define, for $m > 1$,

$$\lambda = \frac{a}{m-1};$$

λ is called the *break-even* concentration. Equation (4.1) has two rest points, $x = 0$ and $x = 1-\lambda$, and the equation can be rewritten as

$$x' = x\left[\frac{m-1}{1+a-x}\right][1-\lambda-x]. \tag{4.2}$$

Clearly, if $m < 1$ or $m > 1$ and $\lambda > 1$, then $\lim_{t \to \infty} x(t) = 0$ ($x'(t)$ is negative and $x(t)$ is bounded below by zero). On the other hand, if $\lambda < 1$ and $m > 1$, then $\lim_{t \to \infty} x(t) = 1-\lambda$ (and hence $\lim_{t \to \infty} S(t) = \lambda$). If $m < 1$, the organism is washing out faster than its maximal growth rate, whereas if $\lambda \ge 1$ there is insufficient nutrient available for the organism to survive. In either case, extinction is not a surprising outcome. The case $m = 1$ is handled by using (4.1) directly.

5. Competition

To study competition in the chemostat, introduce two different microorganisms into the system, labeled x_1 and x_2, with corresponding parameters

a_i and m_i, $i = 1, 2$. We assume that the corresponding lambdas, λ_1 and λ_2, are different. The overall system becomes

$$S' = 1 - S - \frac{m_1 S x_1}{a_1 + S} - \frac{m_2 S x_2}{a_2 + S},$$

$$x_1' = x_1 \left(\frac{m_1 S}{a_1 + S} - 1 \right),$$

$$x_2' = x_2 \left(\frac{m_2 S}{a_2 + S} - 1 \right),$$

$$S(0) \geq 0, \quad x_1(0) > 0, \quad x_2(0) > 0.$$

(5.1)

Again, let $\Sigma(t) = 1 - S(t) - x_1(t) - x_2(t)$, and rewrite the system as

$$\Sigma' = -\Sigma,$$

$$x_1' = x_1 \left(\frac{m_1(1 - \Sigma - x_1 - x_2)}{a_1 + 1 - \Sigma - x_1 - x_2} - 1 \right),$$

$$x_2' = x_2 \left(\frac{m_2(1 - \Sigma - x_1 - x_2)}{a_2 + 1 - \Sigma - x_1 - x_2} - 1 \right),$$

$$\Sigma(0) \leq 1, \quad x_1(0) > 0, \quad x_2(0) > 0.$$

$(5.1')$

In the same manner as before, one has that

$$\lim_{t \to \infty} \Sigma(t) = 0,$$

where the convergence is exponential. Again this shows that the system is dissipative and that, on the set $\Sigma = 0$, trajectories satisfy

$$x_1' = x_1 \left(\frac{m_1(1 - x_1 - x_2)}{a_1 + 1 - x_1 - x_2} - 1 \right),$$

$$x_2' = x_2 \left(\frac{m_2(1 - x_1 - x_2)}{a_2 + 1 - x_1 - x_2} - 1 \right),$$

$$x_1(0) > 0, \quad x_2(0) > 0, \quad x_1 + x_2 \leq 1$$

(5.2)

or

$$x_1' = x_1 \left[\frac{m_1 - 1}{1 + a_1 - x_1 - x_2} \right] [1 - \lambda_1 - x_1 - x_2],$$

$$x_2' = x_2 \left[\frac{m_2 - 1}{1 + a_2 - x_1 - x_2} \right] [1 - \lambda_2 - x_1 - x_2],$$

$$x_1(0) > 0, \quad x_2(0) > 0, \quad x_1 + x_2 \leq 1.$$

The system (5.2) has three rest points:

$$E_0 = (0,0), \quad E_1 = (1-\lambda_1, 0), \quad E_2 = (0, 1-\lambda_2).$$

If λ_1 is different from λ_2 then there is no "interior" rest point, that is, a rest point with both components positive. In view of the results for growth without a competitor, the only interesting cases are where $m_i > 1$ and $0 < \lambda_i < 1$ for $i = 1$ and 2. If not, the corresponding population washes out of the chemostat even without a competitor (and hence is called an inadequate competitor).

The following is the principal theorem for competition between two adequate competitors under Michaelis–Menten dynamics. Proofs (with varying degrees of mathematical rigor) may be found in [AH; HHW; P; SL1]. Extensions will be discussed in the next chapter.

THEOREM 5.1. *Suppose that $m_i > 1$, $i = 1$ and 2, and that $0 < \lambda_1 < \lambda_2 < 1$. Then any solution of the system (5.2) with $x_i(0) > 0$ satisfies*

$$\lim_{t \to \infty} S(t) = \lambda_1,$$

$$\lim_{t \to \infty} x_1(t) = 1 - \lambda_1,$$

$$\lim_{t \to \infty} x_2(t) = 0.$$

Proof. We begin by analyzing (5.2). The first step is to compute the stability of the rest points of system (5.2) by finding the eigenvalues of the Jacobian matrix evaluated at each of these rest points. At $(0,0)$ this matrix takes the form

$$\begin{bmatrix} \dfrac{(m_1-1)(1-\lambda_1)}{1+a_1} & 0 \\ 0 & \dfrac{(m_1-1)(1-\lambda_2)}{1+a_2} \end{bmatrix}.$$

Both eigenvalues are positive and the origin is a repeller. In particular, the origin is not in the omega limit set of any trajectory (other than itself). At $(1-\lambda_1, 0)$, the variational matrix is of the form

$$\begin{bmatrix} \dfrac{(\lambda_1-1)(a_1 m_1)}{(\lambda_1+a_1)^2} & \dfrac{(\lambda_1-1)(a_1 m_1)}{(\lambda_1+a_1)^2} \\ 0 & \dfrac{(m_2-1)(\lambda_1-\lambda_2)}{\lambda_1+a_2} \end{bmatrix}.$$

Since $0 < \lambda_1 < \lambda_2$ and $m_2 > 1$, both eigenvalues are negative. Thus E_1 is (locally) asymptotically stable. At $(0, 1 - \lambda_2)$, the variational matrix takes the form

$$\begin{bmatrix} \dfrac{(m_1 - 1)(\lambda_2 - \lambda_1)}{\lambda_2 + a_1} & 0 \\[2ex] \dfrac{(\lambda_2 - 1)(a_2 m_2)}{(\lambda_2 + a_2)^2} & \dfrac{(\lambda_2 - 1)(a_2 m_2)}{(\lambda_2 + a_2)^2} \end{bmatrix}.$$

One eigenvalue is negative since $\lambda_2 < 1$ and one is positive since $\lambda_1 < \lambda_2$. Thus the stable manifold is one-dimensional and, since E_2 attracts along the $x_1 = 0$ axis, the stable manifold lies there. In particular, the Butler-McGehee theorem (stated in Section 3) allows one to conclude that no trajectory with positive initial conditions can have E_2 as an omega limit point. Since the initial data are positive, the omega limit set cannot equal E_2. If it contained E_2, then it must also contain an entire orbit different from E_2 belonging to the stable manifold of E_2. There are only two possible orbits; one is unbounded, and the other has alpha limit set E_0. But the omega limit set cannot contain an unbounded orbit and it cannot contain E_0 since it is a repeller. Therefore, E_2 is not a limit point.

Since E_1 is a local attractor, to prove the theorem it remains only to show that it is a global attractor. This is taken care of by the Poincaré-Bendixson theorem. As noted previously, stability conditions preclude a trajectory with positive initial conditions from having E_0 or E_2 in its omega limit set. The system is dissipative and the omega limit set is not empty. Thus, by the Poincaré-Bendixson theorem, the omega limit set of any such trajectory must be an interior periodic orbit or a rest point. However, if there were a periodic orbit then it would have to have a rest point in its interior, and there are no such rest points. Hence every orbit with positive initial conditions must tend to E_1. (Actually, two-dimensional competitive systems cannot have periodic orbits.) Figure 5.1 shows the x_1-x_2 plane.

Although this argument resolves the asymptotic behavior on the set $\Sigma = 0$, there remains the question of whether the systems (5.1) and (5.2) have the same asymptotic behavior. This question is answered in considerable generality in Appendix F; however, we give a direct proof of the current case.

Although the stable manifold of E_0 was E_0 in the planar system, and the stable manifold of E_2 was one-dimensional in that system, these manifolds have an extra dimension when one considers the full system (5.1′) involving Σ. Specifically,

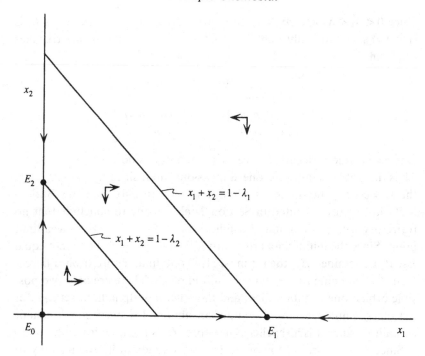

Figure 5.1. The phase plane diagram for the system (5.2).

$$M^+(E_0) = \{(\Sigma, x_1, x_2) \mid x_i = 0, i = 1, 2\},$$

$$M^+(E_2) = \{(\Sigma, 0, x_2) \mid x_2 > 0, x_2 + \Sigma \le 1\}.$$

Let $p = x(0)$ be an arbitrary initial point with $x_i(0) > 0$. Then the initial data do not belong to either stable manifold. Hence $\omega(p)$ is not equal to either E_0 or E_2, but it does lie on $\Sigma = 0$. Since it is invariant and since every solution of (5.2) on $\Sigma = 0$ converges to an equilibrium, $\omega(p)$ contains an equilibrium. By the Butler–McGehee theorem, $E_0 \notin \omega(p)$ since $M^+(E_0)$ is unbounded. If $\omega(p)$ contains E_2, then $\omega(p)$ also contains either E_0 or an unbounded orbit, again by the Butler–McGehee theorem (see Figure 5.2). Since this is impossible, E_1 must be in $\omega(p)$. However, E_1 is a local attractor, so $\omega(p) = E_1$. This completes the proof.   □

If $\lambda_1 = \lambda_2$ then it is not difficult to show that coexistence is possible. This is a "knife-edge" effect – exactly balanced parameters – and cannot be expected to be found in nature.

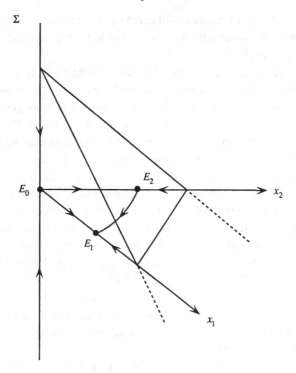

Figure 5.2. The phase space for the system (5.1).

6. The Experiments

The experiments of Hansen and Hubbell [HH] confirm the mathematical result. By working with various microorganisms, Hansen and Hubbell showed that it is the lambda value which determines the outcome of the competition. It is worth noting that this was an example of the mathematics preceding the definitive biological experiment on this type of competition. Although the experiments may be described by a couple of paragraphs, one should not conclude that such experiments were easy. Microbial experiments are always fraught with difficulties. The "proper" organisms and limiting nutrient must be selected so that they can grow under the circumstances where the chemostat is operated. The chemostat must be operated at a turnover rate (washout rate) that allows no growth on the cell walls and no buildup of metabolic products. The feed bottle must not contain substances from whose molecules the organisms can synthesize the limiting nutrient. If the organism mutates during the experiment, the experiment is invalid, so careful checks are run at the end to determine that what is grown is actually what was introduced into the culture vessel. The list could continue.

To motivate the form of the experiments, note first that two parameters are properties of the organism: the m and the a of the chemostat equations. One might postulate that the competitor with the largest m or the one with the smallest a should win the competition. Recall that m is the maximal growth rate and that a (the Michaelis–Menten constant) represents the half-saturation concentration (and so is an indicator of how well an organism thrives at low concentrations). Both of these quantities are obtainable in the laboratory by growing the organism (without a competitor) on the nutrient. (Hansen and Hubbell used a Lineweaver–Burk plot.)

Three experiments were carried out (each reproduced three times) with these postulates in mind. In the first experiment, the organism with the larger m had the smaller lambda value. In the second, the organism with the smaller Michaelis–Menten constant had the larger lambda value. In both cases, the organism with the smaller lambda value won the competition as predicted by the theorem. Finally, organisms with differing m and a values but with (approximately) the same lambda value were shown to coexist (for a reasonably long time).

We reproduce here one table (as Table 6.1) and four graphs (as Figure 6.1) from [HH]. The table shows the organisms used, their parameters, and the run parameters for the chemostat. The limiting nutrient was tryptophan. Hansen and Hubbell used different notation than that presented here, so one should translate as follows: $r = m - D$, $J = \lambda$, $K = a$, $\mu = m$, and $S_0 = S^{(0)}$. The graphs show the predicted time course (with the unscaled variables) in dashed lines and the experimental values with dots connected by solid lines.

An interesting part of the technique was the way in which the equal lambdas were obtained. One strain of *E. coli* had a growth rate that was inhibited by a chemical (Nalidixic acid) while the other strain was essentially immune to the compound (see the third graph in Figure 6.1). By adding the proper amount of the chemical, it was possible to alter the growth rate so as to make the resulting lambda values equal. The chemostat with an inhibitor will be studied in Chapter 4.

7. Discussion

Theorem 5.1 is an example of the principle of competitive exclusion: only one competitor can survive on a single resource. Many of the well-known

Table 6.1

Exper-iment No.	Bacterial strain	Auxotrophic for tryptophan					Other run parameters		
		Yield (cell/g)	K_s (g/liter)	μ (per hour)	r (per hour)	J (g/liter)	S_0 (g/liter)	D (per hour)	Volume (ml)
1	C-8[a]	2.5×10^{10}	3.0×10^{-6}	0.81	0.75	2.40×10^{-7}	1×10^{-4}	6.0×10^{-2}	200
	PAO283[b]	3.8×10^{10}	3.1×10^{-4}	0.91	0.85	2.19×10^{-5}			
2	C-8 nal[r]spec[s]	6.3×10^{10}	1.6×10^{-6}	0.68	0.61	1.98×10^{-7}	5×10^{-6}	7.5×10^{-2}	200
	C-8 nal[s]spec[r]	6.2×10^{10}	1.6×10^{-6}	0.96	0.89	1.35×10^{-7}			
3[c]	C-8 nal[r]spec[s]	6.3×10^{10}	1.6×10^{-6}	0.68	0.61	1.98×10^{-7}	5×10^{-6}	7.5×10^{-2}	200
	C-8 nal[s]spec[r]	6.2×10^{10}	0.9×10^{-6}	0.41	0.34	1.99×10^{-7}			

[a] Escherichia coli.   [b] Pseudomonas aeruginosa.   [c] Nalidixic acid added (0.5 μg/ml).

Source: [HH, p. 1491], Copyright 1980, American Association for the Advancement of Science. Reproduced by permission.

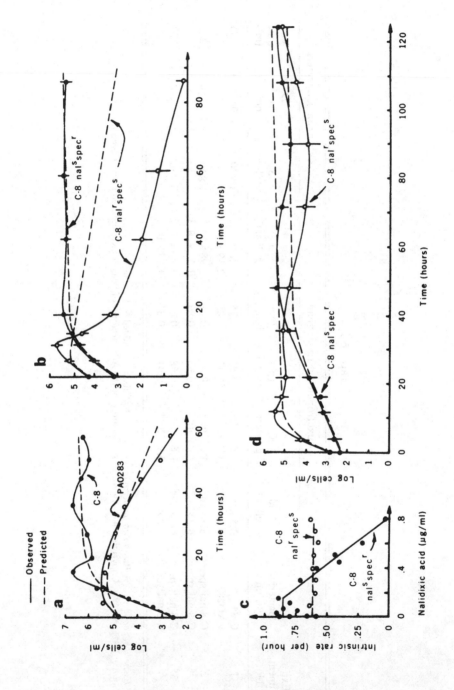

models of competitive systems seem to satisfy this conclusion – for example, the two-dimensional Lotka–Volterra competition model. In the discussion to follow, it is helpful to consider the unscaled version of the system (5.1) (see (2.1) and (2.2) for the unscaled equations (ignoring yield constants) of single population growth):

$$S' = (S^{(0)} - S)D - \frac{m_1 S x_1}{a_1 + S} - \frac{m_2 S x_2}{a_2 + S},$$

$$x_1' = x_1\left(\frac{m_1 S}{a_1 + S} - D\right),$$

$$x_2' = x_2\left(\frac{m_2 S}{a_2 + S} - D\right),$$

(7.1)

$$S(0) \geq 0, \quad x_1(0) > 0, \quad x_2(0) > 0.$$

Three Michaelis–Menten functions are shown in Figure 7.1, where

$$f_i(S) = \frac{m_i S}{a_i + S}.$$

Theorem 5.1 asserts that if the equations $f_i(S) = D$ have unique positive solutions $S = \lambda_i$ ($i = 1, 2$) and if $0 < \lambda_1 < \lambda_2 < S^{(0)}$, then the population x_1 is the winner of the competition; it eliminates population x_2 from the chemostat. It follows that if, as in Figure 7.1a, one function dominates the other for all positive values of S (or at least for $0 < S < S^{(0)}$), then the corresponding population should win the competition. For there to be any hope of coexistence, the two functions must cross before $S = S^{(0)}$. Denote by S^* the point where the functions cross. If $S^* < S^{(0)}$, let $D^* = f_i(S^*)$. If the dilution rate D is set equal to D^*, then $\lambda_1 = \lambda_2 = S^*$ (see Figure 7.1b) and coexistence is possible.

Caption for Figure 6.1
a Experiment 1: Strains differ principally in their half-saturation constants for tryptophan, and PAO283 loses to C-8 as predicted. **b** Experiment 2: Strains differ in their intrinsic rate of increase but not in their half-saturation constants, and C-8 nal[r]spec[s] loses to C-8 nal[s]spec[r] as predicted. **c** Effect of naladixic acid on the intrinsic rate of increase of strains C-8 nal[r]spec[s] and C-8 nal[s]spec[r]. **d** Experiment 3: Strains differ in the half-saturation constants and in their intrinsic rates of increase, but nevertheless have identical J parameters; the strains coexisted for the duration of the experiment, as predicted. In each experiment, the predicted curves were obtained by numerical integration. Bars around points in experiments 2 and 3 indicate ranges of three replicate values. (From [HH, p. 1492], Copyright 1980, American Association for the Advancement of Science. Reproduced by permission.)

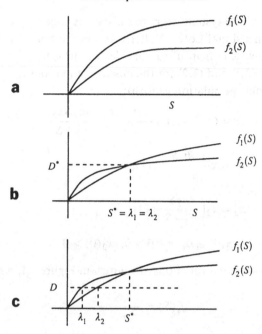

Figure 7.1. Example of Monod-type uptake (or growth) functions: **a** $f_1(S)$ dominates $f_2(S)$ for all values of S, and competitive exclusion holds; **b** $f_1(S)$ and $f_2(S)$ cross exactly at the value of D, producing coexistence, the knife-edge phenomenon; **c** $f_1(S)$ and $f_2(S)$ cross, and competitive exclusion holds.

In the coexistence case there is a line segment of rest points joining E_1 to E_2, each of which can be shown to be stable (but not asymptotically stable), and each (except E_1 and E_2) representing a coexistence rest point. As noted earlier, this occurrence is a knife-edge phenomenon in the sense that it requires exact equality, $D = D^*$. If $D < D^*$ then the population whose growth function dominates in the region $0 < S < S^*$ is the winner, whereas if $D > D^*$ then the population whose growth function dominates in the region $S^* < S < S^{(0)}$ is the winner.

Although this equality was arranged artificially in the experiments of Hansen and Hubbell, it is (as noted previously) extremely unlikely to occur in nature. Since $S(t)$ does not oscillate, if one of the f_is passes through D first (i.e., has D in its range for a lower value of S) then the limiting value of S will fall there, and the corresponding population wins the competition; see Figure 7.1c. Intuitively, if the nutrient equilibrated at a higher value then the first population (the one with the lowest λ) would continue to grow without limit (since $x_i'(t) > 0$ for such S), consuming nutrient and driving its value lower.

From a mathematical point of view, the knife-edge phenomenon is structurally unstable in the sense that a small perturbation of the equations, accounting for some higher-order effect ignored in the original modeling, can be expected to "smooth out" the abrupt transition at $D = D^*$ and change the qualitative structure of the solution set. Instead of a line segment of rest points joining E_1 to E_2, the rest point E_1 could split, at a value $D = D_1 \approx D^*$, into two rest points (the same E_1 and E_c in the interior) – a bifurcation phenomenon. A new rest point appears as D exceeds D_1. The new rest point could be either asymptotically stable or unstable. If it is asymptotically stable, then for D close to D_1 the coordinates of E_c would be close to those of E_1; population 1 would dominate but not exclude population 2. As D is increased, the new rest point could persist as an isolated rest point, moving toward E_2 as D increases toward a value $D = D_2$. As D nears D_2 this rest point would coalesce with E_2 so that, for $D < D_2$, population 2 would dominate in the chemostat but not exclude population 1. If the new rest point were a saddle, then it would lie on the boundary of the basin of attraction of the two stable rest points E_1 and E_2. The winner of the competition would depend on the initial values, on the way the chemostat was originally charged. One would anticipate that the "window" of existence of the new rest point, the interval $D_1 < D < D_2$ of dilution rates, would have size $D_2 - D_1$ of the same order as the size of the perturbation in the model equations. If this perturbation were small then fine tuning of the chemostat would be required in order to observe it. Coexistence would be theoretically possible but difficult to observe.

The very careful experiments of Hansen and Hubbell described in the previous section seem to confirm the knife-edge effect within the accuracy of experimental technique. Where then is the perturbation? Baltzis and Fredrickson [BFr] consider the case where one population is capable of attaching itself to the wall of the chemostat and so escapes the model's implications (the dilution process does not affect the cells attached to the wall). The wall is a refuge. This is a real, practical problem in the operation of the chemostat, one that frequently affects the allowable dilution rates. (When the turnover is too slow, wall growth is a problem.) Another possibility is that there is a small perturbation due to mutual interference between the two populations. A mass action term $-\epsilon x_1 x_2$ in one of the equations would remove the instability. Neither of these situations meets the basic "conservation" principle of the chemostat. G. E. Powell [Po] studies the problem quite generally and considers the case where there is an additional substitutable nutrient, which he calls P. Powell includes a term $\epsilon g_i(P)$ in the equations of growth of the competitors, along with the constant input of the substance P. Although the two-nutrient problem does have the conservation property, it is a totally different phenomenon. An alternate nutrient is not likely in a carefully controlled laboratory experiment but certainly is reasonable in nature, where the organism may be able to synthesize the molecules it needs from many sources.

All of these objections to purely exploitative competition are valid, at least in some settings. However, the types of deviation from purely exploitative assumptions are very diverse, and no single general model is likely to cover all cases and to yield both mathematical results and experimental confirmation. As noted in the Preface, it is the view in this work that studying the chemostat is much like studying the pendulum in physics; the precise assumptions for the pendulum model are not met, but there is much to be learned about oscillations by studying it. Similarly, there is much to be learned about competition by studying it in this simple form. Moreover, like the pendulum in physics, the assumptions are close enough to reality that experiments are meaningful.

In nature, many populations do seem to coexist together, ostensibly on the same resource. It becomes then an interesting problem to modify the chemostat model, taking into account some new aspect that will produce coexistence without destroying the basic assumptions of the chemostat. It is important to note that in doing so we are modeling different phenomena, not introducing small perturbations in the old model. In the following material, modifications that might be expected to produce coexistence are discussed. The modifications add something to the model to take into account some new facet that might be found in nature, while retaining the purely exploitative competition of the chemostat. The modifications lead to mathematically interesting difficulties that require a wide variety of techniques to resolve. The new models also introduce a number of new modeling and mathematical questions whose answers would be of some interest.

The chemostat model could be modified as follows:

(1) introduce more competitors;
(2) modify the functional response (remove the Michaelis–Menten assumption);
(3) introduce an additional trophic level;
(4) introduce inhibitors to growth;
(5) remove the "well stirred" hypothesis (provide for a nutrient gradient);
(6) remove the constant-yield hypothesis;
(7) make either the nutrient concentration or the washout rate time-dependent (to introduce seasonal variations);

(8) allow cell size to be a factor;

(9) allow additional limiting nutrients.

Items (1)–(8) will be discussed in considerable detail in the chapters that follow. Item (9) is a vast subject with many open questions, and the analysis is much more difficult. It is not discussed here; see [BaWo; BWo2; HCH; HsH; LeT; Po; Ti] for more information.

2

The General Chemostat

1. Introduction

In the previous chapter it was shown that the simple chemostat produces competitive exclusion. It could be argued that the result was due to the two-dimensional nature of the limiting problem (and the applicability of the Poincaré–Bendixson theorem) or that this was a result of the particular type of dynamics produced by the Michaelis–Menten hypothesis on the functional response. This last point was the focus of some controversy at one time, inducing the proposal of alternative responses. In this chapter it will be shown that neither additional populations nor the replacement of the Michaelis–Menten hypothesis by a monotone (or even nonmonotone) uptake function is sufficient to produce coexistence of the competitors in a chemostat. This illustrates the robustness of the results of Chapter 1. It will also be shown that the introduction of differing "death rates" (replacing the parameter D by D_i in the equations) does not change the competitive exclusion result.

A brief mathematical digression on Liapunov stability theory will set the stage for the results of this chapter. Those familiar with the LaSalle corollary to Liapunov stability theory, also called the "invariance principle" by some authors, can skip immediately to Section 3.

2. Liapunov Theory

A more sophisticated mathematical approach is required for the results of this chapter because the equations considered cannot be reduced to planar systems. The theorem described in this section and used throughout this chapter is one of great power. When applied to a particular system of differential equations, it generally provides an elegant, simple,

28

and complete global description of the dynamics. In order to apply the theorem, one must find a function that decreases along trajectories of the given system. Herein lies the difficulty in applying the theorem – there are no general methods for finding such functions. Considerable ingenuity is usually required.

Consider the system of differential equations

$$x' = f(x), \qquad (2.1)$$

where f is continuously differentiable on some open subset of $\mathbb{R}^n$. Let G^* be closed and positively invariant for (2.1). A function V is called a *Liapunov function* for (2.1) on a set $G \subset G^*$ if

(i) V is continuously differentiable on G;

(ii) for each $\bar{x} \in \bar{G}$, the closure of G, the limit

$$\lim_{\substack{x \to \bar{x} \\ x \in G}} V(x)$$

exists as either a real number or $+\infty$; and

(iii) $\dot{V} \equiv \nabla V \cdot f \le 0$ on G.

If V is a Liapunov function for (2.1) on G, then we extend the domain of V to $\bar{G}$ by defining V at $x \in \bar{G}$ to be the limiting value of $V(y)$ as y approaches x with $y \in G$. At points $x_0 \in \bar{G}$ where $V(x_0) < \infty$ but where x_0 does not belong to G, we define $\dot{V}(x_0) = \limsup_{h \to 0^+} h^{-1}[V(x(h)) - V(x_0)]$ where $x(t)$ is the solution of (2.1) satisfying $x(0) = x_0$. Set

$$E = \{x \in \bar{G} : V(x) < \infty \text{ and } \dot{V}(x) = 0\}$$

and let M be the largest invariant set for (2.1) in E. Notice that if $x(t)$ is a solution of (2.1) in G then

$$\frac{d}{dt} V(x(t)) = \nabla V(x(t)) \cdot x'(t) = \nabla V(x(t)) \cdot f(x(t)) \le 0 \qquad (2.2)$$

by the chain rule and (iii), so that $V(x(t))$ is nonincreasing along solutions of (2.1).

The following theorem is a minor variation of the LaSalle corollary of Liapunov stability theory, taken from [WLu] (see also [H2], where V is required to be continuous on the closure of G). It also holds under less restrictive hypotheses than are required here.

THEOREM (LaSalle corollary). *Assume that V is a Liapunov function for (2.1) on G and let $\gamma^+ = \{x(t) : t \ge 0\}$ be a bounded orbit of (2.1) belonging to G. Then the omega limit set of γ^+ belongs to M.*

Proof. By (2.2), $V(x(t))$ is nonincreasing so $\lim_{t\to\infty} V(x(t)) = c$ exists. Our hypotheses ensure that c is finite and that $V \equiv c$ on the omega limit set ω of γ^+. Clearly, $\dot V = 0$ on ω, so ω is contained in M since it is an invariant subset of E. $\qquad\square$

For the reader unfamiliar with Liapunov theory, the intuition goes like this: V is some sort of measure of "height." The magnitude of $dV/dt = \nabla V \cdot f \le 0$ is a measure of how fast solutions run "downhill." The downhill slide stops at E. The theory of dynamical systems says that the omega limit set is invariant and therefore contained in M.

3. General Monotone Response and Many Competitors

The equations describing competition between n competitors having concentrations $x_1(t), x_2(t), \ldots, x_n(t)$ for growth-limiting nutrient $S(t)$ in a chemostat are a straightforward generalization of equations (2.1) and (2.2) of Chapter 1. Replacing the Michaelis–Menten response function by general monotone response functions $f_i(S)$ for the ith competitor, $1 \le i \le n$ (precise hypotheses on these functions will be described shortly), one obtains the system

$$S' = D(S^{(0)} - S) - \sum_{i=1}^{n} f_i(S)\frac{x_i}{\gamma_i},$$

$$x_i' = x_i(f_i(S) - D),$$

$$i = 1, 2, \ldots, n,$$

with $S(0) \ge 0$ and $x_i(0) > 0$.

By measuring S and x_i/γ_i in units of $S^{(0)}$ and time in units of D^{-1}, one obtains the nondimensional system

$$S' = 1 - S - \sum_{i=1}^{n} x_i f_i(S),$$

$$x_i' = x_i(f_i(S) - 1), \tag{3.1}$$

$$i = 1, 2, \ldots, n.$$

(Actually, $f_i(S)$ in (3.1) should really be $D^{-1}f_i(S^{(0)}S)$, but we rename the latter to be the former.) Assume that the f_i satisfy the following:

(i) $f_i: \mathbb{R}^+ \to \mathbb{R}^+$;
(ii) $f_i(0) = 0$;
(iii) $f_i(u) < f_i(v)$ if $u < v$; and
(iv) f_i is continuously differentiable.

These assumptions are really quite mild; the f_i need only be increasing and sufficiently smooth. It is not even required that f_i be bounded on $\mathbb{R}^+$. From a biological perspective, of course, the model loses relevance for really large values of S. In addition to the Monod functions, other functions that have been suggested include the exponential kinetics $m(1 - \exp(-S \log 2/a))$, hyperbolic kinetics $m \tanh(S \log 3/2a)$, and piecewise linear kinetics given by $mS/2a$ for $S < 2a$ and by m for $S > 2a$. The piecewise linear kinetics fails to satisfy the strict monotonocity of (iii) and fails to satisfy (iv) at one point, but these assumptions could be weakened so as to include this case.

Let λ_i be the unique solution of $f_i(S) = 1$ if one exists; otherwise let $\lambda_i = +\infty$. We assume the equations are numbered such that

$$0 < \lambda_1 < \lambda_2 \le \cdots \le \lambda_n \le \infty.$$

This assures that competitor x_1 has the least requirements for growth and therefore is favored to win the competition.

As in Chapter 1, set

$$\Sigma = S + \sum_{j=1}^{n} x_j - 1$$

and observe that in the variables $\Sigma, x_1, x_2, \ldots, x_n$, (3.1) takes the form

$$\Sigma' = -\Sigma,$$

$$x_i' = x_i \left(f_i \left(1 + \Sigma - \sum_{j=1}^{n} x_j \right) - 1 \right), \tag{3.2}$$

$$i = 1, 2, \ldots, n.$$

Obviously,

$$\lim_{t \to 0} \Sigma(t) = 0,$$

and so it follows that solutions of (3.1) and (3.2) exist and are bounded for $t \ge 0$. Both $S(t)$ and $x_i(t)$ remain nonnegative from the form of (3.1). Again, as in Chapter 1, we are led to consider the system (3.2) restricted to the invariant hyperplane $\Sigma = 0$, to which all solutions are attracted at an exponential rate. This system is given by

$$x_i' = x_i \left(f_i \left(1 - \sum_{j=1}^{n} x_j \right) - 1 \right), \tag{3.3}$$

$$i = 1, 2, \ldots, n.$$

The relevant domain for (3.3) is the set

$$\Omega = \left\{ x \in \mathbb{R}^n_+ : \sum_{j=1}^{n} x_j \le 1 \right\}.$$

It is easily seen that Ω is positively invariant for (3.3); indeed, the vector field (3.3) points into the interior of Ω on that part of its boundary where $\sum_{j=1}^{n} x_j = 1$. To see this, observe that if $x(0)$ lies on this component of the boundary of Ω then

$$\frac{d}{dt}\Big|_{t=0} \left(\sum_{j=1}^{n} x_j(t) \right) = -1 < 0,$$

so $\sum_{j=1}^{n} x_j(t) < 1$ for $t > 0$.

A competitor x_i for which $\lambda_i \ge 1$ is an inadequate competitor, since its break-even concentration equals or exceeds the reservoir concentration of nutrient. Our first result makes this mathematically precise.

PROPOSITION 3.1. *If $\lambda_i \ge 1$ for some i, then $\lim_{t \to \infty} x_i(t) = 0$.*

Proof. Define $V(x(t)) = x_i(t)$ and observe that

$$\dot{V}(x) = x_i \left(f_i \left(1 - \sum_{j=1}^{n} x_j \right) - 1 \right) \le x_i (f_i(1) - 1) \le 0$$

in $\bar{G} = \Omega$. If $\lambda_i > 1$ then $E = \{x \in \Omega : \dot{V}(x) = 0\} = \{x \in \Omega : x_i = 0\}$, whereas if $\lambda_i = 1$ then $E = \{x \in \Omega : x_i = 0 \text{ or } \sum_{j=1}^{n} x_j = 1\}$. As the vector field (3.3) points into the interior of Ω on $\sum_{j=1}^{n} x_j = 1$, it follows that the largest invariant set M in E is equal to $\{x \in \Omega : x_i = 0\}$. By the LaSalle corollary, $x(t) \to M$ as $t \to \infty$, so the assertion of the proposition holds. $\quad\square$

Now consider the case that the favored competitor x_1 is an adequate one; that is, suppose $0 < \lambda_1 < 1$. Let

$$E_1 = (1 - \lambda_1, 0, 0, \dots, 0)$$

be the rest point corresponding to the survival of only species x_1. System (3.3) may have other rest points besides $E_0 = 0$ and E_1 in Ω if $0 < \lambda_j < 1$ for some $j \ge 2$, but the following analysis does not require their consideration.

Our main result here states that competitive exclusion holds for n competitors in a chemostat provided each competitor possesses a monotone uptake function. The proof follows [AM].

THEOREM 3.2. *Let $x(t)$ be a solution of (3.3) in Ω for which $x_1(0) > 0$. Then*

$$\lim_{t \to \infty} x(t) = E_1.$$

Proof. Define $\Delta = \{x \in \Omega : \sum_j x_j = 1 - \lambda_1\}$, $\mathcal{B} = \{x \in \Omega : \sum_j x_j < 1 - \lambda_1\}$, and $\mathcal{C} = \{x \in \Omega : \sum_j x_j > 1 - \lambda_1\}$. It will be shown that a solution starting in $\mathcal{C}$ either remains in $\mathcal{C}$ and converges to E_1 or enters $\mathcal{B}$ and remains there. Once in $\mathcal{B}$, we show that the solution converges to E_1, completing the proof.

First, observe that

$$\left(\sum_{j=1}^{n} x_j \right)' = \sum_{j=1}^{n} x_j \left(f_j \left(1 - \sum_k x_k \right) - 1 \right) < 0$$

for $x \in (\mathcal{C} \cup \Delta) \backslash E_1$ by the monotonicity of the f_j and the fact that

$$1 - \sum_{k=1}^{n} x_k \le \lambda_1.$$

It follows immediately that if $x(t) \in \Delta \backslash E_1$ for some t then $x(s) \in \mathcal{B}$ for $s > t$. In particular, once a solution enters $\mathcal{B}$, it can never get out.

Suppose the solution $x(t)$ remains in $\mathcal{C}$ for all $t \ge 0$. The previous calculation shows that if $V(x) = \sum_{j=1}^{n} x_j$ in $\mathcal{C}$ then $\dot{V}(x) < 0$. It is easy to see that $\dot{V}(x) = 0$ for $x \in \mathcal{C} \cup \Delta$ if and only if $x = E_1$. By the LaSalle corollary with $G^* = \Omega$ and $G = \mathcal{C}$, any solution remaining in $\mathcal{C}$ for $t \ge 0$ converges to E_1.

It suffices to consider a solution $x(t)$ of (3.3) which belongs to $\mathcal{B}$ for $t \ge 0$ and for which $x_1(0) > 0$. For such a solution,

$$x_1'(t) = x_1(t) \left(f_1 \left(1 - \sum_{j=1}^{n} x_j(t) \right) - 1 \right) > x_1(t)(f_1(\lambda_1) - 1) = 0,$$

so $\lim_{t \to \infty} x_1(t)$ exists and exceeds $x_1(0) > 0$. Obviously, if $V(x) = -x_1$ in $\mathcal{B}$ then $\dot{V}(x) < 0$. Further, $\dot{V}(x) = 0$ for $x \in \mathcal{B} \cup \Delta$ if and only if either $x \in \Delta$ or $x \in \mathcal{B}$ and $x_1 = 0$. By the LaSalle corollary, the solution $x(t)$ converges to the largest invariant set in $\{x \in \mathcal{B} \cup \Delta : \dot{V}(x) = 0\}$. For our solution, obviously $\omega(x(0)) \subset \Delta$ since x_1 increases along $x(t)$. The only invariant subset of Δ is E_1, so necessarily $x(t) \to E_1$ as $t \to \infty$. This completes the proof. $\square$

We hope the reader will appreciate the elegance and simplicity of the arguments supporting Theorem 3.2, which are based on the LaSalle corollary. In particular, a linearized stability analysis about each of the rest points of (3.3), required in Chapter 1, was completely avoided. A careful reading of the proof of Theorem 3.2 reveals that assumption (iii) on f_i is not crucial to the proof; we will have more to say about this later. Finally, it should be noted that the assumption (iv) on f_i can be relaxed somewhat. It can be weakened to requiring only that f_i be locally Lipschitz continuous (e.g., piecewise continuously differentiable), since all that is really required for Theorem 3.2 and Proposition 3.1 is that solutions of initial value problems be unique. However, in order to use Theorem F.1 of Appendix F to relate the dynamics of (3.3) to that of (3.1), it must be assumed that f_i is continuously differentiable.

The results of this section have shown that neither the assumption of Michaelis–Menten functional response nor the assumption of only two competitors is essential for the main results of Chapter 1 to hold. In much of what follows we retain the Michaelis–Menten kinetics, since the parameters are readily measurable in the laboratory.

4. Different Removal Rates

Another question that can be raised concerns the validity of our assumption that all of the removal is accounted for by the washout term. If, for example, a competitor's mortality rate is a significant fraction of the washout rate D, then the assumption is not valid. In this case, the removal rate for that competitor should be the sum of D and the mortality rate. Another possibility is that a filter on the output might slow the washout of an organism but not the nutrient. This could result in that organism's removal rate being less than D. A natural question, then, is whether a species-specific removal rate changes the outcome.

This question was studied by Hsu [Hsu1] in the chemostat with Michaelis–Menten dynamics, and his work is presented here. The equations take the form (ignoring the yield constants)

$$S' = (S^{(0)} - S)D - \sum_{i=1}^{n} \frac{m_i S x_i}{a_i + S},$$

$$x_i' = x_i \left(\frac{m_i S}{a_i + S} - \bar{D}_i \right),$$

$$i = 1, 2, \ldots, n,$$

with $S(0) \geq 0$ and $x_i(0) > 0$. Scaling by D and $S^{(0)}$ as before yields the normalized equations

$$S' = 1 - S - \sum_{i=1}^{n} \frac{m_i S x_i}{a_i + S},$$

$$x_i' = x_i \left(\frac{m_i S}{a_i + S} - D_i \right), \tag{4.1}$$

$$i = 1, 2, \ldots, n,$$

where $D_i = \bar{D}_i/D$. The corresponding λs are defined as in Chapter 1 by

$$\lambda_i = \frac{a_i D_i}{m_i - D_i},$$

where it is assumed that $m_i > D_i$ if λ_i is to be defined. Exactly as in Chapter 1, if $m_i \le D_i$ or if $\lambda_i \ge 1$ then $\lim_{t \to \infty} x_i(t) = 0$; in this case, the entire system merely tends to a lower-order dynamical system, that is, one with fewer competitors. This statement provides necessary conditions for survivability, and one need only consider competitors that satisfy $m_i > D_i$ and $\lambda_i < 1$.

The mathematical difficulties in treating (4.1) are immediately apparent – the conservation principle is lost, and the equations cannot be combined to eliminate one of the variables. Enough of the analysis survives, however, to at least show that (4.1) is dissipative. Adding the equations and replacing D_i by $d = \min\{D_1, D_2, ..., D_n, 1\}$ yields a differential inequality for $\psi = S + \sum_{i=1}^{n} x_i$ of the form

$$\psi' \le 1 - d\psi.$$

If $u(t)$ is the solution of $u' = 1 - du$ satisfying $u(0) = \psi(0)$, then it follows by comparison that

$$S(t) + \sum_{i=1}^{n} x_i(t) \le u(t).$$

As $\lim_{t \to \infty} u(t) = d^{-1}$, we may conclude that

$$\limsup_{t \to \infty} \left[S(t) + \sum_{i=1}^{n} x_i(t) \right] \le d^{-1}.$$

In other words, (4.1) is dissipative. Unfortunately, this argument does not show that $S + \sum_{i=1}^{n} x_i$ has a limiting value, which would allow the reduction of (4.1) to a lower-dimensional system. This means that we are forced to deal with the full system (4.1).

The principal result is the following.

THEOREM 4.1. *Suppose that*

$$0 < \lambda_1 < \lambda_2 \le \lambda_3 \le \cdots \le \lambda_n$$

and $\lambda_1 < 1$. Then

$$\lim_{t \to \infty} x_1(t) = (1 - \lambda_1)/D_1$$

and

$$\lim_{t \to \infty} x_i(t) = 0, \quad i = 2, 3, ..., n.$$

Proof. On the set $G = \{(S, x_1, \ldots, x_n) \in \mathbb{R}_+^{n+1}: S > 0, x_1 > 0\}$, define

$$V(S, x_1, \ldots, x_n) = S - \lambda_1 - \lambda_1 \log\left[\frac{S}{\lambda_1}\right]$$

$$+ c_1\left[x_1 - x_1^* - x_1^* \log\left[\frac{x_1}{x_1^*}\right]\right] + \sum_{i=2}^{n} c_i x_i,$$

where $x_1^* = (1 - \lambda_1)/D_1$ and $c_i = m_i/(m_i - D_i)$. Then it follows that in G

$$\frac{d}{dt} V(S(t), x_1(t), \ldots, x_n(t)) = \nabla V \cdot (S', x_1', \ldots, x_n')^T$$

$$= \begin{bmatrix} 1 - \dfrac{\lambda_1}{S} \\[2ex] c_1\left[1 - \dfrac{x_1^*}{x_1}\right] \\[2ex] c_2 \\[1ex] \vdots \\[1ex] c_n \end{bmatrix} \cdot \begin{bmatrix} 1 - S - \displaystyle\sum_{1}^{n} \dfrac{m_i x_i S}{a_i + S} \\[2ex] \dfrac{m_1 - D_1}{a_1 + S}(S - \lambda_1) x_1 \\[2ex] \dfrac{m_2 - D_2}{a_2 + S}(S - \lambda_2) x_2 \\[1ex] \vdots \\[1ex] \dfrac{m_n - D_n}{a_n + S}(S - \lambda_n) x_n \end{bmatrix}$$

$$= (S - \lambda_1)\left[\frac{1 - S}{S} - \frac{m_1 x_1^*}{a_1 + S}\right] + \sum_{2}^{n}(\lambda_1 - \lambda_i)\frac{m_i x_i}{a_i + S}.$$

The term x_1^* may be rewritten as

$$x_1^* = \frac{1 - \lambda_1}{D_1} = \frac{(1 - \lambda_1)(a_1 + \lambda_1)}{m_1 \lambda_1},$$

so that the term

$$\frac{1 - S}{S} - \frac{m_1 x_1^*}{a_1 + S}$$

may be simplified to

$$-\frac{a_1(S - \lambda_1)}{\lambda_1 S(a_1 + S)}.$$

This in turn may be substituted into the expression for dV/dt to obtain

$$\frac{dV}{dt} = -\frac{(S - \lambda_1)^2 a_1}{(a_1 + S) S \lambda_1} + \sum_{2}^{n} m_i(\lambda_1 - \lambda_i)\frac{x_i}{a_i + S} \leq 0,$$

since $0 < \lambda_1 < \lambda_i$ ($i \geq 2$) and $S > 0$. The set

$$E = \{(S, x_1, \ldots, x_n) \mid dV/dt = 0\}$$

is given by

$$E = \{(\lambda_1, x_1, 0, ..., 0): x_1 > 0\}.$$

Since $\lambda_1 < 1$, the only invariant set in E is

$$S = \lambda_1,$$

$$x_1 = (1 - \lambda_1)/D_1,$$

$$x_i = 0, \quad i = 2, ..., n.$$

An application of the LaSalle corollary yields the desired result. □

The biological conclusion is, of course, that differing removal rates do not alter competitive exclusion in the chemostat. One anticipates that a similar conclusion is true if the Michaelis–Menten dynamics is replaced by the general monotone term $f_i(S)$ used in Section 3. However, the Liapunov calculations depend on this form and the general question is still unresolved.

The reader will have noticed that the Liapunov function used in the proof of the theorem was not obvious on either biological or mathematical grounds. Its discovery by Hsu greatly simplified and extended earlier arguments given in [HHW]. This is typical of applications of the LaSalle corollary. Considerable ingenuity, intuition, and perhaps luck are required to find a Liapunov function.

5. Nonmonotone Uptake Functions

It has been shown that competitive exclusion – that is, the extinction of all but one competitor – holds regardless of the number of competitors or the specific monotone functional response. If one restricts attention to the Michaelis–Menten functional response, then competitive exclusion has been shown even in the case of population-specific removal rates.

There is evidence, however, that a monotone functional response may be inappropriate in some cases. A nutrient which is essential at low concentrations may be inhibiting (or, indeed, even toxic) at higher concentrations. Butler and Wolkowicz [BWo1] consider this possibility; their work has been recently extended in [WuL]. We will describe some special cases of their work in terms of the unscaled system immediately preceding (3.1). Assume that the functional response f_i satisfies (i), (ii), and (iv) of Section 3, but replace the strict monotonicity assumption (iii) by

(iii′) there exist unique, positive, extended real numbers λ_i and μ_i with $\lambda_i < \mu_i$ such that

$f_i(S) < D$ if $S \notin [\lambda_i, \mu_i]$

and

$f_i(S) > D$ if $S \in (\lambda_i, \mu_i)$.

Furthermore, $f_i'(\lambda_i) > 0$ if $\lambda_i < \infty$ and $f_i'(\mu_i) < 0$ if $\mu_i < \infty$.

The interpretation of (iii') is that species x_i increases when $\lambda_i < S < \mu_i$ and decreases if $0 \le S < \lambda_i$ or if $\mu_i < S < \infty$ (if $\mu_i \ne \infty$). In particular, if $0 < \lambda_i < \mu_i < \infty$ then x_i grows at moderate nutrient concentration ($\lambda_i < S < \mu_i$) and declines at low concentration ($S < \lambda_i$) and at high concentration ($S > \mu_i$). See Figure 5.1 for examples. If (iii) holds, then (iii') holds where λ_i is defined as in Section 3 and $\mu_i = +\infty$. Consequently (iii') is more general than (iii) of Section 3.

A further assumption of a generic character is that all the finite λ_i and μ_i are distinct from each other and from $S^{(0)}$.

Assume the equations are numbered such that

$$0 < \lambda_1 < \lambda_2 < \cdots < \lambda_\nu < S^{(0)} < \lambda_j, \quad j = \nu+1, \ldots, n.$$

The case $\nu = 0$ corresponds to $S^{(0)} < \lambda_1$. If $\nu \ne 0$, set

$$Q = \bigcup_{i=1}^{\nu} (\lambda_i, \mu_i),$$

where Q is defined to be the empty set if $\nu = 0$. Observe that Q depends on both D and $S^{(0)}$, in addition to the f_i.

It will be convenient to have notation for the equilibrium solutions $(S, x_1, \ldots, x_n)$. These are $E_0 = (S^{(0)}, 0, \ldots, 0)$, E_i, and E_i^*, where all components of E_i (E_i^*) vanish except for the first and the $(i+1)$th, which are $S = \lambda_i$ ($S = \mu_i$) and $x_i = \gamma_i(S^{(0)} - \lambda_i)$ ($x_i = \gamma_i(S^{(0)} - \mu_i)$). The equilibrium E_i exists for $i = 1, 2, \ldots, \nu$, but E_i^* exists only if $\mu_i < S^{(0)}$ and is unstable if it exists. These are the only equilibria.

The main result in [BWo1] is that every solution converges to one of these equilibria. In particular, since at most one population has a non-zero component at equilibrium, no more than one population can survive. As expected, it is also shown that if $S^{(0)} < \lambda_1$ then E_0 is the global attractor. The following result gives sufficient conditions for population x_1 to be the winner.

THEOREM [BWo1]. *Suppose that Q is a non-empty open interval; that is, suppose Q is a connected set. Then $Q = (\lambda_1, \mu_j)$ for some j.*

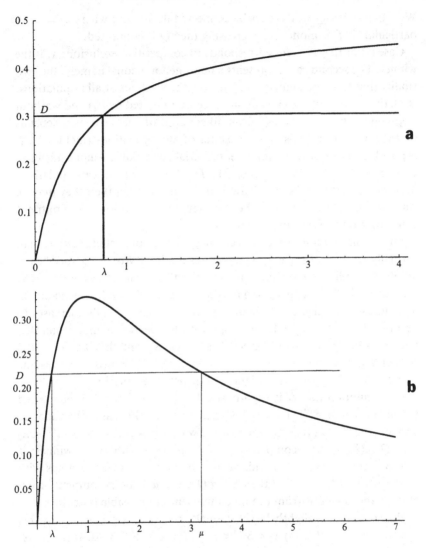

Figure 5.1. **a** A monotone uptake function, where $0 < \lambda < \mu = \infty$. **b** A nonmonotone uptake function, where $0 < \lambda < \mu < \infty$.

(i) *If $\lambda_1 < S^{(0)} < \mu_j$, then E_1 attracts all solutions with $x_k(0) > 0$.*

(ii) *If $S^{(0)} > \mu_j$, then E_0 and E_1 are local attractors, their basins of attraction are non-empty open sets, and the complement of the union of the two basins of attraction has zero Lebesgue measure.*

We remark that Q will always be connected in the case where $\mu_1 = \infty$. In particular, if f_1 is monotone increasing then Q is connected.

Case (i) of the theorem corresponds to competitive exclusion: x_1 is the winner. The second case represents a new phenomenon – namely, the possibility that too much nutrient can lead to the washout of all populations. Both the competitive exclusion outcome (represented by E_1) and washout (represented by E_0) have significant likelihood; which one occurs depends on the initial conditions. That washout of all populations (including x_1) is possible in case (ii), for some set of initial conditions, is not difficult to anticipate. If $S(0) > S^{(0)}$ then $f_i(S) - D < 0$ for all i, so all populations decrease initially. If the populations are initially small then they may be washed out of the chemostat before they are able to lower the nutrient concentration to favorable levels.

Interesting possibilities can occur if Q is not connected. Rather than formally stating a theorem, we apply results in [BWo1] to a specific example. Consider two populations with growth functions as follows: $f_1(S)$ is monotonically increasing to a maximum value $M_1 = f_1(S_1)$, whereupon it monotonically decreases. Similarly, $f_2(S)$ monotonically increases to its maximum $M_2 = f_2(S_2)$, whereupon it also decreases monotonically. Suppose that $f_i'(S) = 0$ only at $S = S_i$ and $S_1 < S_2$, and that $f_1(S) = f_2(S)$ only at $S = 0$ and $S = S^*$, where $S_1 < S^* < S_2$. Let $D^* = f_i(S^*)$ and assume that $f_i(\infty) < D^*$ for $i = 1, 2$. Figure 5.2 depicts the situation.

If the dilution rate D is slightly smaller than D^*, then Q is connected for all values of $S^{(0)}$ since $\lambda_1 < \lambda_2 < \mu_1 < \mu_2$. In this case, the theorem applies. If $S^{(0)} < \lambda_1$ then Q is empty and washout occurs. If $\lambda_1 < S^{(0)} < \lambda_2$ then $Q = (\lambda_1, \mu_1)$ and competitive exclusion holds with x_1 the winner; if $\lambda_2 < S^{(0)}$ then $Q = (\lambda_1, \mu_2)$ and, again, competitive exclusion holds with x_1 the winner. If $\mu_2 < S^{(0)}$ then $S^{(0)} \notin Q = (\lambda_1, \mu_2)$; both competitive exclusion and washout from too much nutrient are possible outcomes.

However, if D is slightly larger than D^* then $\lambda_1 < \mu_1 < \lambda_2 < \mu_2$ and Q is disconnected if either (a) $\lambda_2 < S^{(0)} < \mu_2$ or (b) $\mu_2 < S^{(0)}$ hold. If (a) holds, then E_1 and E_2 are local attractors and the complement of the union of their basins of attraction has zero Lebesgue measure. In this case, the winner depends on how the chemostat is charged at $t = 0$ – that is, on the initial conditions. If (b) holds, then E_0, E_1, E_2 are all local attractors and the complement of the union of their basins has zero Lebesgue measure. In this case, washout of all populations, competitive exclusion of x_2 by x_1, and competitive exclusion of x_1 by x_2 are all possible outcomes, depending on the initial conditions.

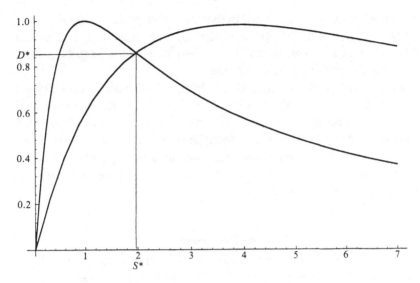

Figure 5.2. Two uptake functions are shown: $f_1(S) = 3S/(S^2 + S + 1)$ and $f_2(S) = 2.5S/(0.2S^2 + S + 3)$; f_1 peaks first, and $f_1(S^*) = f_2(S^*) = D^*$.

These results follow directly from general results in [BWol]. The following intriguing (and fictional) scenario is a slight embellishment of one discussed there. Suppose we would like to use one or both of two populations of microorganisms to remove a contaminant (to humans, a nutrient to the organisms) entering our wastewater treatment plant (chemostat). The contaminant serves as a nutrient for both populations; their uptake rates are depicted in Figure 5.2. Suppose further that the dilution rate is slightly larger than D^* and that it is beyond our control. The Environmental Protection Agency (EPA) sets the maximum acceptable concentration of the contaminant in the water supply at $S = A$. The contaminant concentration entering the plant is $S = S^{(0)}$, and

$$\mu_1 < A < \lambda_2 < S^{(0)} < \mu_2.$$

We are then in an unfortunate situation. Population x_1 alone cannot control the contaminant since $\mu_1 < S^{(0)}$, and therefore it may wash out owing to excess nutrient. Population x_2 alone cannot control the contaminant because λ_2 exceeds the EPA acceptable level A. Both populations together cannot be guaranteed to control the contaminant level either, since population x_2 could competitively exclude x_1 and thereby leave the unacceptable level λ_2 of contaminant in the water. What is needed is another organism,

x_3, for which $\lambda_3 < \mu_1 < \lambda_2 < \mu_3$. In that case, Q is connected ($Q = (\lambda_1, \mu_2)$ or $Q = (\lambda_1, \mu_3)$)) and $S^{(0)} \in Q$, so x_1 competitively excludes x_2 and x_3 and $S(t) \to \lambda_1 < A$ as $t \to \infty$. In practice, x_2 and x_3 would need to be continually added to maintain this situation.

Recent work of Wolkowicz and Lu [WLu] extends the results of [BWo1] described here to include, in some cases, the possibility of population-dependent removal rates. However, at the time of this writing it remains an open problem to describe the global behavior of solutions of the equations modeling n competitors in the chemostat, allowing both for species-specific removal rates and for not necessarily monotone functional responses (e.g., assuming only (iii′)).

6. Discussion

In the previous chapter it was shown that exploitative competition for a single nutrient by two competitors, each with Michaelis–Menten uptake functions, results in the elimination of one competitor; that is, competitive exclusion holds. In the present chapter we have seen that the basic prediction – the elimination of all but one competitor – is unaltered by allowing any number of competitors and allowing quite general, not necessarily monotone, uptake functions. (It can be argued that any biologically realistic uptake function should satisfy the assumptions of Section 5.) The basic prediction is unaltered even if population mortality cannot be neglected – that is, in the case of differing removal rates – provided that Michaelis–Menten uptake functions are used (but see [WLu] for more general uptake functions). More work is needed in the case of differing removal rates to show that they do not affect the prediction for general uptake functions. We can reasonably conclude that coexistence of competitors can only occur under different circumstances than have been considered so far. The remaining chapters address various such possible circumstances.

3

Competition on Three Trophic Levels

1. Introduction

The previous two chapters showed that competitive exclusion holds under a variety of conditions in the chemostat and its modifications. In this chapter it will be shown that if the competition is moved up one level – if the competition occurs among predators of an organism growing on the nutrient – then coexistence may occur. The fact that the competitors are at a higher trophic level allows for oscillations, and the coexistence that occurs is in the form of a stable limit cycle. Along the way it will be necessary to study a three-level food-chain problem which is of interest in its own right; it is the chemostat version of predator–prey equations. The presentation follows that of [BHW1].

Although most of the results can be established with mathematical rigor, there are some elusive problems. These center around the possibility of multiple limit cycles and the difficulty of determining the stability of such limit cycles. At this point one must simply make a hypothesis and resort to numerical evidence in any specific case. Determining the number of limit cycles is a deep mathematical problem, and even in very simple cases the solution is not known. Hilbert's famous sixteenth problem, concerning the number of limit cycles of a second-order system with polynomial right-hand sides, remains basically unresolved. In principle, the stability of a limit cycle can be determined from the Floquet exponents (see Section 4), but this is a notoriously difficult computation – indeed, generally an impossible one.

Throughout the chapter, one assumes that the equilibria and periodic orbits that occur are hyperbolic. This means that local stability is determined from the linearization. Of course one knows this only after making the linearized computations. It is simply that nothing can be said in the

43

case of parameters that yield a nonhyperbolic rest point or cycle. However, since the parameters in the system studied are measured quantities, this is a reasonable assumption; it simply says that certain measured quantities are "unlikely" enough that they may safely be neglected. In these cases stability fails to be determined from the linearization.

The new mathematics that is introduced here is elementary bifurcation theory, in particular, bifurcation from a simple eigenvalue. Although the necessary theorems will not be proved, the material will be discussed in some detail.

2. The Model

We take as the model that of the simple chemostat of Chapter 1, with input nutrient $S(t)$ and organism $x(t)$ growing on that nutrient, and add two predators on x which we label y and z. It is assumed that both the nutrient uptake from the lowest level and the predation from the highest level follow Michaelis–Menten or Monod kinetics. The use of the Monod formulation has already been discussed for the consumption of the nutrient. That the same format should apply in the case of a predator feeding on prey is not immediately clear. This formulation is one of a general class known as a Holling's type-II functional responses [Hol]. A nice discussion can be found in [MD, p. 5], which we repeat here.

The object is to partition the time of an individual predator. Let N denote the number of prey caught during a food-search time period of length T. Let x denote the prey density (units are cells/unit volume), s the search rate (units are volume/time), and b the handling time (units are time/prey). One then has $N = sx(T - bN)$. The functional response (the consumption term) for one predator is N/T or, extending over a unit volume,

$$f(x) = \frac{sx}{1 + sbx}.$$

This is of the same form used for the consumption of nutrient where m is $1/b$ and a is $1/sb$. Modifications and other models are discussed in [CN2, sec. 3.2].

The basic equations with all of the parameters are

$$S' = (S^{(0)} - S)D - \frac{m_1 S x}{\gamma_1(a_1 + S)},$$

$$x' = x\left(\frac{m_1 S}{a_1 + S} - D - \frac{m_2 y}{\gamma_2(a_2 + x)} - \frac{m_3 z}{\gamma_3(a_3 + x)}\right),$$

$$y' = y\left(\frac{m_2 x}{a_2 + x} - D\right),$$ (2.1)

$$z' = z\left(\frac{m_3 x}{a_3 + x} - D\right),$$

$$S(0) = S_0 \geq 0, \quad x(0) = x_0 \geq 0, \quad y(0) = y_0 \geq 0, \quad z(0) = z_0 \geq 0;$$

throughout this section, the prime denotes d/dt.

The scaling is a bit more complicated than before. One makes the following changes:

$$\bar{S} = \frac{S}{S^{(0)}}, \quad \bar{x} = \frac{x}{\gamma_1 S^{(0)}}, \quad \bar{y} = \frac{y}{\gamma_1 \gamma_2 S^{(0)}}, \quad \bar{z} = \frac{z}{\gamma_1 \gamma_3 S^{(0)}},$$

$$\bar{a}_1 = \frac{a_1}{S^{(0)}}, \quad \bar{a}_2 = \frac{a_2}{\gamma_1 S^{(0)}}, \quad \bar{a}_3 = \frac{a_3}{\gamma_1 S^{(0)}},$$

$$\bar{m}_i = \frac{m_i}{D}, \quad i = 1, 2, 3, \quad \bar{t} = Dt.$$

Substituting these into (2.1) and dropping the bars yields

$$S' = 1 - S - \frac{m_1 x S}{a_1 + S},$$

$$x' = x\left(\frac{m_1 S}{a_1 + S} - 1 - \frac{m_2 y}{a_2 + x} - \frac{m_3 z}{a_3 + x}\right),$$

$$y' = y\left(\frac{m_2 x}{a_2 + x} - 1\right),$$ (2.2)

$$z' = z\left(\frac{m_3 x}{a_3 + x} - 1\right),$$

$$S(0) = S_0 \geq 0, \quad x(0) = x_0 \geq 0, \quad y(0) = y_0 \geq 0, \quad z(0) = z_0 \geq 0.$$

The effect of such scaling is to produce a "standardized" environment. The operating parameters – those under the control of the experimenter – have been scaled away by measuring everything relative to them. This has the effect of uncluttering the system by reducing the number of relevant parameters that must be considered.

Following the same approach as in Chapter 1, we seek to limit the size of the system by restricting ourselves to a set containing the omega limit set (see Appendix F). Let $\Sigma = 1 - S - x - y - z$. Then system (2.2) may be written as

$$\Sigma' = -\Sigma,$$

$$x' = x\left(\frac{m_1[1-\Sigma-x-y-z]}{1+a_1-\Sigma-x-y-z} - 1 - \frac{m_2 y}{a_2+x} - \frac{m_3 z}{a_3+x}\right),$$

$$y' = y\left(\frac{m_2 x}{a_2+x} - 1\right),$$

$$z' = z\left(\frac{m_3 x}{a_3+x} - 1\right),$$

$$x(0) = x_0 \geq 0, \quad y(0) = y_0 \geq 0, \quad z(0) = z_0 \geq 0,$$

$$\Sigma(0) = \Sigma_0, \quad \Sigma_0 + x_0 + y_0 + z_0 \leq 1.$$

Clearly, $\lim_{t\to\infty} \Sigma(t) = 0$ and hence the omega limit set of any trajectory lies in the set $\Sigma = 0$. (Alternatively, one could appeal to the theory discussed in Appendix F.) Trajectories in the omega limit set are solutions of the following system:

$$x' = x\left(\frac{m_1(1-x-y-z)}{1+a_1-x-y-z} - 1 - \frac{m_2 y}{a_2+x} - \frac{m_3 z}{a_3+x}\right),$$

$$y' = y\left(\frac{m_2 x}{a_2+x} - 1\right), \tag{2.3}$$

$$z' = z\left(\frac{m_3 x}{a_3+x} - 1\right),$$

$$x(0) = x_0 \geq 0, \quad y(0) = y_0 \geq 0, \quad z(0) = z_0 \geq 0.$$

The first step in the analysis is to eliminate the cases of inadequate competitors and inadequate prey – "inadequate" in the sense that they could not survive in the chemostat without predators, or survive on this prey even at the prey's maximum possible level. Define

$$\lambda_i = \frac{a_i}{m_i - 1}, \quad i = 1, 2, 3,$$

for $m_i > 1$.

LEMMA 2.1. *If $m_1 \leq 1$ or if $m_1 > 1$ and $\lambda_1 \geq 1$, then $\lim_{t\to\infty} x(t) = 0$, $\lim_{t\to\infty} y(t) = 0$, and $\lim_{t\to\infty} z(t) = 0$. If $m_2 \leq 1$ or if $m_2 > 1$ and $\lambda_2 \geq 1$, then $\lim_{t\to\infty} y(t) = 0$. If $m_3 \leq 1$ or if $m_3 > 1$ and $\lambda_3 \geq 1$, then $\lim_{t\to\infty} z(t) = 0$.*

Proof. If $m_1 \leq 1$ then

$$x' < x\left[\frac{1}{1+a_1} - 1\right].$$

From this it follows that

$$x(t) < x(0) \exp\left[-\frac{a_1}{1+a_1}t\right]$$

or

$$\lim_{t \to \infty} x(t) = 0.$$

For t large, $y' < -dy$ and $z' < -dz$ for some $d > 0$, and hence

$$\lim_{t \to \infty} y(t) = 0 \quad \text{and} \quad \lim_{t \to \infty} z(t) = 0.$$

For $m_2 \leq 1$ or $m_3 \leq 1$, the argument is similar except that only the relevant predator population tends to zero. The idea of the proof is the same for the λs. Suppose that $m_2 > 1$ and $\lambda_2 \geq 1$. Since $x \leq 1$, it follows that

$$y' = y\frac{(m_2-1)}{a_2+x}(x-\lambda_2) \leq 0, \tag{2.4}$$

and consequently $\lim_{t \to \infty} y(t)$ exists. If $y(t)$ converged to γ for some $\gamma > 0$, then $\lim_{t \to \infty} x(t) = \lambda_2 \geq 1$. If $\lambda_2 > 1$, this is an immediate contradiction. If $\lambda_2 = 1$, then since $\lim_{t \to \infty} \Sigma(t) = 0$ this implies that $\gamma = 0$, which is the desired contradiction. Similar arguments work for the other dependent variables. $\qquad\square$

The predators represented by $y(t)$ and $z(t)$ may be labeled (assuming $\lambda_2 \neq \lambda_3$) so that $\lambda_2 < \lambda_3$. Thus the only conditions of interest are

$$m_i > 1 \text{ and } \lambda_i < 1, \ i = 1, 2, 3; \ \lambda_2 < \lambda_3, \tag{H1}$$

which we have denoted collectively as hypothesis (H1).

3. A Simple Food Chain

The argument for coexistence will be based on a bifurcation theorem to be discussed shortly. The "base" from which the bifurcation will occur is the system obtained from (2.1) by deleting one of the high-level competitors – in this case, eliminating z (or setting the initial condition for z equal to zero). The result is a food chain wherein x consumes the nutrient and y consumes x but y cannot consume the nutrient. This problem is interesting both mathematically and practically. In a waste treatment process, the bacteria, represented by x, live on the waste (or nutrient) while other organisms such as ciliates feed on the bacteria.

The equations of interest take the form

$$S' = 1 - S - \frac{m_1 x S}{a_1 + S},$$

$$x' = \frac{m_1 x S}{a_1 + S} - x - \frac{m_2 x y}{a_2 + x},$$

$$y' = \frac{m_2 x y}{a_2 + x} - y,$$

$$S(0) = S_0 \geq 0, \quad x(0) = x_0 \geq 0, \quad y(0) = y_0 \geq 0.$$

(These equations appear in [Ca1; Ca2; CN2; JDFT; Se; TDJF] and no doubt in many other papers.) The same simplification effected previously may be applied to this system, letting $\Sigma = 1 - S - x - y$. Then $\lim_{t \to \infty} \Sigma(t) = 0$. The resulting system of interest is

$$x' = x\left(\frac{m_1(1 - x - y)}{1 + a_1 - x - y} - 1 - \frac{m_2 y}{a_2 + x}\right),$$

$$y' = y\left(\frac{m_2 x}{a_2 + x} - 1\right), \tag{3.1}$$

$$x \geq 0, \quad y \geq 0, \quad x + y \leq 1.$$

This is a two-dimensional dynamical system, and we proceed in the standard way: identify the rest points, determine their local stability, and look for limit cycles. The interesting behavior will be in the interior of the positive quadrant, but the analysis begins with the boundary.

There are two rest points on the boundary, given by

$$E_1 = (0, 0) \quad \text{and} \quad E_2 = (1 - \lambda_1, 0).$$

The local stability is determined by the eigenvalues of the variational matrix

$$M = \begin{bmatrix} m_{11} & m_{12} \\ m_{21} & m_{22} \end{bmatrix},$$

where

$$m_{11} = \frac{m_1(1 - x - y)}{1 + a_1 - x - y} - \frac{m_2 y}{a_2 + x} - 1 + x\left(-\frac{m_1 a_1}{(1 + a_1 - x - y)^2} + \frac{m_2 y}{(a_2 + x)^2}\right),$$

$$m_{12} = \frac{-m_1 a_1 x}{(1+a_1-x-y)^2} - \frac{m_2 x}{a_2+x},$$

$$m_{21} = \frac{m_2 a_2 y}{(a_2+x)^2}, \quad m_{22} = \frac{m_2 x}{a_2+x} - 1.$$

The matrix is to be evaluated at the coordinates of each rest point. At E_1 this takes the particularly simple form

$$M = \begin{bmatrix} \dfrac{(m_1-1)(1-\lambda_1)}{1+a_1} & 0 \\ 0 & -1 \end{bmatrix}.$$

The eigenvalues are on the diagonal; one is negative and the other, under our hypothesis (H1), is positive, so E_1 is a saddle point. This is expected because (3.1) is a predator–prey system; without the prey, the predator cannot be expected to survive. The eigenvector corresponding to the negative eigenvalue lies along the y axis, reflecting this. The eigenvector corresponding to the positive eigenvalue lies along the x axis, reflecting the prey's ability to grow in the absence of a predator.

Evaluated at E_2, M takes the form

$$M = \begin{bmatrix} -(1-\lambda_1)\dfrac{a_1 m_1}{(a_1+\lambda_1)^2} & -(1-\lambda_1)\left[\dfrac{m_1 a_1}{(a_1+\lambda_1)^2} + \dfrac{m_2}{1+a_2-\lambda_1}\right] \\ 0 & \dfrac{(m_2-1)(1-\lambda_1-\lambda_2)}{1+a_2-\lambda_1} \end{bmatrix}.$$

Again the eigenvalues are on the diagonal, and one is clearly negative; its eigenvector lies along the x axis. This corresponds to an attracting steady state when the predator is absent. The rest point will be a local attractor for the full system if the remaining eigenvalue is negative. Clearly, this can happen only if $\lambda_1+\lambda_2 > 1$, since we are assuming (H1). This corresponds to the (local) extinction of the predator and the survival of the prey. If $\lambda_1+\lambda_2 < 1$ then E_2 is a saddle point. In this case the eigenvector points into the positive quadrant, since both terms in the first row have the same sign and the eigenvalue is positive. The Butler–McGehee theorem, discussed in Section 3 of Chapter 1, may be applied at both E_1 and E_2 to conclude that no trajectory starting in the interior of the positive quadrant may have E_1 or E_2 in its omega limit set.

We turn now to the question of an interior rest point. Such a point, which we label $E_c = (x_c, y_c)$, exists if there is a solution in the positive quadrant of the algebraic system

$$\frac{m_1(1-x_c-y_c)}{1+a_1-x_c-y_c} - \frac{m_2 y_c}{a_2+x_c} = 1,$$

$$\frac{m_2 x_c}{a_2+x_c} = 1.$$

Clearly, x_c is given by

$$x_c = \lambda_2 = \frac{a_2}{m_2-1},$$

which is positive by our basic hypothesis (H1). Substituting this value into the first equations gives an equation for y_c of the form

$$\frac{m_1(1-\lambda_2-y_c)}{1+a_1-\lambda_2-y_c} - \frac{m_2 y_c}{a_2+\lambda_2} = 1, \tag{3.2}$$

which (using the fact that $a_2+\lambda_2 = m_2\lambda_2$) simplifies to the form

$$\frac{m_1(1-\lambda_2-y_c)}{1+a_1-\lambda_2-y_c} - 1 = \frac{y_c}{\lambda_2}$$

or

$$(m_1-1)(1-\lambda_2-\lambda_1-y_c) = \frac{y_c}{\lambda_2}(1+a_1-\lambda_2-y_c). \tag{3.3}$$

Since λ_2+y_c must be less than 1, by (3.2) one can see from this form of the equation that if $\lambda_1+\lambda_2 > 1$ then there is no positive solution y_c, since one side of the equation would be positive while the other side is negative.

On the other hand, if $\lambda_1+\lambda_2 < 1$ then the left-hand side of (3.3) is a line with positive y intercept and a zero at $1-\lambda_1-\lambda_2$. The right-hand side of (3.3) is a parabola with zeros at 0 and at $1+a_1-\lambda_2$. Since $1-\lambda_1-\lambda_2 < 1+a_1-\lambda_2$ there is a unique value of y_c, $0 < y_c < 1-\lambda_1-\lambda_2$, that satisfies (3.3). The value of x_c is already unique. We summarize this in the following remark.

REMARK. If $\lambda_1+\lambda_2 > 1$ then there is no positive solution of (3.2) and hence no interior equilibrium. In this case E_2 is a globally asymptotically stable rest point. If $\lambda_1+\lambda_2 < 1$, there exists a unique interior rest point and E_2 is unstable.

Everything in the statement has been established except the claim of global asymptotic stability of E_2 in the first case. Since the system is two-dimensional, the Poincaré-Bendixson theorem provides a proof of the global claim.

The next step is to analyze the stability of the interior rest point. To do this one considers the variational matrix at E_c

$$
M = \begin{bmatrix} \dfrac{-m_1\lambda_2 a_1}{(1+a_1-\lambda_2-y_c)^2} + \dfrac{m_2 y_c \lambda_2}{(a_2+\lambda_2)^2} & \dfrac{-m_1 a_1 \lambda_2}{(1+a_1-\lambda_2-y_c)^2} - 1 \\ \dfrac{(m_2-1)y_c}{\lambda_2 + a_2} & 0 \end{bmatrix}.
$$

The determinant of M is positive; hence the real parts of the eigenvalues have the same sign, and stability depends on the trace (the sum of the eigenvalues of M). This is just the term in the upper left-hand corner. The rest point will be locally asymptotically stable if

$$
\frac{m_2 y_c}{(a_2+\lambda_2)^2} < \frac{m_1 a_1}{(1+a_1-\lambda_2-y_c)^2}.
$$

This simplifies to

$$
\frac{y_c}{m_2 \lambda_2^2} < \frac{m_1 a_1}{(1+a_1-\lambda_2-y_c)^2}. \tag{3.4}
$$

If the inequality is reversed then the rest point E_c is unstable – a repeller. The Poincaré–Bendixson theorem then allows one to conclude that there exists a limit cycle. Unfortunately, there may (theoretically) be several limit cycles. If all limit cycles are hyperbolic then there is at least one asymptotically stable one, for if there are multiple limit cycles the innermost one must be asymptotically stable. Moreover, since all trajectories eventually lie in a compact set, there are only a finite number of limit cycles and the outermost one must be asymptotically stable. Since the system is (real) analytic, one could also appeal to results for such systems. For example, Erle, Mayer, and Plesser [EMP] and Zhu and Smith [ZS] show that if E_c is unstable then there exists at least one limit cycle that is asymptotically stable. Stability of limit cycles will be discussed in the next section. We make a brief digression to outline the principal parts of this theory, and then return to the food-chain problem.

4. Elementary Floquet Theory

A standard reference for the material in this section is [CL]. Here, the basic definitions and theorems are given but no proofs are presented.

Floquet theory deals with the structure of linear systems of the form

$$
x' = A(t)x, \tag{4.1}
$$

where $x(t) \in \mathbb{R}^n$ and $A(t)$ is an $n \times n$ matrix of functions of period T. Associated with the system (4.1) is the corresponding matrix system

$$Y' = A(t)Y, \tag{4.2}$$

where Y is an $n \times n$ matrix. A nonsingular matrix solution of (4.2) is called a *fundamental* matrix. Given a fundamental matrix $\Phi(t)$, every solution of (4.1) can be written as $\Phi(t)c$ for a constant vector c. It is usual to take (4.2) with the initial condition $\Phi(0) = I$, where I is the $n \times n$ identity matrix. The principal theorem may be stated as follows.

THEOREM 4.1. *Let $A(t)$ be periodic of period T. Then if $\Phi(t)$ is a fundamental matrix, so is $\Psi(t) = \Phi(t+T)$. Corresponding to any fundamental matrix $\Phi(t)$ there exists a periodic nonsingular matrix $P(t)$ of period T and a constant matrix B such that*

$$\Phi(t) = P(t)e^{Bt}.$$

As customary, the exponential of a matrix means the sum of the matrix series corresponding to the exponential function. The eigenvalues of $\Phi(T) = e^{BT}$ are called the Floquet *multipliers*. The eigenvalues of B are called the Floquet *exponents*. (There is some delicacy about the uniqueness of B which we will ignore because it is not relevant to our use.) Usually it is not possible to compute the Floquet exponents or multipliers. However, for low-dimensional systems of the kind we will investigate, there is a general theorem about the determinant of a fundamental matrix which is helpful. Let $\Phi(t)$ be a fundamental matrix for (4.1) with $\Phi(0) = I$. Then

$$\det \Phi(t) = \exp\left[\int_0^t \operatorname{tr} A(s)\, ds \right], \tag{4.3}$$

where det indicates determinant and tr indicates trace of a matrix. Using Theorem 4.1 yields

$$\det \Phi(T) = \det[e^{BT}] = \exp\left[\int_0^T \operatorname{tr} A(s)\, ds \right]. \tag{4.4}$$

Thus the product of the Floquet multipliers is the determinant of $\Phi(T)$. Equation (4.4) will be useful in some stability calculations.

Suppose now that

$$x' = f(x), \tag{4.5}$$

where f is continuously differentiable, and suppose (4.5) has a periodic trajectory γ that can be thought of as being parameterized by time, giving a solution $\gamma(t)$ which is periodic of period T. (We are using γ to denote the point set or trajectory and $\gamma(t)$ to denote the solution; γ is a closed curve, $\gamma(t)$ is a periodic function.) The linearization about this solution gives a system of the form

$$y' = f_x(\gamma(t))y, \qquad (4.6)$$

where $f_x(\gamma(t))$ is the Jacobian matrix evaluated at the periodic solution. This is a system of the form (4.1). The Jacobian matrix evaluated at $\gamma(t)$ is periodic of period T. A solution $\gamma(t)$ is said to be *asymptotically orbitally stable* if, for every $\epsilon > 0$, there is a $\delta > 0$ such that if a trajectory of (4.5) comes within δ of the orbit γ then that trajectory remains within ϵ of γ and has γ as its omega limit set.

THEOREM 4.2. *Let* $n-1$ *of the Floquet multipliers of* (4.6) *lie inside the unit circle in the complex plane. Then* γ *is an asymptotically orbitally stable trajectory of* (4.5).

A trajectory γ is unstable if one of the multipliers is outside the unit circle. Hyperbolicity, in this case, is taken to mean that exactly one multiplier is on the unit circle. When a periodic solution of (4.5) exists, one multiplier of the linearization is always equal to 1. This is where (4.4) is useful, particularly when the system is two-dimensional.

5. The Food Chain Continued

We return now to the discussion of Section 3. A criterion for the stability or instability of the rest point E_c had been obtained in the form (3.4). If the rest point E_c is locally asymptotically stable, it is possible that there could still be limit cycles surrounding it. The following arguments show that this cannot happen. These arguments are very detailed and very tedious; the reader who is not interested in the technique might be well advised to skip to the statement of the main theorem.

In the proofs that follow, the quantity $1-x(t)-y(t)$ turns up so frequently that we use the equivalent (in terms of the original system) $S(t)$ to keep the computations simpler. Similarly, we use $S_c = 1-x_c-y_c$ in order to shorten the notation. We begin with the following computation.

REMARK.

$$S(t) - S_c = \cfrac{\dfrac{1}{x}\dfrac{dx}{dt} - \dfrac{m_2}{m_2-1}\left(1 + \dfrac{y_c}{a_2+\lambda_2}\right)\dfrac{1}{y}\dfrac{dy}{dt}}{\dfrac{m_1 a_1}{(a_1+S_c)(a_1+S(t))} + \dfrac{m_2}{a_2+x}}.$$

Proof.

$$\frac{x'}{x} = \frac{m_1 S}{a_1 + S} - \frac{m_2 y}{a_2 + x} - 1$$

$$= \frac{m_1 S}{a_1 + S} - \frac{m_1 S_c}{a_1 + S_c} + \frac{m_2 y_c}{a_2 + \lambda_2} - \frac{m_2 y}{a_2 + x}$$

$$= \frac{m_1 a_1 (S - S_c)}{(a_1 + S)(a_1 + S_c)} - \frac{m_2}{a_2 + x}(y - y_c) + \frac{m_2 y_c}{(a_2 + \lambda_2)(a_2 + x)}(x - x_c)$$

$$= \left(\frac{m_1 a_1}{(a_1 + S)(a_1 + S_c)} + \frac{m_2}{a_2 + x} \right)(S - S_c)$$

$$+ \left(\frac{m_2}{a_2 + x} + \frac{m_2 y_c}{(a_2 + x)(a_2 + \lambda_2)} \right)(x - \lambda_2)$$

$$= \left(\frac{m_1 a_1}{(a_1 + S)(a_1 + S_c)} + \frac{m_2}{a_2 + x} \right)(S - S_c)$$

$$+ \frac{y'}{y} \frac{m_2}{m_2 - 1} \left(1 + \frac{y_c}{a_2 + \lambda_2} \right),$$

from which the remark follows by solving for $S - S_c$. (Note that use has been made of (3.2) and various arrangements of the fact that $a_2 + \lambda_2 = m_2 \lambda_2$.) $\qquad\square$

To avoid excessive mathematical notation the next lemma is stated in terms of the system

$$x' = f(x, y), \qquad y' = g(x, y), \tag{5.1}$$

where we have in mind that the functions f and g are given by the right-hand side of (3.1).

LEMMA 5.1. *Let $\Gamma(t) = (x(t), y(t))$ be an arbitrary periodic trajectory of (3.1) with period T. Let R denote the set of points in the plane that are interior to Γ and let*

$$\Delta = \int_0^T \left(\frac{\partial f}{\partial x}(x(t), y(t)) + \frac{\partial g}{\partial y}(x(t), y(t)) \right) dt. \tag{5.2}$$

Then one can write Δ as a sum of the form

$$\Delta = \left(\frac{y_c}{m_2 x_c} - \frac{m_1 a_1 x_c}{(a_1 + S_c)^2} \right) T + \iint_R Q(x, y) \, dx \, dy, \tag{5.3}$$

where $Q(x, y) < 0$.

Since $S_c = 1 - x_c - y_c$ and $x_c = \lambda_2$, if the quantity in brackets in (5.3) is negative then (3.4) is satisfied (and conversely). The use of this complicated lemma has the following consequence for the stability of a periodic orbit.

COROLLARY 5.2. *If (3.4)* holds then Γ is asymptotically (orbitally) stable.

Proof. The quantity under the integral sign in the definition of Δ in (5.2) is the trace of the Jacobian matrix for the system (5.1) evaluated along the periodic orbit. Theorem 4.2 then applies. A periodic orbit for an autonomous system has one Floquet multiplier equal to 1. Since there are only two multipliers and one of them is 1, e^Δ is the remaining one. The periodic orbit is asymptotically orbitally stable because, in view of Lemma 5.1, $\Delta < 0$. $\qquad\square$

Proof of Lemma 5.1. Differentiation and substitution yield

$$\Delta = \int_0^T \left\{ \left[\frac{m_1(1-x-y)}{1+a_1-x-y} - \frac{m_2 y}{a_2+x} - 1 \right] \right. $$
$$\left. + x \left[\frac{-m_1 a_1}{(1+a_2-x-y)^2} + \frac{m_2 y}{(a_2+x)^2} \right] + \left[\frac{m_2 x}{a_2+x} - 1 \right] \right\} dt. \qquad (5.4)$$

The quantity in the first square bracket is just $x'(t)/x(t)$ and hence integrates to 0 since $x(t)$ is periodic of period T. Similarly, the third square bracket is just $y'(t)/y(t)$ and integrates to 0 for similar reasons. Thus one obtains that

$$\Delta = \int_0^T \left(\frac{m_2 y(t)}{(a_2+x(t))^2} - \frac{a_1 m_1}{(1+a_1-x(t)-y(t))^2} \right) x(t)\, dt. \qquad (5.5)$$

The proof reduces to the application of Green's theorem to yield a function $Q(x, y)$ with the alleged sign. Unfortunately, the details are somewhat involved (although not atypical of the use of Green's theorem). Note that

$$\int_0^T \frac{m_2 x(t) y(t)}{(a_2+x(t))^2}\, dt = \int_0^T \left(\frac{x(t)}{a_2+x(t)} \right) \left(\frac{m_2 y(t)}{a_2+x(t)} \right) dt$$

$$= \int_0^T \frac{x(t)}{a_2+x(t)} \left(\frac{m_1 S(t)}{a_1+S(t)} - 1 - \frac{x'(t)}{x(t)} \right) dt$$

$$= \int_0^T \frac{x(t)}{a_2+x(t)} \left(\frac{m_1 S_c}{a_1+S_c} - 1 \right) dt$$

$$+ \int_0^T \frac{x(t)}{a_2+x(t)} \left(\frac{m_1 S(t)}{a_1+S(t)} - \frac{m_1 S_c}{a_1+S_c} \right) dt,$$

where we have used Equation (3.1) and the fact that $x'/(a_2+x)$ integrates to 0. Thus one has that $\Delta = I_1 + I_2$. We investigate each integral separately, beginning with I_2. First the terms are combined to give

$$I_2 = \int_0^T \frac{x(t)}{a_2+x(t)} \left(\frac{m_1 S(t)(a_1+S_c) - m_1 S_c(a_1+S(t))}{(a_1+S(t))(a_1+S_c)} \right) dt$$

$$= \frac{m_1 a_1}{a_1+S_c} \int_0^T \left(\frac{x(t)}{a_2+x(t)} \frac{S(t)-S_c}{a_1+S(t)} \right) \cdot dt.$$

The remark preceding Lemma 5.1 can be applied to convert I_2 to a line integral of the form

$$I_2 = \frac{m_1 a_1}{a_1+S_c} \int_\Gamma \frac{x}{(a_2+x)(a_1+S)} \left(\frac{m_1 a_1}{(a_1+S_c)(a_1+S)} + \frac{m_2}{a_2+x} \right)^{-1}$$

$$\times \left\{ \frac{dx}{x} - \frac{m_2}{m_2-1} \left(1 + \frac{y_c}{a_2+\lambda_2} \right) \frac{dy}{y} \right\}.$$

This is of the form

$$\int_\Gamma P_1 \, dx + Q_1 \, dy.$$

Green's theorem may now be applied to deduce that

$$I_2 = \frac{m_1 a_1}{a_1+S_c} \iint_R \left(\frac{\partial Q_1}{\partial x} - \frac{\partial P_1}{\partial y} \right) dx \, dy,$$

where

$$Q_1(x,y) = -\frac{m_2}{m_2-1} \left(1 + \frac{y_c}{m_2\lambda_2} \right) \frac{x}{y} P(x,y),$$

$$P_1(x,y) = (a_2+x)^{-1} P(x,y),$$

$$P(x,y) = \left(\frac{m_1 a_1}{a_1+S_c} + \frac{m_2(a_1+1-x-y)}{a_2+x} \right)^{-1}.$$

Differentiation shows that

$$\frac{\partial Q_1}{\partial x} < 0, \qquad \frac{\partial P_1}{\partial y} > 0$$

and hence that the integrand is negative.

The integral I_1 is much easier. First write

$$I_1 = \left(\frac{m_1 S_c}{a_1+S_c} - 1 \right) \int_0^T \frac{x(t)}{a_2+x(t)} \, dt.$$

Then use the differential equation for y and (3.2) to obtain

$$I_1 = \frac{1}{m_2}\left(\frac{m_1 S_c}{a_1 + S_c} - 1\right)\int_0^T \left(\frac{y'}{y} + 1\right)dt$$

$$= \frac{T}{m_2}\left(\frac{m_1 S_c}{a_1 + S_c} - 1\right) = T\frac{y_c}{a_2 + \lambda_2} = \frac{y_c T}{m_2 \lambda_2}.$$

The same technique will be applied to the second integral in (5.5), although it is much more complicated. First of all, write

$$-\int_0^T \frac{a_1 m_1 x(t)}{(a_1 + S(t))^2}\,dt$$

$$= -\int_0^T \frac{x(t)}{a_2 + x(t)}\left\{\left[\frac{m_1 a_1(a_2 + x(t))}{(a_1 + S(t))^2} - \frac{m_1 a_1}{(a_1 + S_c)^2}(a_2 + x(t))\right]\right.$$

$$+ \left[\frac{m_1 a_1(a_2 + x(t))}{(a_1 + S_c)^2} - \frac{m_1 a_1(a_2 + \lambda_2)}{(a_1 + S_c)^2}\right]$$

$$\left.+ \left[\frac{m_1 a_1(a_2 + \lambda_2)}{(a_1 + S_c)^2}\right]\right\}dt$$

$$= I_3 + I_4 + I_5.$$

We begin with I_3. As before (suppressing the notation for the dependence on t in the integrand),

$$I_3 = -\int_0^T \frac{x}{a_2 + x}\left[\frac{m_1 a_1(a_2 + x)}{(a_1 + S)^2} - \frac{m_1 a_1(a_2 + x)}{(a_1 + S_c)^2}\right]dt$$

$$= a_1 m_1 \int_0^T \frac{x(S - S_c)(2a_1 + S + S_c)}{(a_1 + S_c)^2(a_1 + S)^2}\,dt.$$

This can be broken into two integrals:

$$I_3 = \frac{a_1 m_1}{(a_1 + S_c)^2}\int_0^T \frac{x(S - S_c)}{a_1 + S}\,dt + \frac{a_1 m_1}{a_1 + S_c}\int_0^T \frac{x(S - S_c)}{(a_1 + S)^2}\,dt.$$

Each of these integrals can be converted to a line integral (as before) by using the remark preceding Lemma 5.1. The procedure is exactly the same and, in the interest of cutting the presentation short, we note only that the end result is

$$\int_0^T \frac{x(S - S_c)}{a_1 + S}\,dt = \iint_R \left(\frac{\partial Q_2}{\partial x} - \frac{\partial P_2}{\partial y}\right)dx\,dy,$$

where

$$P_2 = P(x, y), \qquad Q_2 = -\frac{m_2}{m_2 - 1}\left(1 + \frac{y_c}{m_2 \lambda_2}\right)\frac{x}{y}P(x, y),$$

and

$$\int_0^T \frac{x(S-S_c)}{(a_1+S)^2}\,dt = \iint_R \left(\frac{\partial Q_3}{\partial x} - \frac{\partial P_3}{\partial y} \right) dx\,dy,$$

$$P_3 = \frac{P(x,y)}{1+a_1-x-y},$$

$$Q_3 = \frac{m_2}{m_2-1}\left(1 + \frac{y_c}{m_2\lambda_2} \frac{xP(x,y)}{y(1+a_1-x-y)} \right).$$

Note that

$$\frac{\partial Q_2}{\partial x} < 0, \quad \frac{\partial Q_3}{\partial x} < 0, \quad \frac{\partial P_2}{\partial y} > 0, \quad \frac{\partial P_3}{\partial y} > 0.$$

Thus I_3 can be written as an integral over R which has a negative integrand. The remaining two integrals are easy. By (2.4), one has

$$I_4 = -\frac{m_1 a_1}{(a_1+S_c)^2} \int_0^T \frac{x(x-\lambda_2)}{a_2+x}\,dt$$

$$= -\frac{m_1 a_1}{(m_2-1)(a_1+S_c)^2} \int_0^T \frac{xy'}{y}\,dt$$

$$= -\frac{m_1 a_1}{(m_2-1)(a_1+S_c)^2} \int_\Gamma \frac{x}{y}\,dy$$

$$= -\frac{m_1 a_1}{(m_2-1)(a_1+S_c)^2} \iint_R \frac{1}{y}\,dx\,dy$$

and

$$I_5 = -\frac{m_1 a_1(a_2+\lambda_2)}{(a_1+S_c)^2} \int_0^T \frac{x}{a_2+x}\,dt$$

$$= -\frac{m_1 a_1(a_2+\lambda_2)}{m_2(a_1+S_c)^2} \int_0^T \left(\frac{y'}{y} + 1 \right) dt$$

$$= \frac{-m_1 a_1(a_2+\lambda_2)}{m_2(a_1+S_c)^2} T = \frac{-m_1 a_1 \lambda_2}{(a_1+S_c)^2} T.$$

Collecting all of the above integrals yields

$$\Delta = \left(\frac{-m_1 a_1 \lambda_2}{(a_1+S_c)^2} + \frac{y_c}{m_2\lambda_2} \right) T + \iint_R Q(x,y)\,dx\,dy. \qquad \square$$

THEOREM 5.3. *If (3.4) holds then all trajectories of (3.1) tend to E_c as t tends to infinity.*

Before giving a proof, note that the components of E_c are readily obtainable: $x_c = \lambda_2$, and y_c is the root of a quadratic. The condition is stated in the form (3.4) to avoid the complicated expression that would result from using the quadratic formula.

Proof of Theorem 5.3. Condition (3.4) makes E_c locally asymptotically stable. By the Poincaré–Bendixson theorem, it is necessary only to show that with condition (3.4) there are no limit cycles. Suppose there were a limit cycle. However, there is at most a finite number of limit cycles and each must contain E_c in its interior. Hence there is a periodic trajectory Γ that contains no other periodic trajectory in its interior. Intuitively speaking, Γ is the trajectory "closest" to the rest point. The constant term in the formula given in Lemma 5.1 is negative. The corollary shows that Γ is asymptotically stable. This is a contradiction, since the rest point is asymptotically stable – that is, between the two there must be an unstable periodic orbit. $\qquad\qquad\square$

When the inequality in (3.4) is reversed, there will be a periodic orbit (by an application of the Poincaré–Bendixson theorem). By our assumption of hyperbolicity, this orbit must be asymptotically orbitally stable since it is so from the "inside." These comments establish the next result.

THEOREM 5.4. *If E_c exists in the positive quadrant and if*

$$\frac{y_c}{m_2 \lambda_2^2} > \frac{m_1 a_1}{(1 + a_1 - \lambda_2 - y_c)^2}, \qquad (5.6)$$

then there exists an asymptotically orbitally stable periodic orbit for (3.1).

Kuang [K1] has shown that if the parameters are such that

$$\frac{y_c}{m_2 \lambda_2^2} - \frac{m_1 a_1}{(1 + a_1 - \lambda_2 - y_c)^2}$$

is small and positive, then the limit cycle is unique and asymptotically stable.

6. Bifurcation from a Simple Eigenvalue

In Section 4 it was shown that a food chain depending on the nutrient S, a first-level consumer x, and a predator y could possess a periodic solution. From the standpoint of the full system (2.2) or the simplified system

(2.3), this food chain corresponds to the absence of a predator z competing with y. After the simplification, one can view (3.1) as (2.3) with the initial condition $z(0) = 0$, since the x-y subsystem represents an invariant set. A natural question, then, is: If one chooses the parameters so that the food chain has a periodic limit cycle, can the full system have a limit cycle in the positive cone in $\mathbb{R}^3$? The tools to answer questions of this type come from bifurcation theory. The kind of theorem needed is of the simplest sort, bifurcation from a simple eigenvalue. Although no theorems will be proved, this section attempts to sketch the basic ideas needed. Some familiarity is required with the fundamental theory of ordinary differential equations, particularly the Floquet theory, so the reader who is interested in only the results should skip ahead to Section 7 for application to the system under study.

A good explanation of the theory can be found in Smoller [Smo] (see also [MM]). All forms of the basic result are essentially equivalent. Bifurcation theory is not restricted to differential equations but is actually concerned with mappings or functions. A principal tool in developing the theory is the implicit function theorem. When the theory is used in infinite-dimensional spaces, quite sophisticated mathematics is required. However, the problem here can be dealt with in a finite-dimensional setting.

Suppose one has an equation

$$f(x, \lambda) = 0, \tag{6.1}$$

where f is a smooth function from $\mathbb{R}^n \times \mathbb{R}$ into $\mathbb{R}^n$ and where we think of λ as a parameter. One seeks the set of solutions – that is, the set

$$f^{-1}(0) = \{(x, \lambda) \in \mathbb{R}^n \times \mathbb{R} \mid f(x, \lambda) = 0\}.$$

Further, we are interested in curves of such solutions (i.e., a solution for each value of λ in some interval).

Suppose that (x_0, λ_0) is a solution of (6.1) belonging to some smooth curve Γ of solutions that can be parameterized by λ. That is, suppose there exists an interval I with λ_0 as an interior point and a smooth function $x: I \to \mathbb{R}^n$ such that $x(\lambda_0) = x_0$ and

$$\Gamma = \{(x(\lambda), \lambda) \mid \lambda \in I\} \subset f^{-1}(0).$$

Typically, Γ is a curve of obvious (usually trivial) solutions. The point (x_0, λ_0) will be called a *bifurcation point* of (6.1) if every neighborhood of (x_0, λ_0) contains a solution of (6.1) which does not belong to Γ. Figure 6.1 crudely pictures a bifurcation point with $x_0 = 0$.

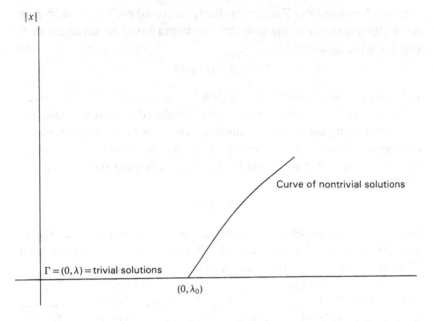

Figure 6.1. A bifurcation diagram.

If the Jacobian of f with respect to x, the matrix

$$J = \left[\frac{\partial f_i}{\partial x_j}\right],$$

is nonsingular at (x_0, λ_0), then the implicit function theorem implies that there exists an interval J containing λ_0 as an interior point and a smooth function $\bar{x}: J \to \mathbb{R}^n$ such that $\bar{x}(\lambda_0) = x_0$, and

$$\bar{\Gamma} = \{(\bar{x}(\lambda), \lambda) \mid \lambda \in J\}$$

consists of solutions of (6.1). Furthermore, these are the only solutions of (6.1) in a neighborhood of (x_0, λ_0). Obviously, then, Γ and $\bar{\Gamma}$ overlap and

$$x(\lambda) = \bar{x}(\lambda)$$

for $\lambda \in J \cap I$. Consequently, we see that a necessary condition for (x_0, λ_0) to be a bifurcation point is that J be singular at (x_0, λ_0).

Fortunately, although the implicit function theorem would appear to be inapplicable as a tool to discover the structure of solutions of (6.1) in a neighborhood of a bifurcation point, it can be successfully applied once the branch of solutions Γ is appropriately "factored out" of the map f. In most applications it is the case that, or it can easily be arranged that, $(x_0, \lambda_0) = (0, \lambda_0)$ and that

$$\Gamma = \{(0, \lambda) \mid \lambda \in I\}.$$

In stating the result used in this chapter, we assume this to be the case.

The ideas of dynamical systems were introduced in Chapter 1 (Section 3). Two mappings appear in the discussion here, both based on ideas best expressed in terms of dynamical systems. The discussion is limited to dimension 3 because that is what we need, but it is valid for $\mathbb{R}^n$. Suppose one has a differential equation

$$x' = f(x), \qquad (6.2)$$

where $x \in \mathbb{R}^3$. Given an initial point x_0, there is a trajectory through it, $\pi(x_0, t)$. If the time t is fixed, say at $t = T$, then $\pi(x, T)$ is a function from $\mathbb{R}^3$ to $\mathbb{R}^3$. This is called the *solution mapping*. Now suppose that x_0 is a point on a periodic orbit γ of period T. Let P be a plane through x_0 and orthogonal to the tangent vector $f(x_0)$. Let N be a sufficiently (in a sense to be made clear in what follows) small neighborhood of x_0 in P. By continuity, solutions corresponding to initial conditions near x_0 will remain "close" to the periodic orbit and return to a point "near" x_0 at time T. Indeed, if we are close enough, the orbit will reach the plane P although not necessarily at time T. By using the implicit function theorem, one can show that given x close enough to x_0 there is a time $\tau(x)$ such that $\pi(x, \tau(x)) \in P$. (There is a good discussion of this in the textbook of Hartman [Har].) Hence, for $x \in N$, $\pi(x, \tau(x))$ defines a mapping from $\mathbb{R}^2$ into $\mathbb{R}^2$. This mapping is called the *Poincaré map* associated with γ. The two mappings are related and we shall make use of this relation. The Poincaré map is what we need, but the solution map makes the computations tractable. The principal approach may be summarized in the following statement.

REMARK 6.1. Let γ and x_0 be as before. A fixed point of the Poincaré mapping (different from x_0) gives the initial conditions for a periodic orbit of equation (6.2) in a neighborhood of x_0.

Since each of these mappings is from $\mathbb{R}^n$ to $\mathbb{R}^n$, where $n = 3$ for the solution map and $n = 2$ for the Poincaré map, their linearizations are given by a matrix. The following is a basic result connecting the linearization of the two maps.

LEMMA 6.2. *Let γ be as before, and consider the solution map and the Poincaré map in a neighborhood of a point of γ. Then the eigenvalues of the linearization of the Poincaré map together with $\{1\}$ are the eigenvalues of the linearization of the solution map.*

A proof of this lemma is given in Appendix E.

Now replace the differential equation by one with a parameter,

$$x' = f(x, \lambda). \tag{6.3}$$

Suppose now that γ is a periodic trajectory for every value of the parameter. This corresponds to a curve of fixed points – a trivial curve – of the Poincaré map in the terms discussed for equation (6.1). We can take coordinates in the plane P such that x_0 corresponds to the origin, and then look for conditions that will yield a bifurcation point of the Poincaré mapping. The following theorem provides such conditions in very general terms.

THEOREM 6.3. *Let W be an open neighborhood of $0 \in R^n$ and let I be an open interval containing λ_0 in $\mathbb{R}$. Let $P(x, \lambda)$ be a twice continuously differentiable mapping of $W \times I$ into $\mathbb{R}^n$ satisfying $P(0, \lambda) = 0$ for all $\lambda \in I$. Let $L(\lambda)$ be the Jacobian of P with respect to x evaluated at $(x, \lambda) = (0, \lambda)$. Suppose that $l(\lambda)$ is a real, simple eigenvalue of $L(\lambda)$ satisfying $l(\lambda_0) = 1$ with $dl(\lambda_0)/d\lambda \neq 0$ and that v_0 spans the null space of $L(\lambda_0) - \text{Id}$. Then there exist $\delta > 0$ and a continuously differentiable map $(\phi, \bar{\lambda}) : (-\delta, \delta) \rightarrow v_0^\perp \times \mathbb{R}$ such that*

  (i) $(\phi(0), \bar{\lambda}(0)) = (0, 0)$ *and*
 (ii) $P(s(v_0 + \phi(s)), \bar{\lambda}(s) + \lambda_0) = s(v_0 + \phi(s)), |s| < \delta.$

Furthermore, there is a neighborhood of $(0, \lambda_0) \in W \times I$ such that the only solutions of $P(x, \lambda) - x = 0$ in the neighborhood are $(0, \lambda)$ and $(x, \lambda) = (s(v_0 + \phi(s)), \lambda_0 + \bar{\lambda}(s)).$

A *simple* eigenvalue is one where the algebraic multiplicity as a root of the characteristic polynomial is 1. Theorem 6.3 follows from [Smo, thm. 13.4] and the remarks following it. The difficulty in applying the theorem is in computing the eigenvalues of the linearization of the Poincaré map. The term $v_0^\perp$ denotes the $(n-1)$-dimensional subspace orthogonal to v_0, and Id is the identity mapping.

In the linearization of the two-dimensional system (3.1) around the stable limit cycle, one multiplier is equal to 1 and, if hyperbolic, the other must be inside the unit circle because of the stability. For the Poincaré map corresponding to (2.3), then, one eigenvalue will be inside the unit circle; the object of applying this theorem is to have the remaining one move across the unit circle as prescribed in the theorem's hypotheses. The linearization of (2.3) about the periodic orbit in the $z = 0$ plane takes a special form; the following useful lemma is used to compute the needed eigenvalues.

LEMMA 6.4. *Let $A(t)$ be a 2×2 continuous, periodic matrix of period T, and let $Y(t)$ be the fundamental matrix of $y' = A(t)y$. Let $B(t)$ be a 3×3 periodic matrix of the form*

$$B(t) = \begin{bmatrix} A(t) & \begin{matrix} b_1(t) \\ b_2(t) \end{matrix} \\ 0 \quad 0 & b_3(t) \end{bmatrix}.$$

Then the fundamental matrix $\Phi(t)$ of $z' = B(t)z$ is given by

$$\Phi(t) = \begin{bmatrix} Y(t) & \begin{matrix} z_1 \\ z_2 \end{matrix} \\ 0 \quad 0 & z_3 \end{bmatrix},$$

where

$$z_3(t) = \exp\left[\int_0^t b_3(s)\,ds\right]$$

and $z = (z_1, z_2)$ is given by

$$z(t) = \int_0^t Y(t)Y^{-1}(s)\,b(s)\,z_3(s)\,ds$$

with $b(s) = \text{col}(b_1(s), b_2(s))$. In particular, if ρ_1, ρ_2 are the Floquet multipliers associated with the 2×2 system (eigenvalues of $Y(T)$) then the Floquet multipliers of the 3×3 system are ρ_1, ρ_2, and $\rho_3 = z_3(T)$.

Proof. The proof is a straightforward computation. The $z_3(t)$ term decouples from the system and can be obtained by a quadrature. The formulas for $z_i(t)$, $i = 1, 2$, can be obtained by substituting $z_3(t)$ into the two-dimensional system

$$z_1'(t) = a_{11}z_1 + a_{12}z_2 + b_1(t)z_3(t),$$

$$z_2'(t) = a_{21}z_1 + a_{22}z_2 + b_2(t)z_3(t),$$

and applying the variation-of-constants formula. The formulas for the multipliers follow by expanding the determinant of $\Phi(T)$. $\square$

7. Competing Predators

Armed with the preceding discussion, we can now state the main theorem of this chapter. It shows that, in contrast to the basic chemostat, coexistence can occur if the competition is at a higher trophic level. (We remind the reader of the general assumption of hyperbolicity of limit cycles.)

THEOREM 7.1. *Let a_i and m_i $(i = 1, 2)$ be fixed so that $m_i > 1$, let $\lambda_i < 1$, and assume that the parameters are fixed so that (5.6) holds. Then there exists a number a_3^* such that for some values of a_3 with $|a_3 - a_3^*|$ sufficiently small, (2.3) has a periodic orbit in the positive octant near the x–y plane.*

Proof. Let $\gamma = (x(t), y(t), 0)$ be the orbitally asymptotically stable periodic orbit of period T given by Theorem 5.4. (We have already noted that if there are several orbits then one must be asymptotically stable, by our assumption of hyperbolicity.) Let the Floquet multipliers of γ, viewed as a solution of (3.1), be 1 and ρ_1, where $0 < \rho_1 < 1$. This trajectory remains a solution of the system (2.3) for all values of m_3 and a_3. Fix m_3. For this periodic orbit, define $\mu(a_3)$ by

$$\mu(a_3) = \frac{m_3}{T} \int_0^T \frac{x(\xi)}{a_3 + x(\xi)} \, d\xi.$$

Note that $\mu(0) = m_3 > 1$ and $\mu(a_3)$ is decreasing in a_3. In fact, it follows that

$$\frac{\partial \mu}{\partial a_3} = \frac{-m_3}{T} \int_0^T \frac{x(\xi)}{(a_3 + x(\xi))^2} \, d\xi < 0. \tag{7.1}$$

It also follows that

$$\mu(a_3) \le \frac{m_3}{a_3 T} \int_0^T x(\xi) \, d\xi. \tag{7.2}$$

Therefore, as a_3 tends to infinity, $\mu(a_3)$ tends to zero. Consequently, there exists a unique value of a_3 such that $\mu(a_3) = 1$. Denote this value of a_3 by a_3^*.

Fix a point $p_0 = (x(t_0), y(t_0), 0)$ on γ such that $y'(t_0) = 0$. Then the plane $x = x(t_0) = \lambda_2$ cuts the periodic orbit γ at the point p_0. See Figure 7.1.

The intersection of the plane with a sufficiently small ball centered at p_0 is transversal to γ at p_0 for all a_3 in a neighborhood I of a_3^*. Denote this transversal section by Ω. We use coordinates (Y, Z) on Ω given by

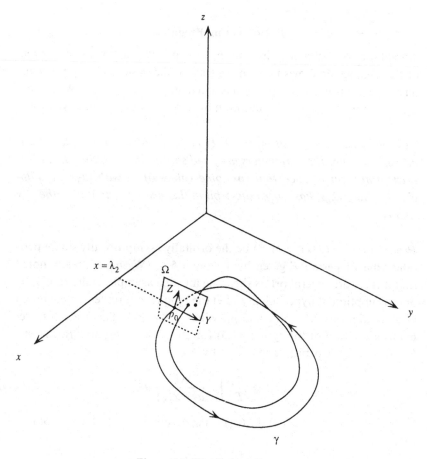

Figure 7.1. The Poincaré map.

$Y = y - y(t_0)$ and $Z = z - z(t_0)$, so that p_0 corresponds to $(0, 0)$ in Ω. The Poincaré map

$$P = P(Y, Z; a_3) = (P_1(Y, Z; a_3), P_2(Y, Z; a_3))$$

is defined for (Y, Z) in a neighborhood W of $(0, 0)$ in Ω and for $a_3 \in I$. Here P_1 denotes the Y coordinate and P_2 denotes the Z coordinate of the point $P(Y, Z; a_3) \in \Omega$ (see Figure 7.1). Since γ is a periodic solution of (2.3) for all values of a_3, it follows that

$$P(0, 0; a_3) = (0, 0)$$

for all values of $a_3 \in I$.

Since the x-y plane is an invariant set for (2.3), one has

$$P_2(Y, 0; a_3) \equiv 0$$

for all $(Y, 0) \in W$ and $a_3 \in I$. Consequently the Jacobian of P with respect to (Y, Z) at $(0, 0, a_3)$ is given by

$$J = \begin{bmatrix} \dfrac{\partial P_1}{\partial Y} & \dfrac{\partial P_1}{\partial Z} \\ 0 & \dfrac{\partial P_2}{\partial Z} \end{bmatrix}.$$

Because the periodic solution $(x(t), y(t))$ of (3.1) is assumed to have Floquet multipliers 1 and ρ_1, where $0 < \rho_1 < 1$, Lemma 6.2 implies that

$$\rho_1 = \frac{\partial P_1}{\partial Y}(0, 0; a_3)$$

is independent of a_3.

To compute the remaining eigenvalue of P, one computes the eigenvalues of the solution map. The linearization of (2.3) about γ takes the form

$$\begin{bmatrix} A(t) & \begin{matrix} f_z \\ 0 \end{matrix} \\ 0 \quad 0 & \dfrac{m_3 x(t)}{a_3 + x(t)} - 1 \end{bmatrix},$$

where f_z denotes the partial derivative of f, the right-hand side of the first equation in (2.3), with respect to z. The matrix $A(t)$ is just the linearization of (3.1) about $(x(t), y(t))$. Its fundamental matrix possesses Floquet multipliers 1 and $\rho_1 \in (0, 1)$ as noted previously. By Lemma 6.4, the other multiplier is $\rho_2 = e^{T(\mu(a_3)-1)}$. Consequently, from the upper triangular form of J and Lemma 6.2,

$$\rho_2 = \frac{\partial P_2}{\partial Z}(0, 0; a_3) = e^{T(\mu(a_3)-1)}.$$

This eigenvalue, which we denote by $l(a_3)$ (in anticipation of verifying Theorem 6.3), clearly satisfies $l(a_3^*) = 1$ and $dl(a_3^*)/da_3 < 0$, since $d\mu/da_3 < 0$.

An eigenvector v_0 corresponding to the eigenvalue $l(a_3^*) = 1$ of J is easily computed to be

$$v_0 = \left(\frac{\partial P_1}{\partial Z}, 1 - \rho_1 \right).$$

Note that its second Z component is positive. The hypotheses of Theorem 6.3 have now been verified. Therefore, the Poincaré map has a fixed point given in (x, y, z) coordinates by

$$(x, y, z) = p_0 + s(0, v_0) + o(s)$$

corresponding to a parameter a_3 given by

$$a_3 = a_3^* + \bar{\lambda}(s),$$

where $o(s)$ represents a term satisfying $\lim_{s \to 0} o(s)/s = 0$, $\bar{\lambda}(0) = 0$, and $\bar{\lambda}(s)$ is continuously differentiable. This completes the proof of the theorem.   □

The question of which values of a_3 yield the periodic orbit is not resolved by the theorem. Generically, the derivative of $\bar{\lambda}$ at $s = 0$ is not expected to be zero. In that case there would be an interval of values on one side of a_3^* for which there would be a periodic solution in the positive cone; see Figure 7.2. This affects the stability as well. A sample computation is given in the next section.

8. Numerical Example

The limit cycle found in the previous section holds only for $|a_3 - a_3^*|$ small. Obviously, once the limit cycle exists, it can be continued, either globally or until certain "bad" things happen such as the period tending to infinity or the orbit collapsing to a point. It is very difficult to show analytically that these events do not occur. Moreover, the computations necessary to actually prove the asymptotic stability of the bifurcating orbit are very difficult. We discuss briefly some numerical computations, shown in Figure 8.1, which suggest answers to both these problems.

The differential equations were solved for a variety of values of a_3 less than a_3^*. The program was run for considerable time and the last 100 points saved. If the limiting periodic orbit were asymptotically stable, these points would be near the periodic orbit - equal as well as the eye can determine. These periodic orbits, corresponding to different parameters and hence to different systems of differential equations, were then plotted on a single three-dimensional graph (Figure 8.1). This illustrates the stability.

At first the orbits were in the x-y plane, and then out of it but close to it, as the theorem states. However, as the parameter was varied, the periodic orbits continued to exist (as one might expect). Moreover, as the parameter was continued, the periodic orbits moved near the x-z plane and collapsed into it. The figure also illustrates this point.

Figure 8.1 gives a pictorial view of bifurcation. The periodic orbits in the respective planes correspond to the periodic orbits of the food chains studied in Sections 3 and 5. The periodic orbits in the interior of the cone correspond to periodic orbits of two predators feeding on a single prey growing on the nutrient.

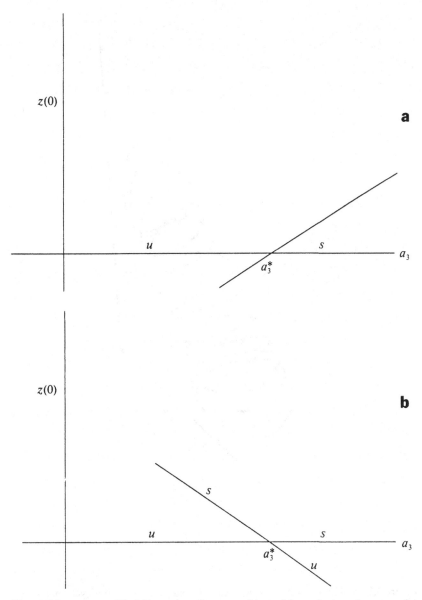

Figure 7.2. **a** A possible bifurcation diagram with positive solutions for $a_3 > a_3^*$.
b A possible bifurcation diagram with positive solutions for $a_3 < a_3^*$.

9. A Long Food Chain

It is expected that more complicated dynamic behavior would be possible in a longer food chain. Indeed, we will illustrate (numerically) that adding

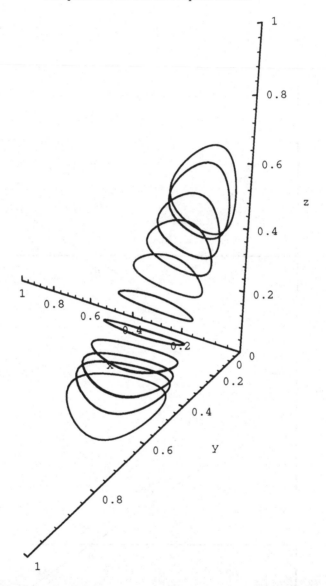

Figure 8.1. Numerical solutions showing bifurcation from a limit cycle in a plane into a limit cycle in the interior. The fixed parameters are $a_1 = 0.3$, $a_2 = 0.4$, $m_1 = 10.0$, $m_2 = 4.5$, and $m_3 = 5.0$. The parameter a_3 was varied between 0.46 and 0.48. The limit cycle moves from one plane (a face of $\mathbb{R}_+^3$) through the interior of $\mathbb{R}_+^3$ (showing coexistence) and collapses into the opposite face of $\mathbb{R}_+^3$.

another trophic level to the model of Section 3 can produce complex dynamics. This was suggested by recent (unpublished) work of Wolkowicz on a food chain with a reproducing prey at the lowest level, and by the results of Hastings and Powell [HP].

If one adds another predator to the model of Section 3, then the system is governed by the equations (in scaled form)

$$S' = 1 - S - xf_1(S),$$

$$x' = x(f_1(S) - 1) - yf_2(x),$$

$$y' = y(f_2(x) - 1) - zf_3(y),$$

$$z' = z(f_3(y) - 1),$$

where

$$f_i(u) = \frac{m_i u}{a_i + u}.$$

As in the earlier work, one has immediately that

$$\lim_{t \to \infty} S(t) + x(t) + y(t) + z(t) = 1.$$

On the omega limit set, then, one may eliminate the S variable by using $S = 1 - x - y - z$ to obtain the system

$$x' = x(f_1(1 - x - y - z) - 1) - yf_2(x),$$

$$y' = y(f_2(x) - 1) - zf_3(y),$$

$$z' = z(f_3(y) - 1).$$

The parameters of the first two levels were fixed: $m_1 = 10.0$, $m_2 = 4.0$, $a_1 = 0.08$, and $a_2 = 0.23$. The values are consistent with those suggested in [CN2]. The growth rate of the top predator was fixed at $m_3 = 3.5$, and the value of a_3 varied to obtain Figures 9.1-4. Figure 9.1 shows the attractor as a limit cycle; Figure 9.2 indicates the appearance of a period-doubling bifurcation. Figure 9.3 shows the attractor after two such bifurcations, while Figure 9.4 suggests a chaotic attractor.

10. Discussion

The final conclusion of Section 7 was that two predators could survive on a common prey. It is significant that this can happen only as an oscillatory phenomenon. This is easy to understand intuitively. If the prey

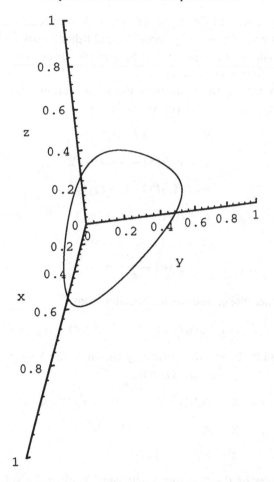

Figure 9.1. A periodic orbit in the long food chain. Parameters are $a_1 = 0.08$, $a_2 = 0.23$, $a_3 = 0.40$, $m_1 = 10.0$, $m_2 = 4.0$, and $m_3 = 3.5$.

level came to a steady state, it would favor one predator or the other. While it is oscillating, sometimes the concentration favors one competitor and sometimes the other. Each manages to grow enough while it has the advantage to survive. One does well at higher concentrations and then crashes rapidly; the other is steady at lower concentrations and improves while the first is "crashing." This is a manifestation of the common r-K strategist discussion in the ecology literature.

The "platform" from which the (four-dimensional) oscillations were launched was the (three-dimensional) food chain, which was already oscillatory. Experiments confirming the behavior of the food chain exist in

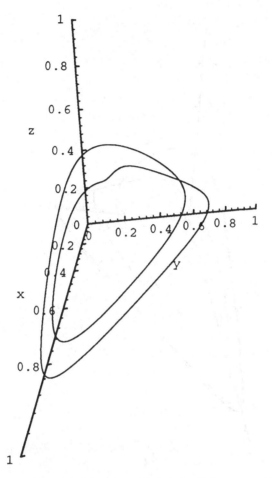

Figure 9.2. Period doubling in the long food chain. Parameters are as in Figure 9.1 except that $a_3 = 0.3$.

the literature. Figure 10.1 shows the results of the experiments of Jost et al. [JDFT]. A food chain was constructed in a chemostat consisting of a nutrient (glucose), a bacteria living on that nutrient (*Azotobacter vinelandii*), and a ciliate (*Tetrahymena pyriformis*) feeding on the bacteria. Both sustained oscillations (Figure 10.1a) and convergence to a steady state (Figure 10.1b) were observed, depending on the washout rate. A discussion of these and other experiments can be found in Jannasch and Mateles [JM]. Although we are unaware of any definitive experiments, it is clear from the mathematics that if the food chain can oscillate then competing predators at the top can also oscillate.

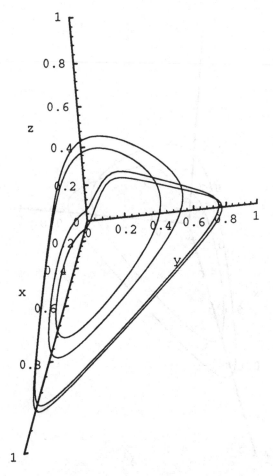

Figure 9.3. Two period doublings in the long food chain. Parameters are as in Figure 9.1 except that $a_3 = 0.24$.

If coexistence is possible because of the oscillation, then might two competitors also coexist in an oscillatory environment – that is, with oscillatory inputs but with only two trophic levels? The answer is clearly Yes. This problem, however, requires very different techniques, since it is a forced oscillation and does not generate a dynamical system. Chapter 7 is devoted to one approach to such problems.

Delays can also introduce oscillations. These problems are infinite-dimensional ones and much more difficult to analyze rigorously. Comments on the delayed chemostat can be found in Chapter 10, along with

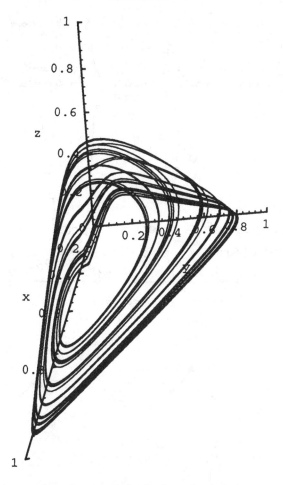

Figure 9.4. A complicated attractor in the long food chain, possibly chaotic. Parameters are as in Figure 9.1 except that $a_3 = 0.20$.

references. However, the problem will not be considered in detail in this work.

Although Theorem 7.1 can be interpreted to show the possibility of coexistence, many questions of a mathematical nature remain unanswered. One would like to know the direction of bifurcation (i.e., determining the range of the function $\bar{\lambda}(s)$) of the bifurcating solutions, the extent of their continuability, and so forth. The numerical results answer these questions for a particular case, but this is mathematically unsatisfactory. The questions are answerable in principle, but the required calculations

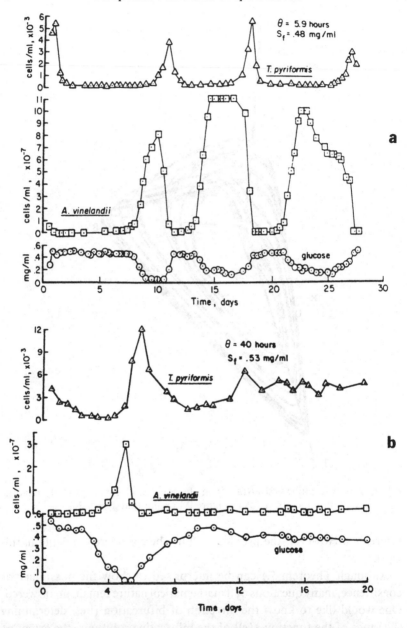

Figure 10.1. A food chain constructed of a nutrient (glucose), a bacteria (*A. vinelandii*) living on that nutrient, and a ciliate (*T. pyriformis*) feeding on the bacteria: **a** sustained oscillations; **b** convergence to a steady state. (From [JDFT], Copyright 1973, American Society for Microbiology. Reproduced by permission.)

are formidable. Generically, of course, the derivative of $\bar{\lambda}(s)$ at $s = 0$ is nonzero. Hence there will be periodic orbits in the positive cone for a_3 on one side of a_3^*. Keener [Ke1; Ke2] has a technique for answering such questions under a further assumption. Essentially, the idea is to make all of the bifurcations occur at one point and then "peel" them apart (unfold them) as the parameter changes.

Pavlou and Kevrekidis [PK] consider periodically forced food chains and investigate the effect of forcing on the periodic orbit. Using numerical bifurcation techniques, they exhibit quasiperiodicity and entrainment of frequencies.

4

The Chemostat with an Inhibitor

1. Introduction

In the first two chapters the general theory of the chemostat was developed, and it was shown that competitive exclusion is the expected outcome. In Chapter 3, coexistence was shown to occur when the competition was at a higher trophic level; the mechanism was simply the oscillation of the object of the competition – the prey in the case being considered. In this chapter, we return to the basic chemostat model but add another factor, the presence of an inhibitor. The inhibitor affects the nutrient uptake rate of one of the competitors but is taken up by the other without ill effect. The use of Nalidixic acid in the experiments of Hansen and Hubbell [HH], discussed in Chapter 1, is an example. Its effect on one strain of *E. coli* was essentially nil while the growth rate of the other was severely diminished.

The interest in this subject goes far beyond laboratory examples. It is common for one strain of bacteria to be affected by an antibiotic while another is resistant. The antibiotic acts as the inhibitor in our discussion, and we want to know whether the resistant strain will eliminate the nonresistant strain. If it can, the antibiotic will not be effective in treating the disease. In commercial applications, genetically altered organisms are used to generate "products." Resistance to an antibiotic is often introduced with the genetic material as well, in order to prevent reversion of the organism to the "wild" (i.e. the unaltered) type which presumably is a better competitor. A similar situation occurs in the use of bacteria to eliminate pollutants. One strain may be able to take up the pollutant without ill effect while another may be inhibited by it. One wants to know whether the inhibited strain can eliminate the uninhibited strain; if it can, then the pollutant cannot be removed. Intuitively, if the inhibited strain is the weaker competitor without the presence of the inhibiting agent, then one expects it to be eliminated. The focus is thus on the case where the superior competitor is the one sensitive to the inhibitor. The question remains the same: Does competitive exclusion still hold?

Taking the chemostat as the basic model, the inquiry focuses on determining the outcome as a function of the basic parameters of the organisms and the operating parameters of the chemostat. In the Michaelis–Menten formulation, the parameters of the organisms are the a_is and m_is of Chapter 1; the operating parameters of the chemostat are $S^{(0)}$ and D, the input concentration of nutrient and the washout rate. If the inhibitor is assumed to be input into the chemostat at a constant rate, then new parameters are introduced. One new operating parameter would be this constant input concentration. The uptake characteristic of the uninhibited organism and the effect of the inhibitor on the growth of the inhibited organism are the other new elements that can be expected to play key roles. The former will be treated in Michaelis–Menten form whereas the latter will be treated in a quite general fashion.

The basic model is that of two competitors in the chemostat, and the new element is the effect of the inhibitor. The model was first proposed by Lenski and Hattingh [LH], who investigated the various outcomes with a computer simulation. The mathematical analysis here follows that contained in Hsu and Waltman [HW1].

2. The Model

As noted in the introduction, the model is that of a standard chemostat with two competitors, but with the added feature that an inhibitor is also input from an external source. The nutrient (and inhibitor) uptake and conversion (in the case of nutrient) are assumed to follow Michaelis–Menten dynamics. The results are probably valid for general monotone dynamics, although this has not been established.

Let $S(t)$ denote the nutrient concentration at time t in the culture vessel, $x_1(t), x_2(t)$ the concentration of the competitors, and $p(t)$ the concentration of the inhibitor. The equations of the model take the form [LH]:

$$S' = (S^{(0)} - S)D - \frac{m_1 x_1 S}{a_1 + S} f(p) - \frac{m_2 x_2 S}{a_2 + S},$$

$$x_1' = x_1 \left(\frac{m_1 S}{a_1 + S} f(p) - D \right),$$

$$x_2' = x_2\left(\frac{m_2 S}{a_2 + s} - D\right), \tag{2.1}$$

$$p' = (p^{(0)} - p)D - \frac{\delta x_2 p}{K + p},$$

$$S(0) \geq 0, \quad x_i(0) > 0, \quad (i = 1, 2), \quad p(0) \geq 0,$$

where $S^{(0)}$ is the input concentration of the nutrient and $p^{(0)}$ that of the inhibitor, both of which are assumed constant; D is the dilution rate of the chemostat. The terms m_i, a_i $(i = 1, 2)$ are the usual parameters, the maximal growth rates of the competitors (without an inhibitor) and the Michaelis–Menten (or half-saturation) constants, respectively. The parameters δ and K play similar roles for the inhibitor, with δ the maximal uptake rate by x_2 and K a half-saturation parameter. The function $f(p)$ represents the degree of inhibition of p on the growth rate of x_1.

To reduce the number of parameters – and to provide a standard environment so that comparisons can be made in terms of the parameters of the competing populations – the equations will be scaled, much as was done in the first three chapters. First, scale the units of concentration of S, x_1, x_2, by the input concentration $S^{(0)}$. This includes the parameters $a_i, i = 1, 2$. (We have already tacitly scaled out the yield parameters, which scale the conversion of nutrient to organism.) Then scale time by the dilution rate (with units 1/time). This reduces D to unity and replaces m_i by m_i/D $(i = 1, 2)$ and δ by δ/D. Finally, scale p by $p^{(0)}$, which has the effect of scaling $p^{(0)}$ to unity. In [LH],

$$f(p) = e^{-\lambda p},$$

so this would now be written

$$f(p) = e^{-\lambda p^{(0)}(p/p^{(0)})}.$$

The new variable is $p/p^{(0)}$ and the new parameter is $\lambda p^{(0)}$. If one makes these changes and then returns to the "old" names (e.g., using m_1 as the new "maximal growth rate," the "old" m_1/D), the system (2.1) takes the form

$$S' = 1 - S - \frac{m_1 x_1 S}{a_1 + S} f(p) - \frac{m_2 x_2 S}{a_2 + S},$$

$$x_1' = x_1\left(\frac{m_1 S}{a_1 + S} f(p) - 1\right),$$

$$x_2' = x_2\left(\frac{m_2 S}{a_2+S}-1\right), \tag{2.2}$$

$$p' = 1-p-\frac{\delta x_2 p}{K+p},$$

$$S(0) \geq 0, \quad x_i(0) > 0 \;\; (i=1,2), \quad p(0) \geq 0.$$

The analysis of this set of nonlinear ordinary differential equations forms the content of this chapter.

Concerning the function $f(p)$, we shall assume that

(i) $f(p) \geq 0$, $f(0) = 1$;
(ii) $f'(p) < 0$, $p > 0$.

The function $f(p)$ adjusts the effective value of the parameter m_1; the quantity $m_1 f(p)$ represents the maximal growth rate of this population if the concentration of the inhibitor is p.

As noted in [LH], the ability of x_2 to consume the inhibitor ($\delta > 0$) is of crucial importance. Lenski and Hattingh refer to this ability of x_2 to "detoxify" the environment and note that without it (i.e., with $\delta = 0$ in (2.2)) $p(t)$ tends to unity as t tends to infinity. Therefore, the limiting system obtained by dropping the p equation and replacing p by 1 describes the dynamics of (2.2) on the omega limit set. This limiting system is just the equations for competition in the chemostat without an inhibitor and where m_1 is replaced by $m_1 f(1)$. Competitive exclusion must then result.

3. The Conservation Principle

The method of procedure is much like that in the previous chapters. The equations will be rearranged to make clear that the omega limit set lies in a smaller set, one governed by three ordinary differential equations. The analysis will then proceed on this reduced set of differential equations. Since every trajectory is asymptotic to its omega limit set, knowledge of the global behavior of this set of equations is essential for determining the asymptotic behavior of (2.2). (See Appendix F for the general result on returning to the asymptotic behavior of (2.2).)

Let

$$\Sigma = 1-S-x_1-x_2.$$

Then

$$\Sigma' = -S'-x_1'-x_2' = -1+S+x_1+x_2 = -\Sigma.$$

The system (2.2) may then be replaced by

$$\Sigma' = -\Sigma,$$

$$x_1' = x_1\left(\frac{m_1(1-\Sigma-x_1-x_2)}{a_1+1-\Sigma-x_1-x_2}f(p)-1\right),$$

$$x_2' = x_2\left(\frac{m_2(1-\Sigma-x_1-x_2)}{a_2+1-\Sigma-x_1-x_2}-1\right), \qquad (3.1)$$

$$p' = 1-p-\delta\frac{x_2 p}{K+p}.$$

Clearly, $\lim_{t\to\infty}\Sigma(t) = 0$. Hence, the solutions in the omega limit set of (3.1) must satisfy

$$x_1' = x_1\left(\frac{m_1(1-x_1-x_2)}{1+a_1-x_1-x_2}f(p)-1\right),$$

$$x_2' = x_2\left(\frac{m_2(1-x_1-x_2)}{1+a_2-x_1-x_2}-1\right), \qquad (3.2)$$

$$p' = 1-p-\delta\frac{x_2 p}{K+p},$$

$$x_i(0) > 0 \ (i=1,2), \quad x_1(0)+x_2(0) < 1, \quad p(0) \ge 0.$$

Let

$$\lambda_1 = \frac{a_1}{m_1-1}, \qquad \lambda_2 = \frac{a_2}{m_2-1}. \qquad (3.3)$$

These are the determining parameters for the simple chemostat, as discussed in Chapter 1, and would determine the outcome if the inhibitor p were not present. The form of the equations (3.2) guarantees that if $x_i(0) > 0 \ (i=1,2)$ then $x_i(t) > 0$ for $t > 0$. Moreover, $p'|_{p=0} = 1 > 0$, so if $p(0) \ge 0$ then $p(t) > 0$ for $t > 0$. The terms $x_1(t)$ and $x_2(t)$ satisfy

$$x_1' \le x_1\left(\frac{m_1(1-x_1-x_2)}{1+a_1-x_1-x_2}-1\right),$$

$$x_2' = x_2\left(\frac{m_2(1-x_1-x_2)}{1+a_2-x_1-x_2}-1\right). \qquad (3.4)$$

Taking advantage of the monotonicity (in the variable $1-x_1-x_2$) in the right-hand side of (3.4) yields a set of two scalar differential inequalities of the form

$$x_1' \le x_1\left(\frac{m_1(1-x_1)}{1+a_1-x_1}-1\right),$$

$$x_2' \le x_2\left(\frac{m_2(1-x_2)}{1+a_2-x_2}-1\right).$$

The comparison with the growth equations for the chemostat (equation (4.1) or (4.2) of Chapter 1) establishes the following proposition.

PROPOSITION 3.1. *If $m_i \le 1$ or if $m_i > 1$ and $\lambda_i \ge 1$, $\lim_{t\to\infty} x_i(t) = 0$. If $m_i > 1$ and $0 < \lambda_i < 1$ then $\limsup_{t\to\infty} x_i(t) \le 1-\lambda_i$ with $i = 1$ or 2.*

This simply states the biologically intuitive fact that if one of the competitors could not survive in the simple chemostat, that competitor will not survive in the chemostat with an inhibitor. Thus we may assume $m_i > 1$ and $0 < \lambda_i < 1$, $i = 1, 2$. There are some other simple cases that can also be easily dispatched.

PROPOSITION 3.2. *If $0 < \lambda_2 < \lambda_1 < 1$, then*

$$\lim_{t\to\infty} x_1(t) = 0,$$

$$\lim_{t\to\infty} x_2(t) = 1-\lambda_2,$$

$$\lim_{t\to\infty} p(t) = p_2^* < 1,$$

where p_2^* is the positive root of the quadratic

$$(1-p)(K+p)-\delta(1-\lambda_2)p = 0. \tag{3.5}$$

(The reason for labeling the root p_2^* will become clear in the next section.) Biologically, the proposition states that if x_2 eliminates x_1 in a chemostat without x_1 being inhibited, x_2 eliminates x_1 when x_1 is inhibited.

Proof of Proposition 3.2. That $p_2^* < 1$ follows from the fact that $p(t)$ satisfies

$$p' < 1-p$$

and the basic comparison theorem. The boundedness of p and the fact that Σ tends to zero also show that all solutions of (3.1) are bounded. The other assertions in the proof will follow from the more general comparison theorem (see Appendix B, Theorem B.4) and a knowledge of the behavior of solutions of the basic chemostat equations, system (5.2) of Chapter 1. Since "=" is also "≥," (3.4) may be written

$$x_1' \le x_1 \left(\frac{m_1(1-x_1-x_2)}{1+a_1-x_1-x_2} - 1 \right),$$

$$x_2' \ge x_2 \left(\frac{m_2(1-x_1-x_2)}{1+a_2-x_1-x_2} - 1 \right).$$

Let u_1 and u_2 be solutions of the system (5.2) in Chapter 1 satisfying $u_i(0) = x_i(0)$, $i = 1, 2$. It follows from Theorem B.4 that

$$x_1(t) \le u_1(t) \quad \text{and} \quad x_2(t) \ge u_2(t).$$

If $\lambda_2 < \lambda_1$, then $\lim_{t \to \infty} u_1(t) = 0$ and $\lim_{t \to \infty} u_2(t) = 1 - \lambda_2$ (by Theorem 5.1 of Chapter 1). Thus one can conclude that $\lim_{t \to \infty} x_1(t) = 0$ and $\liminf_{t \to \infty} x_2(t) \ge 1 - \lambda_2$ (Theorem 5.1 of Chapter 1). Since we already have (by Proposition 3.1) that $\limsup_{t \to \infty} x_2(t) \le 1 - \lambda_2$, the proof is complete. □

The hypothesis of Proposition 3.2 excludes the case $\lambda_1 = \lambda_2$. This is ordinarily not biologically important because the λ_i are computed from measured quantities; it is unlikely that they would be exactly the same (or the same with respect to this environment). However, an interesting potential application is the case where the organisms are indeed the same (mutants of the same organism) except for their sensitivity to the antibiotic. Intuitively, if the organisms are the same except for sensitivity to the inhibitor, one expects the x_1 population to lose the competition when the inhibitor is present. However, establishing this mathematically cannot be done directly from the comparison theorem as used before, since if $\lambda_1 = \lambda_2$ then coexistence occurs with the chemostat equations used for comparison purposes. In order to use the comparison principle, one needs a better estimate than $f(p) \le 1$, $p \ge 0$. This is the purpose of the following lemma.

LEMMA 3.3. *There exists a number $\gamma > 0$ such that $p(t) > \gamma$ for t sufficiently large.*

Proof. Suppose $\liminf_{t \to \infty} p(t) = 0$. If $p(t)$ decreased to zero monotonically then there would be a point t_0 such that, for $t > t_0$, $p(t) + \delta p(t)/(K+p(t)) < 1$. For such values, $p'(t) > 0$, which contradicts $p(t)$ decreasing. Hence there exists a set of points t_n, $t_n \to \infty$, such that $p'(t_n) = 0$ and $p(t_n) \to 0$ as $t_n \to \infty$. For such values of t_n,

$$0 = 1 - p(t_n) - \frac{\delta p(t_n) x_2(t_n)}{K + p(t_n)}$$

$$> 1 - p(t_n) - \frac{\delta p(t_n)}{K + p(t_n)}$$

$$> 0$$

for n large. This establishes the lemma.   □

THEOREM 3.4. *If* $0 < \lambda_2 \le \lambda_1 < 1$, *then*

$$\lim_{t \to \infty} x_1(t) = 0,$$

$$\lim_{t \to \infty} x_2(t) = x_2^* > 0,$$

$$\lim_{t \to \infty} p(t) = p_2^* > 0,$$

where x_2^* *and* p_2^* *are as before.*

Proof. In view of Lemma 3.3, the inequalities (3.4) can be replaced by

$$x_1' \le x_1 \left(\frac{m_1(1 - x_1 - x_2) f(\gamma)}{1 + a_1 - x_1 - x_2} - 1 \right),$$

$$x_2' \ge x_2 \left(\frac{m_2(1 - x_1 - x_2)}{1 + a_2 - x_1 - x_2} - 1 \right),$$

where γ is given by Lemma 3.3 and t is sufficiently large. This system of inequalities can be compared to the equations for the chemostat, where λ_1 is replaced by $a_1/(m_1 f(\gamma) - 1)$, so that the first component of the comparison system tends to zero as t tends to infinity, as in the proof of Proposition 3.2. Hence, so does $x_1(t)$.   □

Thus, for the remainder of this chapter, one may assume

$$m_i > 1, \quad i = 1, 2,$$

$$0 < \lambda_1 < \lambda_2 < 1. \tag{3.6}$$

The results in this section provide conditions for one or both of the competitors to wash out of the chemostat; these are the uninteresting cases. To avoid "unlikely" cases in the analysis to follow, we shall tacitly assume that all rest points are hyperbolic – that is, their stability is determined by their linearization.

It is easy to anticipate that oscillations in the concentrations of x_1, x_2, and p may be possible when $0 < \lambda_1 < \lambda_2 < 1$. If $p(0)$ is small then the system (3.2) behaves as if there were essentially no inhibitor. The superior competitor x_1 will begin to prevail over its rival x_2, driving its concentration to a low level. When x_2 becomes small it is no longer detoxifying the chemostat by consuming p, so the concentration of p increases toward $p = 1$. Once p attains an appreciable concentration, it begins to inhibit the growth of x_1. This may allow x_2 to recover and begin to detoxify the chemostat by consuming the inhibitor. Once again, we may arrive at the situation where p is small. Repeating this scenario results in the oscillation of all concentrations. Therefore, the possibility of coexistence in the form of oscillatory behavior is plausible. Section 7 will clarify the conditions for a specific $f(p)$.

4. Rest Points and Stability

There are three potential rest points of (3.2) on the boundary, which we label

$$E_0 = (0, 0, 1), \quad E_1 = (x_1^*, 0, 1), \quad E_2 = (0, x_2^*, p_2^*).$$

These correspond to one or both competitors becoming extinct.

The rest point E_0 always exists, and E_2 exists with $x_2^* = 1 - \lambda_2$ and p_2^* the root of (3.5) if $0 < \lambda_2 < 1$, which is contained in our basic assumption (3.6). The existence of E_1 is a bit more delicate. In keeping with the definitions in (3.3), define $\lambda_0 = a_1/(m_1 f(1) - 1)$. Then $0 < \lambda_0 < 1$ corresponds to the survivability of the first population in a chemostat under maximal levels of the inhibitor. Easy computations show that $E_1 = (1 - \lambda_0, 0, 1)$ will exist if $\lambda_0 > 0$ and will have positive coordinates and be asymptotically stable in the x_1-p plane if $0 < \lambda_0 < 1$. If $1 - \lambda_0$ is negative then E_1 is neither meaningful nor accessible from the given initial conditions, since the x_2-p plane is an invariant set. The stability of either E_1 or E_2 will depend on comparisons between the subscripted λs. The local stability of each rest point depends on the eigenvalues of the linearization around those points. The Jacobian matrix for the linearization of (3.2) at E_i, $i = 1, 2$, takes the form

$$J = \begin{pmatrix} m_{11} & m_{12} & m_{13} \\ m_{21} & m_{22} & 0 \\ 0 & m_{32} & m_{33} \end{pmatrix}. \tag{4.1}$$

At E_0,

$$J = \begin{pmatrix} \dfrac{m_1 f(1)}{1 + a_1} - 1 & 0 & 0 \\ 0 & \dfrac{m_2}{1 + a_2} - 1 & 0 \\ 0 & -\dfrac{\delta}{1 + K} & -1 \end{pmatrix}.$$

The eigenvalues are the diagonal elements. One eigenvalue is -1, and the associated eigenvector lies along the p axis. This corresponds to the growth of the inhibitor to its limiting value in the absence of a consumer. The set $\{(0, 0, p) \mid p > 0\}$ is positively invariant and is part of the stable manifold of E_0. Because $\lambda_2 < 1$, $m_{22} = m_2/(1+a_2) - 1$ is positive. Similarly, the remaining diagonal term, m_{11}, is positive if $0 < \lambda_0 < 1$ and negative otherwise. When this eigenvalue is negative, the stable manifold of E_0 is the entire x_1-p plane.

At E_1, $m_{21} = 0$, which means (since $m_{23} = m_{31} = 0$) that the eigenvalues are just the diagonal elements of J. Thus

$$\mu_1 = -\frac{m_1 a_1 (1 - \lambda_0)}{(a_1 + \lambda_0)^2} f(1),$$

$$\mu_2 = \frac{m_2 a_2 (\lambda_0 - \lambda_2)}{(a_2 + \lambda_0)(a_2 + \lambda_2)},$$

$$\mu_3 = -1.$$

If $0 < \lambda_0 < \lambda_2 < 1$, then E_1 is asymptotically stable. This reflects the fact that x_1, in the presence of the maximal inhibitor concentration, is still a better competitor than x_2. If $\lambda_0 > \lambda_2$ then E_1 is unstable and, of course, if $\lambda_0 > 1$ then E_1 does not exist.

At E_2, $m_{12} = m_{13} = m_{23} = 0$, so again the eigenvalues are just the diagonal elements

$$\mu_1 = \frac{m_1 \lambda_2 f(p_2^*)}{a_1 + \lambda_2} - 1,$$

$$\mu_2 = -\frac{m_2 a_2 (1 - \lambda_2)}{(a_2 + \lambda_2)^2},$$

$$\mu_3 = -1 - \frac{\delta K (1 - \lambda_2)}{(K + p_2^*)^2}.$$

Clearly, μ_2 and μ_3 are negative so that E_2 always has a two-dimensional stable manifold; E_2 is asymptotically stable (unstable) if $\mu_1 < 0$ ($\mu_1 > 0$).

The local behavior of the rest point set on the boundary is summarized in Table 4.1, where $0 < \lambda_1 < \lambda_2 < 1$ is assumed. The more interesting case is that of an interior rest point. Let $E_c = (x_{1c}^*, x_{2c}^*, p_c^*)$ denote the coordinates of a possible interior rest point. First of all, it must be the case that

$$1 - x_{1c}^* - x_{2c}^* = \lambda_2, \tag{4.2}$$

for this is the only nontrivial zero of the derivative of x_2 in (3.2). Using this, one sets the derivative of x_1 equal to zero to find

Table 4.1

Point	Existence	Stability
E_0	Always	1- or 2-dimensional stable manifold
E_1	$0 < \lambda_0 < 1$	Asymptotically stable if $0 < \lambda_0 < \lambda_2$
E_2	Always	Asymptotically stable if $m_1\lambda_2 f(p_2^*) < a_1 + \lambda_2$

$$\frac{m_1\lambda_2}{a_1+\lambda_2} f(p) = 1$$

or that one needs $(a_1+\lambda_2)/m_1\lambda_2$ to be in the range of f. The inequality $\lambda_1 < \lambda_2$ implies $(a_1+\lambda_2)/m_1\lambda_2 < 1$. If

$$\frac{a_1+\lambda_2}{m_1\lambda_2} > \lim_{p \to \infty} f(p)$$

then $(a_1+\lambda_2)/m_1\lambda_2$ will be in the range of f. If this inequality holds, set

$$p_c^* = f^{-1}\left(\frac{a_1+\lambda_2}{m_1\lambda_2}\right). \tag{4.3}$$

Since f is monotone, if p_c^* exists then it is unique. If $p_c^* < 1$ then x_{2c}^* can be determined by setting $p'(t)$ equal to zero, yielding

$$1 - p_c^* - \frac{\delta x_{2c}^* p_c^*}{K + p_c^*} = 0$$

or

$$x_{2c}^* = \frac{(1-p_c^*)(K+p_c^*)}{\delta p_c^*}. \tag{4.4}$$

Hence, we see that $p_c^* < 1$ is a necessary condition for E_c to exist. The coordinate x_{2c}^* is unique because p_c^* is unique. If $x_{2c}^* < 1 - \lambda_2$, then x_{1c}^* is uniquely determined from (4.2) as

$$x_{1c}^* = 1 - x_{2c}^* - \lambda_2. \tag{4.5}$$

Thus E_c exists if and only if $f(1) \leq (a_1+\lambda_2)/m_1\lambda_2$ and $x_{2c}^* < 1-\lambda_2$. Since $1-\lambda_2 = x_2^*$, it follows that if x_{2c}^* exists then $x_{2c}^* < x_2^*$. This last inequality is the biologically expected statement that x_2 will do less well in the coexistent steady state than in the steady state where it is the sole survivor. The inequality is true if and only if (see (3.5))

$$x_{2c}^* = \frac{(1-p_c^*)(K+p_c^*)}{\delta p_c^*} < \frac{(1-p_2^*)(K+p_2^*)}{\delta p_2^*} = 1-\lambda_2 = x_2^*.$$

In view of the monotonicity of this expression in p, this occurs if and only if $p_2^* < p_c^*$. From (4.3) one has that the preceding inequality is equivalent to

$$f(p_2^*) > \frac{a_1+\lambda_2}{m_1\lambda_2} \tag{4.6}$$

or to the instability of E_2 if E_2 is nondegenerate. This establishes that a necessary condition for the existence of an interior equilibrium for (3.2) is that E_2 is unstable. The next proposition improves on this result.

PROPOSITION 4.1. *The rest point E_c exists (and is positive) if and only if (i) the Jacobian matrix at E_2 has a positive eigenvalue and (ii) if E_1 exists, then the Jacobian matrix at E_1 has a positive eigenvalue.*

Proof. To see the sufficiency of the conditions, note that if the Jacobian of E_2 has a positive eigenvalue then

$$f(p_2^*) > \frac{a_1+\lambda_2}{m_1\lambda_2}.$$

If E_1 exists and its Jacobian has a positive eigenvalue, then $\lambda_2 < \lambda_0$. Consequently, one has $f(1) < (a_1+\lambda_2)/m_1\lambda_2$. It is easily seen from the monotonicity of $\lambda/(a_1+\lambda)$ that this inequality also holds if E_1 does not exist. There are two cases. If $\lambda_0 < 0$ then

$$f(1) < m_1 f(1) < 1 = \frac{a_1+\lambda_1}{m_1\lambda_1} < \frac{a_1+\lambda_2}{m_1\lambda_2}.$$

Similarly, if $\lambda_0 > 1$ then

$$f(1) < \frac{a_1+1}{m_1} < 1 = \frac{a_1+\lambda_1}{m_1\lambda_1} < \frac{a_1+\lambda_2}{m_1\lambda_2}.$$

Thus the inequality holds even if E_1 does not exist.
From

$$f(1) < \frac{a_1+\lambda_2}{m_1\lambda_2} < f(p_2^*)$$

one can conclude from the mean value theorem that there exists a unique number p_c^* with $f(p_c^*) = (a_1+\lambda_2)/m_1\lambda_2$ and $p_2^* < p_c^* < 1$.

Since

$$1 = p_2^* + \frac{\delta x_2^* p_2^*}{K + p_2^*},$$

it follows from (4.4) that $x_{2c}^* < x_2^* = 1 + \lambda_2$. Therefore E_c exists and is positive.

Conversely, if E_c exists then $p_2^* < p_c^* < 1$ and $x_{2c}^* < x_2^*$, so

$$f(p_2^*) > f(p_c^*) = \frac{a_1 + \lambda_2}{m_1 \lambda_2}.$$

Hence, the Jacobian evaluated at E_2 has a positive eigenvalue (denoted by μ_1 in the previous discussion of the eigenvalues of this Jacobian). Since $f(p_2^*) > f(p_c^*) > f(1)$,

$$\frac{m_1 \lambda_2 f(1)}{a_1 + \lambda_2} < \frac{m_1 \lambda_2 f(p_c^*)}{a_1 + \lambda_2} = 1.$$

If E_1 exists $(0 < \lambda_0 < 1)$, then $\lambda_2 < \lambda_0$. Therefore the Jacobian at E_1 has a positive eigenvalue (denoted by μ_2 in the previous discussion of the eigenvalues of this Jacobian). This completes the proof of the proposition.
□

There remains the question of the stability of E_c. The matrix J in (4.1) takes the form

$$J = \begin{bmatrix} -\dfrac{m_1 a_1}{(a_1 + \lambda_2)^2} f(p_c^*) x_{1c}^* & -\dfrac{m_1 a_1}{(a_1 + \lambda_2)^2} f(p_c^*) x_{1c}^* & \dfrac{m_1 \lambda_2}{a_1 + \lambda_2} x_{1c}^* f'(p_c^*) \\[3ex] -\dfrac{m_2 a_2}{(a_2 + \lambda_2)^2} x_{2c}^* & -\dfrac{m_2 a_2}{(a_2 + \lambda_2)^2} x_{2c}^* & 0 \\[3ex] 0 & -\dfrac{\delta p_c^*}{K + p_c^*} & -1 - \dfrac{\delta K x_{2c}^*}{(K + p_c^*)^2} \end{bmatrix}.$$

By expanding the determinant of J in elements of the last row, one sees that it is negative; thus the dimension of the stable manifold is 1 or 3.

The characteristic roots of J satisfy

$$\mu^3 + \mu^2 \left(1 + \frac{\delta K x_{2c}^*}{(K + p_c^*)^2} + \frac{a_1 x_{1c}^*}{(a_1 + \lambda_2)\lambda_2} + \frac{a_2 x_{2c}^*}{(a_2 + \lambda_2)\lambda_2} \right)$$

$$+ \mu \left(1 + \frac{\delta K x_{2c}^*}{(K + p_c^*)^2} \right) \left(\frac{a_1 x_{1c}^*}{(a_1 + \lambda_2)\lambda_2} + \frac{a_2 x_{2c}^*}{(a_2 + \lambda_2)\lambda_2} \right)$$

$$- \frac{f'(p_c^*)}{f(p_c^*)} \frac{a_2}{(a_2 + \lambda_2)\lambda_2} \frac{\delta p_c^*}{K + p_c^*} x_{1c}^* x_{2c}^* = 0. \qquad (4.7)$$

Since $f'(p) < 0$, the constant term is positive, so the Routh–Hurwitz criterion (Appendix A) says that E_c will be asymptotically stable if and only if

$$\left(1 + \frac{\delta K x_{2c}^*}{(K + p_c^*)^2} + \frac{a_1 x_{1c}^*}{(a_1 + \lambda_2)\lambda_2} + \frac{a_2 x_{2c}^*}{(a_2 + \lambda_2)\lambda_2}\right)$$

$$\times \left(1 + \frac{\delta K x_{2c}^*}{(K + p_c^*)^2}\right)\left(\frac{a_1 x_{1c}^*}{(a_1 + \lambda_2)\lambda_2} + \frac{a_2 x_{2c}^*}{(a_2 + \lambda_2)\lambda_2}\right)$$

$$> -\frac{f'(p_c^*)}{f(p_c^*)} \frac{a_2}{(a_2 + \lambda_2)\lambda_2} \frac{\delta p_c^*}{K + p_c^*} x_{1c}^* x_{2c}^*. \tag{4.8}$$

This condition is obviously difficult to verify in general. It will be used later for the special case $f(p) = e^{-\eta p}$.

5. Competition without an Interior Equilibrium

With our standing hypotheses that all rest points are hyperbolic, Proposition 4.1 implies that if E_c does not exist then either E_2 is asymptotically stable (all eigenvalues of the Jacobian are negative) or E_1 exists and is asymptotically stable (all eigenvalues are negative). The results to follow establish that if E_i, $i = 1, 2$, is asymptotically stable then it attracts all solutions (is globally asymptotically stable for positive initial conditions). Therefore, when E_c does not exist, one of the rest points E_1 or E_2 attracts all solutions of (3.2).

THEOREM 5.1. *If* $0 < \lambda_0 < \lambda_2$, *then* E_1 *attracts all solutions of* (3.2) *with* $x_1(0) > 0$; *that is,*

$$\lim_{t \to \infty} x_1(t) = x_1^*,$$

$$\lim_{t \to \infty} x_2(t) = 0,$$

$$\lim_{t \to \infty} p(t) = 1.$$

Proof. Since

$$0 < \frac{a_1}{m_1 f(1) - 1} < \lambda_2,$$

there exists an $\epsilon > 0$ such that

$$0 < \frac{a_1}{m_1 f(1 + \epsilon) - 1} < \lambda_2.$$

Since $p(t) < 1 + \epsilon$ for all large t, for such t one has that

$$x_1' \geq x_1 \left(\frac{m_1(1-x_1-x_2)}{1+a_1-x_1-x_2} f(1+\epsilon)-1 \right),$$

$$x_2' = x_2 \left(\frac{m_2(1-x_1-x_2)}{1+a_2-x_1-x_2} -1 \right).$$

By Theorem B.4, $x_1(t)$ and $x_2(t)$ can be compared to the solutions of the chemostat system (5.2) of Chapter 1. (See the proof of Proposition 3.2 with the inequalities reversed.) The break-even concentrations are $a_1/(m_1 f(1+\epsilon)-1)$ and λ_2. From the comparison, one may conclude that $\lim_{t \to \infty} x_2(t) = 0$ and

$$\liminf_{t \to \infty} x_1(t) \geq 1 - \frac{a_1}{m_1 f(1+\epsilon)-1}.$$

Since ϵ may be chosen arbitrarily small, it follows that

$$\liminf_{t \to \infty} x_1(t) \geq 1 - \lambda_0.$$

Using $\lim_{t \to \infty} x_2(t) = 0$ in the equation for p, it is easy to see that

$$\lim_{t \to \infty} p(t) = 1.$$

Then, for ϵ small,

$$x_1' \leq x_1 \left(\frac{m_1(1-x_1)}{1+a_1-x_1} f(1-\epsilon)-1 \right)$$

holds for all large t (where t depends on ϵ). By comparison with the growth equations (4.1) of Chapter 1, it follows that

$$\limsup_{t \to \infty} x_1(t) \leq 1 - \frac{a_1}{m_1 f(1-\epsilon)-1}.$$

Since ϵ may be chosen arbitrarily small,

$$\limsup_{t \to \infty} x_1(t) \leq 1 - \lambda_0.$$

This completes the proof. $\qquad\square$

THEOREM 5.2. *If* $m_1 \lambda_2 f(p_2^*)/(a_1+\lambda_2) < 1$, *then* E_2 *attracts all solutions of* (3.2) *with* $x_2(0) > 0$; *that is,*

$$\lim_{t \to \infty} x_1(t) = 0,$$

$$\lim_{t \to \infty} x_2(t) = x_2^*,$$

$$\lim_{t \to \infty} p(t) = p_2^*.$$

Proof. For $\epsilon > 0$, let $p = p^*(\epsilon)$ be the root of

$$1 - p - \frac{\delta(1 - \lambda_2 + \epsilon)p}{K + p} = 0.$$

Clearly,

$$\lim_{\epsilon \to 0} p^*(\epsilon) = p^*(0) = p_2^*.$$

Since $\limsup_{t \to \infty} x_2(t) \leq 1 - \lambda_2$ (by Proposition 3.1), for each $\epsilon > 0$ there exists $T_1(\epsilon) > 0$ such that $x_2(t) \leq 1 - \lambda_2 + \epsilon$ for $t > T_1(\epsilon)$. It follows that

$$p' \geq 1 - p - \frac{\delta(1 - \lambda_2 + \epsilon)p}{K + p},$$

and consequently there exists $T_2(\epsilon) \geq T_1(\epsilon)$ such that $p(t) \geq p^*(\epsilon) - \epsilon$ for $t \geq T_2(\epsilon)$. One therefore has that

$$x_1' \leq x_1\left(\frac{m_1(1 - x_1 - x_2)}{1 + a_1 - x_1 - x_2} f(p^*(\epsilon) - \epsilon) - 1\right)$$

$$\leq x_1\left(\frac{m_1(1 - x_1)}{1 + a_1 - x_1} f(p^*(\epsilon) - \epsilon) - 1\right)$$

and

$$x_2' = x_2\left(\frac{m_2(1 - x_1 - x_2)}{1 + a_2 - x_1 - x_2} - 1\right)$$

for $t \geq T_2(\epsilon)$. If $m_1 f(p^*(\epsilon) - \epsilon) \leq 1$, then the second differential inequality for x_1 implies that

$$\lim_{t \to \infty} x_1(t) = 0.$$

If $m_1 f(p_2^*) > 1$, then $m_1 f(p^*(\epsilon) - \epsilon) > 1$ for sufficiently small ϵ. Since the hypotheses imply that $\lambda_2 < a_1/(m_1 f(p_2^*) - 1)$, it follows that $\lambda_2 < a_1/(m_1 f(p^*(\epsilon) - \epsilon) - 1)$. Using the first comparison for x_1, the equations for x_2, and Theorem B.4, one can again conclude that $\lim_{t \to \infty} x_1(t) = 0$. The remainder of the proof is similar to that of the previous theorem.

□

6. Three-Dimensional Competitive Systems

We interrupt the analysis of the inhibitor model in order to present some theorems on competitive systems which are needed in the analysis. Consider the system

$$x' = f(x), \quad x \in \mathbb{R}^3, \tag{6.1}$$

where one assumes that

$$\frac{\partial f_i}{\partial x_j} \leq 0, \quad i \neq j. \tag{6.2}$$

Such systems are said to be *competitive* (as noted in Chapter 1). When (6.1) represents a population growth equation, (6.2) indicates that an increase in the size of one component inhibits the growth of the others. Such a system is not necessarily order-preserving, so the theory of monotone systems does not apply. However, if solutions exist for all time and if one runs time "backwards" (more correctly, if one makes the change of variables $t = -\tau$ and regards τ as "time"), then the corresponding dynamical system is monotone. More formally, the system

$$x' = -f(x) \tag{6.3}$$

is monotone and the results of Appendix C apply. Thus the limit properties in the theory of monotone dynamical systems (discussed in Appendix C) apply to the alpha limit sets, not the omega limit sets, of competitive systems. Observe that periodic orbits are both alpha and omega limit sets. The results on monotone systems may be reinterpreted for (6.1) satisfying (6.2).

REMARK 6.1. If two points are ordered at time t, then the trajectories through them have been ordered for all previous time. That is, $x \leq y$ implies that $\pi(x, t) \leq \pi(y, t)$ for all $t < 0$.

An alternative statement is as follows.

REMARK 6.2. If two points are unordered at time t, then the trajectories through them remain unordered for all future time. That is, x not comparable to y implies that $\pi(x, t)$ is not comparable to $\pi(y, t)$ for all $t > 0$.

In Remark 6.1 we have used $\leq$ rather than $\leq_K$ because K is the usual positive cone. A consequence of these remarks is that planar cooperative or competitive systems do not have periodic orbits. For example, consider a planar cooperative system and let P be an arbitrary point on a periodic orbit. Impose the standard two-dimensional coordinate system at P. The orbit cannot be tangent to both the x and the y axes at P and so must have points in both quadrants II and IV (the sets unordered with respect to P), since points in quadrant I or III would be ordered. The orbit cannot pass through P again and cannot have points in quadrants I and III, and so is not a closed orbit. This argument fails in three dimensions, which can support periodic orbits for competitive systems.

A major difference between competitive and cooperative systems is that cycles may occur as attractors in competitive systems. However, three-dimensional systems behave like two-dimensional general autonomous equations in that the possible omega limit sets are similarly restricted. Two important results are given next. These allow the Poincaré–Bendixson conclusions to be used in determining asymptotic behavior of three-dimensional competitive systems in the same manner used previously for two-dimensional autonomous systems. The following theorem of Hirsch is our Theorem C.7 (see Appendix C, where it is stated for cooperative systems).

THEOREM 6.3 [Hi4]. *Let L be a compact omega limit set of a competitive system in* $\mathbb{R}^3$. *If L contains no equilibria, then L is a closed orbit.*

The system (6.1) is said to be *competitive and irreducible* if it is competitive and if the Jacobian of $f(x)$ in (6.1) is an irreducible matrix (see Appendix A) for all x.

PROPOSITION 6.4. *Let* x_0 *be a hyperbolic rest point of a continuously differentiable, competitive and irreducible dynamical system generated by* (6.1) *in* $\mathbb{R}^3$. *Suppose that* $M^+(x_0)$ *is one-dimensional and* $M^-(x_0)$ *is two-dimensional. Then:*

(1) $M^-(x_0)$ *is unordered; that is, if* $x_i \in M^-(x_0)$ *for* $i = 1, 2$ *and* $x_1 \leq x_2$ *or* $x_2 \leq x_1$, *then* $x_1 = x_2$.
(2) $M^+(x_0)$ *is totally ordered by* $<$; *that is, if* x_1 *and* x_2 *are distinct points of* $M^+(x_0)$, *then either* $x_1 < x_2$ *or* $x_2 < x_1$.

Proof. Assertion (1) is just Theorem C.4. The assertion concerning $M^+(x_0)$ follows from the Perron–Frobenius theory (Theorem A.5) and the monotonicity of the time-reversed system (6.3). If J is the Jacobian matrix of f at x_0, then (6.2) implies that $-J$ satisfies the hypotheses of Theorem A.5. It follows that $r = -s(-J) < 0$ is an eigenvalue of J corresponding to an eigenvector $v > 0$. Because $M^+(x_0)$ is tangent at x_0 to the line through x_0 in the direction v, the local stable manifold of x_0 is totally ordered. Since $M^+(x_0)$ is the extension of the local stable manifold by the order-preserving "backward" (or time-reversed) system, it follows that $M^+(x_0)$ is totally ordered. $\square$

7. Competition with an Interior Equilibrium

By Proposition 4.1, if E_c exists then E_2 is unstable and E_1, if it exists, is also unstable. We begin by showing that if E_c exists, the omega limit set of every solution for which $x_i(0) > 0$ ($i = 1, 2$) remains a positive distance away from the boundary of $\mathbb{R}^3_+$; in the terminology of Appendix D, the system (3.2) is *persistent*.

THEOREM 7.1. *Let E_c exist and let $(x_1(t), x_2(t), p(t))$ be a solution of (3.2) with $x_i(0) > 0$, $i = 1, 2$. Then*

$$\liminf_{t \to \infty} x_1(t) > 0 \quad and \quad \liminf_{t \to \infty} x_2(t) > 0.$$

The omega limit set of any trajectory lies interior to the positive cone.

Proof. Note that $M^+(E_0)$, the stable manifold of E_0, is either the p axis if E_1 exists or the x_1-p plane if E_1 does not exist. The manifold $M^+(E_2)$ is the x_2-p plane less the p axis; if E_1 exists, $M^+(E_1)$ is the x_1-p plane less the p axis. Since $(x_1(0), x_2(0), p(0))$ does not belong to any of these stable manifolds, its omega limit set (denoted by ω) cannot be any of the three rest points. Moreover, ω cannot contain any of these rest points by the Butler–McGehee theorem (see Chapter 1). (By arguments that we have used several times before, if ω did then it would have to contain E_0 or an unbounded orbit.) If ω contains a point of the boundary of $\mathbb{R}^3_+$ then, by the invariance of ω, it must contain one of the rest points E_0, E_1, E_2 or an unbounded trajectory. Since none of these alternatives are possible, ω must lie in the interior of the positive cone. This completes the proof. $\square$

One can work a little harder, using the full theory of Appendix D, and conclude uniform persistence. However, uniform persistence will not be needed in this chapter.

Theorem 7.1 guarantees the coexistence of both the x_1 and x_2 populations when E_c exists. However, it does not give the global asymptotic behavior. The further analysis of the system is complicated by the possibility of multiple limit cycles. Since this is a common difficulty in general two-dimensional systems, it is not surprising that such difficulties occur in the analysis of three-dimensional competitive systems.

THEOREM 7.2. *Suppose that system (3.2) has no limit cycles. Then E_c is globally asymptotically stable.*

Proof. In view of Theorem 7.1, the omega limit set of any trajectory cannot be on the boundary $x_1 = 0$ or $x_2 = 0$. Away from the boundary, the system is irreducible. Since there are no limit cycles, all trajectories must tend to E_c, by Theorem 6.3.   □

THEOREM 7.3. *Suppose that E_c is unstable and hyperbolic, and let $(x_1(t)$, $x_2(t), p(t))$ be a solution of (3.2) satisfying $x_i(0) > 0$, $i = 1, 2$. If $q = (x_1(0), x_2(0), p(0)) \notin M^+(E_c)$, then $\omega(q)$ is a nonconstant periodic orbit.*

Proof. Theorem 7.1 guarantees that none of the boundary rest points belong to its limit set, $\omega(q)$. It must be shown that E_c does not belong to $\omega(q)$. The result will then follow from Theorem 6.3.

Suppose $E_c \in \omega(q)$. Since $q \notin M^+(E_c)$, it follows that $\omega(q) \neq E_c$. The Butler–McGehee theorem implies that $\omega(q)$ contains a point r of $M^+(E_c)$ distinct from E_c. By Proposition 6.4, either $r > E_c$ or $r < E_c$. In either case, $\omega(q)$ contains the two ordered points E_c and r, contradicting the fact that $\omega(q)$ is unordered (Theorem C.5). This contradiction establishes the theorem.   □

To show the instability of E_c, one must show that the inequality (4.8) is violated. This, however, is a very formidable task without further information about the function $f(p)$, which gives the effect of the inhibitor. In the next remark, a specific function is chosen, the function used originally by Lenski and Hattingh [LH]. The proof then reduces to the application of the Routh–Hurwitz condition, still a computationally formidable task. However, the point of interest is deriving any oscillation at all, given the tendency of the chemostat to approach a steady state.

REMARK 7.4. Let $f(p) = e^{-\eta p}$ in (3.2). Then for η sufficiently large there exists a $\delta_0 > 0$ and a K_0 such that, for $\delta > \delta_0$ and $K < K_0$, (3.2) has a limit cycle.

Proof. As noted before, the remark follows if E_c is unstable. To show this, one must show that (4.8) is violated. In the case under consideration, $-f'(p)/f(p) = \eta$; this simplifies the computation. Define c by

$$c = \ln\left(\frac{m_1\lambda_2}{a_1 + \lambda_2}\right)$$

and note that $c = \eta p_c$. It follows that

$$1 - p_c^* = 1 - \frac{c}{\eta} = \frac{(\eta - c)}{\eta}.$$

Note that when η is fixed, p_c is fixed. From the definition of x_{2c}^* (see (4.4)), it follows that

$$\frac{\delta K x_{2c}^*}{(K + p_c^*)^2} = \frac{K}{K + p_c^*} \frac{1 - p_c^*}{p_c^*} = \left(\frac{K}{K + p_c^*}\right)\left(\frac{\eta}{c} - 1\right) \qquad (7.1)$$

for any choice of η and the corresponding p_c. Fix η satisfying

$$\eta > c + 2\left(3 + \frac{a_1(1 - \lambda_2)}{(a_1 + \lambda_2)\lambda_2}\right)\left(\frac{a_1(1 - \lambda_2)}{(a_1 + \lambda_2)\lambda_2} + 1\right)\left(\frac{2(a_2 + \lambda_2)\lambda_2}{(1 - \lambda_2)a_2}\right) \qquad (7.2)$$

and

$$\frac{m_1}{a_1 + 1} e^{-\eta} < 1.$$

Let K_0 be so small that the expression in (7.1) is less than unity. To show that (4.8) is violated, we estimate both sides. Note that, since $x_{1c}^* = 1 + \lambda_2 - x_{2c}^*$, $\lim_{\delta \to \infty} x_{2c}^* = 0$ and $\lim_{\delta \to \infty} x_{1c}^* = 1 + \lambda_2$. Hence, for δ sufficiently large,

$$\frac{a_2 x_{2c}^*}{(a_2 + \lambda_2)\lambda_2} < 1 \quad \text{and} \quad 1 - \lambda_2 > x_{1c}^* > \frac{1 - \lambda_2}{2}.$$

The right-hand side of (4.8) is bounded below by (using (3.1))

$$\frac{a_2}{2(a_2 + \lambda_2)\lambda_2}(1 - \lambda_2)(\eta - c).$$

The left-hand side of (4.8) has three factors, which we denote by F_1, F_2, F_3 in the order given. By the discussion so far, for δ sufficiently large we have

$$F_1 < \left(3 + \frac{a_1(1 - \lambda_2)}{(a_1 + \lambda_2)\lambda_2}\right),$$

$$F_2 < 2,$$

$$F_3 < \frac{a_1(1 - \lambda_2)}{(a_1 + \lambda_2)\lambda_2} + 1.$$

Using (7.2), it follows that

$$F_1 F_2 F_3 < \frac{(\eta - c)a_2}{(a_2 + \lambda_2)\lambda_2} \frac{(1 - \lambda_2)}{2},$$

which contradicts (4.8). Hence, E_c is unstable and the conclusion of Remark 7.4 follows from Theorem 7.3, completing the proof. $\qquad \square$

If, in addition, the vector field is analytic, then it can be shown (see [ZS]) that there must be an attracting periodic orbit.

8. Discussion

The material in this chapter makes several interesting points. First of all, the need for the more sophisticated tools is clear; without them, there can be no global results. It also illustrates that three- and four-dimensional systems can be analyzed. Better mathematical tools might make even more complex systems tractable. Although coexistence can occur either as an attracting rest point or as a stable periodic orbit, the interesting case is the oscillatory one. The time course is shown in Figure 8.1 for a sample problem; Figure 8.2 shows a three-dimensional phase plot of the limit cycle. Experiments involving the chemostat with an inhibitor should be no more difficult than those of ordinary chemostat experiments (which, however, are not easy). The oscillatory case suggests an interesting experiment, which is hinted at in the work of Hansen and Hubbell [HH] discussed in Chapter 1: they used an inhibitor (in a very limited way) as an alternate explanation of the coexistence case. A definitive experiment remains to be done to see if the oscillations can occur for parameters in a realistic region.

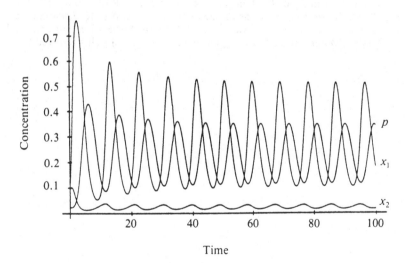

Figure 8.1. Plot of 100 time steps in the case of oscillatory coexistence. Parameters are $a_1 = 0.5$, $a_2 = 3.4$, $m_1 = 5.0$, $m_2 = 6.0$, $K = 0.1$, $\eta = 5.0$. (From [HW1], reprinted with permission from the *SIAM Journal on Applied Mathematics*, volume 52, number 2, pp. 528–40. Copyright 1992 by the Society for Industrial and Applied Mathematics, Philadelphia, Pennsylvania. All rights reserved.)

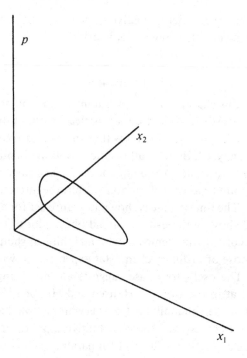

Figure 8.2. Plot in E^3 of the limit cycle given in Figure 8.1. (From [HW1], reprinted with permission from the *SIAM Journal on Applied Mathematics,* volume 52, number 2, pp. 528–40. Copyright 1992 by the Society for Industrial and Applied Mathematics, Philadelphia, Pennsylvania. All rights reserved.)

5

The Simple Gradostat

1. Introduction

In the preceding chapter we saw how the chemostat could be modified to account for a new phenomenon – the presence of an inhibitor. In this chapter we extend the idea behind the simple chemostat to a new apparatus in order to model a property of ecological systems that it is not possible to model in the simple chemostat. The idea is to capture the essentials of the new phenomenon without destroying the tractability of the chemostat either as a mathematical model or as an experimental one. A very simple situation will be described here; a more complicated one – with a less explicit (in the sense of less computable) analysis – will be discussed in the next chapter. Just as the chemostat is a basic model for competition in the simplest situation, the apparatus here shows promise of being the model for competition along a nutrient gradient.

The "well mixed" hypothesis for the chemostat does not allow a nutrient gradient to be generated. A basic tenet is that the nutrient concentration is the same everywhere; hence any advantage in nutrient consumption is present everywhere. The model that incorporates a true gradient would be one involving partial differential equations; a new variable, space, must be accommodated. Systems of nonlinear partial differential equations are difficult mathematical objects to understand and analyze. Even numerical solutions pose added and significant difficulties. Moreover, even if an experimental gradient is devised, measurements that do not disturb the local environment take on new difficulties.

A piece of laboratory apparatus was devised by Lovitt and Wimpenny [LW1; LW2; WL] for experiments along a nutrient gradient. It is a concatenation of chemostats in which the adjacent vessels are connected in both directions. Output occurs at the first and last vessels, and those in between exchange their contents – nutrient and organisms. The flow rates in, out, and between vessels are constant and equal. The apparatus was named a *gradostat*. It does not occur in nature, at least in this form. Indeed, although we shall think of the apparatus as connected horizontally, the closest approximation in nature may be vertically, as in a water column. In Chapter 6, we shall see that much more imaginative connection patterns are possible. Growth along nutrient gradients does occur in abundance in nature. For example, the surface films in dental plaque represent growth along such a gradient, as does growth along the banks of a stream or along a seacoast. In a water column, sunlight replaces the nutrient as an essential source needed for growth.

In a loose sense the apparatus generates a "discrete" gradient; the nutrient concentrations will vary from vessel to vessel, so the "parameters" of competition change from vessel to vessel. If there is no consumption, the nutrient concentrations arrange themselves as discrete points along a linear gradient. The effect of a nutrient gradient on growth and competition can be studied with such a device.

Figure 1.1 shows the device used by Lovitt and Wimpenny, and Figure 1.2 is a schematic of the device to be analyzed in this chapter. The mathematical analysis is restricted to two vessels, two competitors, and one

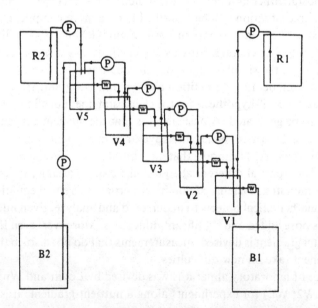

Figure 1.1. The device of Lovitt and Wimpenny. (From [LW2], Copyright 1981, Society for General Microbiology. Reproduced by permission.)

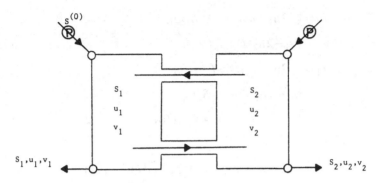

Figure 1.2. A schematic of a two-vessel gradostat. (From [JSTW], Copyright 1987, *Journal of Mathematical Biology*. Reproduced by permission.)

nutrient. The analysis in this special case takes a particularly elegant form, the relevant conditions being expressed as a set of basic inequalities. As we shall see in the next chapter, much more general formulations can cover many more interesting situations (more interesting gradients), but at the expense of a more complex analysis. The analysis will make extensive use of the theory of monotone systems discussed in Appendix C. The presentation closely follows that of [JSTW]. Since there are only two vessels, the model can be completely analyzed.

To keep the volume in each vessel constant, it is necessary to have medium (without nutrient) input at the right-hand end (as shown in Figure 1.2).

2. The Model

The competitors will be labeled as u and v with a subscript denoting the vessel: u_1 is the first competitor in the first vessel, v_1 the other competitor in the first vessel, and so on. The nutrient concentration in each vessel is also labeled this way. The equations will differ from that of the basic chemostat only in the flow of nutrient between vessels; for example, there are two outputs from each vessel, one out of the system and one into the adjoining vessel. If there were more vessels in the gradostat, then the middle ones would have two connections as well – that is, connections to the two adjoining vessels.

The equations take the form

$$S_1' = (S^{(0)} - 2S_1 + S_2)D - f_u(S_1)u_1/\gamma_u - f_v(S_1)v_1/\gamma_v,$$
$$S_2' = (S_1 - 2S_2)D - f_u(S_2)u_2/\gamma_u - f_v(S_2)v_2/\gamma_v,$$

$$u_1' = (-2u_1 + u_2)D + f_u(S_1)u_1,$$

$$u_2' = (u_1 - 2u_2)D + f_u(S_2)u_2, \tag{2.1}$$

$$v_1' = (-2v_1 + v_2)D + f_v(S_1)v_1,$$

$$v_2' = (v_1 - 2v_2)D + f_v(S_2)v_2,$$

$$S_i(0) \geq 0, \ u_i(0) \geq 0, \ v_i(0) \geq 0, \ i = 1, 2.$$

The functions

$$f_u(S_i) = \frac{m_u S_i}{a_u + S_i}, \quad f_v(S_i) = \frac{m_v S_i}{a_v + S_i}, \quad i = 1, 2,$$

are the usual Michaelis–Menten response terms. The constants m_u and m_v are the maximal growth rates for the u and v populations, while a_u and a_v are the corresponding Michaelis–Menten (or half-saturation) constants and γ_u and γ_v are the conversion factors. The term $S^{(0)}$ denotes the input concentration of the limiting nutrient, and D is called the washout rate. These terms are used exactly as in Chapter 1. As noted there, the last two quantities are under the control of the experimenter, while m, a, and γ depend on the population being cultured. Since the focus is on competition, it is reasonable to assume that the functions f_u and f_v are different. Specifically, it is assumed that either $m_u \neq m_v$ or $a_u \neq a_v$.

The quantities a_u and a_v have units of concentration, so if all concentrations (nutrients, organisms, and Michaelis–Menten constants) are measured in units of $S^{(0)}$ then $S^{(0)}$ may be scaled out of system (2.1). Similarly, the units of m_u, m_v, and D are reciprocal time, so with an appropriate change of time scale D may also be scaled out of the system. Moreover, the conversion factors γ_u and γ_v can be incorporated into u_i and v_i. This is essentially the scaling that has been used in all of the previous models. With these changes of scale, the new system takes the form

$$S_1' = 1 - 2S_1 + S_2 - f_u(S_1)u_1 - f_v(S_1)v_1,$$

$$S_2' = S_1 - 2S_2 - f_u(S_2)u_2 - f_v(S_2)v_2,$$

$$u_1' = -2u_1 + u_2 + f_u(S_1)u_1,$$

$$u_2' = u_1 - 2u_2 + f_u(S_2)u_2, \tag{2.2}$$

$$v_1' = -2v_1 + v_2 + f_v(S_1)v_1,$$

$$v_2' = v_1 - 2v_2 + f_v(S_2)v_2,$$

$$S_i(0) \geq 0, \ u_i(0) \geq 0, \ v_i(0) \geq 0, \ i = 1, 2.$$

The *a*s and *m*s have changed their biological meaning, although the new *f*s are formally the same as those in (2.1).

Now define

$$\Sigma_1(t) = \tfrac{2}{3} - S_1(t) - u_1(t) - v_1(t)$$

and

$$\Sigma_2(t) = \tfrac{1}{3} - S_2(t) - u_2(t) - v_2(t).$$

Adding the equations that correspond to a dependent variable with a subscript 1 and those with a subscript 2 yields

$$\begin{aligned}
\Sigma_1' &= -2\Sigma_1 + \Sigma_2, \\
\Sigma_2' &= \Sigma_1 - 2\Sigma_2, \\
u_1' &= -2u_1 + u_2 + f_u(\tfrac{2}{3} - \Sigma_1 - u_1 - v_1)u_1, \\
u_2' &= u_1 - 2u_2 + f_u(\tfrac{1}{3} - \Sigma_2 - u_2 - v_2)u_2, \\
v_1' &= -2v_1 + v_2 + f_v(\tfrac{2}{3} - \Sigma_1 - u_1 - v_1)v_1, \\
v_2' &= v_1 - 2v_2 + f_v(\tfrac{1}{3} - \Sigma_2 - u_2 - v_2)v_2.
\end{aligned} \tag{2.3}$$

The form of (2.2) and the uniqueness of initial value problems guarantee that the nonnegative cone (in $\mathbb{R}^6$) is positively invariant (see Appendix B, Proposition B.7). The following "conservation" result parallels that for the simple chemostat.

LEMMA 2.1. *The system* (2.2) *is dissipative. Moreover,*

$$\lim_{t\to\infty} \Sigma_1(t) = 0 \quad \text{and} \quad \lim_{t\to\infty} \Sigma_2(t) = 0,$$

where the convergence is exponential. The four-dimensional set

$$\Gamma = \{(\Sigma_1, \Sigma_2, u_1, u_2, v_1, v_2) \in \mathbb{R}_+^6 \mid \Sigma_1 = 0, \Sigma_2 = 0\}$$

is positively invariant.

Proof. Σ_1 and Σ_2 satisfy the linear system

$$\begin{aligned}
z_1' &= -2z_1 + z_2, \\
z_2' &= z_1 - 2z_2,
\end{aligned}$$

and the assertions about limits follow by solving this constant coefficient system. $\qquad\square$

The difference between this and the previous conservation results is that it holds in each vessel rather than for the entire "biomass." As a consequence of this lemma, the omega limit set of any trajectory is nonempty, compact, connected, and contained in Γ. Since every trajectory is asymptotic to its omega limit set, it is important to analyze the system on this set. (See Appendix F for a rigorous justification.) Trajectories in the omega limit set satisfy

$$
\begin{aligned}
u_1' &= -2u_1 + u_2 + f_u(\tfrac{2}{3} - u_1 - v_1)u_1, \\
u_2' &= u_1 - 2u_2 + f_u(\tfrac{1}{3} - u_2 - v_2)u_2, \\
v_1' &= -2v_1 + v_2 + f_v(\tfrac{2}{3} - u_1 - v_1)v_1, \\
v_2' &= v_1 - 2v_2 + f_v(\tfrac{1}{3} - u_2 - v_2)v_2
\end{aligned}
\tag{2.4}
$$

on the region that we again call Γ to conserve notation. This region Γ is defined by

$$
\Gamma = \{(u_1, u_2, v_1, v_2) \in \mathbb{R}_+^4 \mid u_1 + v_1 \leq \tfrac{2}{3} \text{ and } u_2 + v_2 \leq \tfrac{1}{3}\}.
$$

System (2.4) is the one that will receive most of the analysis. Several of the results in the appendices will be used; the theory of monotone systems and the persistence results will be particularly useful. It is generally not possible to analyze a four-dimensional system such as (2.4) because the dynamics can be very complicated; indeed, they can be chaotic. One must work very hard, using the theory developed, to show rigorously that the dynamics are, in fact, very simple. From the standpoint of dynamical systems, this is extraordinary "luck"; from the standpoint of the biology, it is expected. What is new, biologically, is that coexistence is possible and the competition uncomplicated.

3. The Set of Rest Points

As with the chemostat, the basic approach is to locate the rest points, analyze their local stability, and determine the global properties of the dynamical system. The first lemma gives some estimates of quantities that will be important in the analysis. It turns out to be easier to state the results in terms of rest points of the system (2.2) rather than those of the system (2.4). These results will be interpreted as needed for the system (2.4).

LEMMA 3.1. *Let* $(S_1^*, S_2^*, u_1^*, u_2^*, v_1^*, v_2^*)$ *be an equilibrium point of* (2.2). *Then:*

(a) $u_1^* > 0$ *if and only if* $u_2^* > 0$,
   $v_1^* > 0$ *if and only if* $v_2^* > 0$, *and*
   $S_1^* > S_2^* > 0$;
(b) $f_u(S_1^*) > f_u(S_2^*)$ *and*
   $f_v(S_1^*) > f_v(S_2^*)$;
(c) *if* $u_1^* > 0$ *then*
   $\frac{1}{2} < 2 - f_u(S_1^*) < 1$,
   $1 < 2 - f_u(S_2^*) < 2$, *and*
   $(2 - f_u(S_1^*))(2 - f_u(S_2^*)) = 1$;
(d) *if* $v_1^* > 0$ *then*
   $\frac{1}{2} < 2 - f_v(S_1^*) < 1$,
   $1 < 2 - f_v(S_2^*) < 2$, *and*
   $(2 - f_v(S_1^*))(2 - f_v(S_2^*)) = 1$.

Before proving this technical lemma, we make some observations. One might think that coexistence in the gradostat is possible by having one competitor win in one vessel and the other in the second vessel. Statement (a) shows this is not possible (at least, not as a steady state). The last statements in (c) and (d) provide a crucial key to actual computation of the coordinates of the rest points. The other inequalities provide estimates that will be useful in stability considerations.

Proof. The proofs of these assertions are all relatively straightforward. The results will be established for u_i^*; the others follow similarly. The equations for the rest point are just the equations with the left-hand side of (2.2) set equal to zero. The equations for u_i^* yield the first statement in (a) directly. The second equation in (2.2) at equilibrium becomes

$$S_1^* - 2S_2^* = u_2^* f_u(S_2^*) + v_2^* f_v(S_2^*) \geq 0.$$

It follows that

$$S_1^* \geq 2S_2^* > S_2^*.$$

The S_2^* term cannot be zero. Since f_u is monotone increasing, (b) follows.
   To prove (c), note that

$$u_2^* = (2 - f_u(S_1^*))u_1^* \quad \text{and} \quad u_1^* = (2 - f_u(S_2^*))u_2^*.$$

From (a), it follows that

$$2 - f_u(S_1^*) > 0 \quad \text{and} \quad 2 - f_u(S_2^*) > 0$$

and hence, combining these two equations, that

$$(2 - f_u(S_1^*))(2 - f_u(S_2^*)) = 1.$$

The inequalities then follow from (b). This completes the proof of the lemma.   □

There is an equilibrium point of the form $(\frac{2}{3}, \frac{1}{3}, 0, 0, 0, 0)$. This corresponds to both organisms washing out of the gradostat. In view of Lemma 3.1(a), there are potential equilibrium points of the form $(\hat{S}_1, \hat{S}_2, \hat{u}_1, \hat{u}_2, 0, 0)$ and $(\tilde{S}_1, \tilde{S}_2, 0, 0, \tilde{v}_1, \tilde{v}_2)$, with all nonzero entries positive. These correspond to one of the competitors washing out of the gradostat. For coexistence to occur as a steady state, it must be shown that there also is an equilibrium point with all components strictly positive. The existence and the stability of these rest points are closely related, as the following sections show. To determine the stability of rest points it turns out that certain functions of the coordinates of the rest points must be evaluated, so it is important to be able to compute them explicitly. Our inability to make these computations when the number of vessels is more than two restricts the analysis, as the next chapter will show.

Both $\hat{u}_1$ and $\hat{u}_2$, if they exist, are solutions of

$$-2u_1 + u_2 + f_u(\tfrac{2}{3} - u_1)u_1 = 0,$$
$$u_1 - 2u_2 + f_u(\tfrac{1}{3} - u_2)u_2 = 0. \qquad (3.1)$$

A straightforward analysis reduces the question of solutions to finding the roots of a cubic equation. Of course, these quantities must be positive for the rest points to be meaningful. Conditions for this, based on stability considerations, will be given in the next section. Numerically solving a cubic is simple; in fact, there is an unlovely formula that does it explicitly. The important point is that since solving the cubic is possible, conditions based on the numerical coordinates are testable. A similar computation gives the coordinates of $\tilde{v}_1$ and $\tilde{v}_2$.

The more difficult question is the existence of an interior rest point – that is, where S_i, u_i, and v_i are all positive. (Stability considerations, or the arguments that follow here, show that there is at most one interior rest point.) From Lemma 3.1 one has

$$(2 - f_u(S_1^*))(2 - f_u(S_2^*)) = 1,$$
$$(2 - f_v(S_1^*))(2 - f_v(S_2^*)) = 1. \qquad (3.2)$$

Although (3.2) is nonlinear in S_1^* and S_2^*, it is linear in $r_1 = S_1^* + S_2^*$ and $r_2 = S_1^* S_2^*$. Rewriting (3.2) gives

$$3a_u^2 - A_{11}a_u r_1 + A_{12}r_2 = 0,$$
$$3a_v^2 - A_{21}a_v r_1 + A_{22}r_2 = 0,$$

(3.3)

where

$$A_{11} = (2m_u - 3), \quad A_{12} = (m_u - 1)(m_u - 3),$$
$$A_{21} = (2m_v - 3), \quad A_{22} = (m_v - 1)(m_v - 3).$$

The condition that the determinant in (3.3),

$$\begin{bmatrix} -A_{11}a_u & A_{12} \\ -A_{21}a_v & A_{22} \end{bmatrix},$$

not vanish is

$$a_u(2m_u - 3)(m_v - 1)(m_v - 3) \neq a_v(2m_v - 3)(m_u - 1)(m_u - 3).$$

If this is the case, then

$$r_1 = c_2/c_1 \quad \text{and} \quad r_2 = c_3/c_1,$$

where

$$c_1 = A_{12}A_{21}a_v - A_{11}A_{22}a_u,$$
$$c_2 = 3A_{12}a_v^2 - 3A_{22}a_u^2,$$
$$c_3 = 3A_{11}a_v^2 - 3A_{21}a_u^2.$$

If either r_1 or r_2 is nonpositive then there is no interior rest point in $\mathbb{R}_+^6$. Given r_1 and r_2, S_1^* and S_2^* can now be recovered from

$$S_1^* S_2^* = c_3/c_1,$$
$$S_1^* + S_2^* = c_2/c_1,$$

or, upon substitution, from

$$c_1(S_1^*)^2 - c_2 S_1^* + c_3 = 0.$$

Hence, one has that

$$S_1^* = \frac{c_2 \pm \sqrt{c_2^2 - 4c_1 c_3}}{2c_1},$$
$$S_2^* = \frac{c_2 \mp \sqrt{c_2^2 - 4c_1 c_3}}{2c_1}.$$

(3.4)

Since $S_1^* > S_2^*$, there is only one choice of signs. Hence S_1^* and S_2^*, if they exist, are uniquely determined by (3.4). If S_1^* and S_2^* are determined from (3.4), then the components of u^* and v^* can be determined from the basic equations (2.2).

Table 3.1

Point	Coordinates
E_0	$(0, 0, 0, 0)$
E_1	$(\hat{u}_1, \hat{u}_2, 0, 0)$, $\quad \hat{u}_i > 0, \ i = 1, 2$
E_2	$(0, 0, \bar{v}_1, \bar{v}_2)$, $\quad \bar{v}_i > 0, \ i = 1, 2$
E_*	$(u_1^*, u_2^*, v_1^*, v_2^*)$, $\quad u_i^* > 0, \ v_i^* > 0, \ i = 1, 2$

We seek to use these values in (2.4) to determine the coordinates of the rest point. We make use of the fact that, at an equilibrium, Lemma 3.1 applies to the coordinates. Hence one has that

$$
\begin{aligned}
f_u(S_1^*)u_1^* &= 2u_1^* - u_2^* \\
&= \tfrac{4}{3} - 2S_1^* + S_2^* - 2v_1^* - \tfrac{1}{3} + S_2^* + v_2^* \\
&= 1 - 2S_1^* + S_2^* - 2v_1^* + v_2^* \\
&= 1 - 2S_1^* + S_2^* - f_v(S_1^*)(\tfrac{2}{3} - S_1^* - u_1^*).
\end{aligned}
$$

Thus the coordinate of u_1^* is

$$
u_1^* = \frac{2S_1^* + f_v(S_1^*)(\tfrac{2}{3} - S_1^*) - S_2^* - 1}{f_v(S_1^*) - f_u(S_1^*)}.
$$

Similarly,

$$
u_2^* = \frac{2S_2^* + f_v(S_2^*)(\tfrac{1}{3} - S_2^*) - S_1^*}{f_v(S_2^*) - f_u(S_2^*)}.
$$

The remaining two coordinates follow directly as

$$
v_1^* = \tfrac{2}{3} - S_1^* - u_1^*,
$$

$$
v_2^* = \tfrac{1}{3} - S_2^* - u_2^*.
$$

The analysis is valid provided that the denominators in u_1^* or u_2^* are not zero. Note that they cannot both be zero since then $f_u \equiv f_v$. If one of them is zero, we can first determine one variable and then solve for the other by direct substitution into the appropriate equation in (2.2).

If the determinant condition does not hold (i.e., if equality prevails), then the solution of (3.2) represents two parallel lines. If they are coincident, there is a continuum of solutions; if they are not coincident, there are no solutions. The possibility of a continuum of solutions will be eliminated on stability grounds (provided, as we are assuming, that either $m_u \neq m_v$ or $a_u \neq a_v$). The proof will be given in Lemma 5.2.

The generic assumption of hyperbolicity is made throughout this chapter. In fact (as we will show), all eigenvalues are real, so the hyperbolicity assumption means that none of them vanish. This section is summarized in Table 3.1, which lists the possible rest points for the system (2.4).

4. Growth without Competition

Before analyzing the full gradostat system, the system without competition will be considered. We arbitrarily eliminate the v variable; similar results hold when u is eliminated. The defining equations are:

$$S_1' = 1 - 2S_1 + S_2 - f_u(S_1)u_1 - f_v(S_1)v_1,$$

$$S_2' = S_1 - 2S_2 - f_u(S_2)u_2 - f_v(S_2)v_2,$$

$$u_1' = -2u_1 + u_2 + f_u(S_1)u_1, \tag{4.1}$$

$$u_2' = u_1 - 2u_2 + f_u(S_2)u_2,$$

$$S_i(0) \geq 0, \quad u_i(0) \geq 0, \quad i = 1, 2.$$

The conservation principle established previously applies here, so one may deal (on the omega limit set) with the system obtained by setting $S_1(t) = \frac{2}{3} - u_1(t)$ and $S_2(t) = \frac{1}{3} - u_2(t)$:

$$u_1' = -2u_1 + u_2 + f_u(\tfrac{2}{3} - u_1)u_1,$$

$$u_2' = u_1 - 2u_2 + f_u(\tfrac{1}{3} - u_2)u_2, \tag{4.2}$$

$$u_i(0) \geq 0, \quad i = 1, 2.$$

Of course, the equations are restricted to the region defined by $0 < u_1(t) < \frac{2}{3}$, $0 < u_2(t) < \frac{1}{3}$. This region may be seen to be positively invariant by checking the derivatives along the boundary. Equation (4.2) will be analyzed by determining the stability of rest points and eliminating limit cycles. Since the system is two-dimensional and dissipative, the Poincaré–Bendixson theory applies, and this program is sufficient for determining the asymptotic behavior.

The off-diagonal entries in the Jacobian matrix of the right-hand side of system (4.2) satisfy

$$\frac{\partial}{\partial u_2}(-2u_1+u_2+f_u(\tfrac{2}{3}-u_1)u_1) = 1,$$

$$\frac{\partial}{\partial u_1}(u_1-2u_2+f_u(\tfrac{1}{3}-u_2)u_2) = 1.$$

Thus the system is cooperative, which eliminates the possibility of limit cycles (see the comments following Remark 6.2 of Chapter 4). The Poincaré–Bendixson theorem then guarantees convergence of each bounded trajectory to a rest point.

Define

$$\alpha_1(z) = 2-f_u(\tfrac{2}{3}-z) \quad \text{and} \quad \alpha_2(z) = 2-f_u(\tfrac{1}{3}-z).$$

The variational matrix for (4.2) at the origin is of the form

$$J = \begin{bmatrix} -\alpha_1(0) & 1 \\ 1 & -\alpha_2(0) \end{bmatrix}$$

with eigenvalues

$$\lambda = \frac{-(\alpha_1(0)+\alpha_2(0)) \pm \sqrt{(\alpha_1(0)-\alpha_2(0))^2+4}}{2}.$$

The eigenvalues are real, and if $\alpha_1(0)\alpha_2(0) < 1$ then the origin is a saddle point.

LEMMA 4.1. *If $\alpha_1(0)\alpha_2(0) > 1$, the origin is an attractor or repeller as $\alpha_1(0)+\alpha_2(0)$ is positive or negative. If $\alpha_1(0)\alpha_2(0) < 1$, the origin is not in the omega limit set of any trajectory with positive initial conditions.*

Proof. If $\alpha_1(0)\alpha_2(0) > 1$ then both eigenvalues have the same sign, so the origin is an attractor or a repeller according to whether $\alpha_1(0)+\alpha_2(0)$ is positive or negative.

Note that for (4.2) the nonnegative quadrant is positively invariant. When $\alpha_1(0)\alpha_2(0) < 1$ (i.e., when the origin is a saddle point), we will show that the stable manifold of $(0,0)$ does not intersect the interior of the positive quadrant. Let λ^+ and λ^- denote the positive and negative eigenvalues of J_u, respectively. One has that

$$\alpha_1(0)+\lambda^+ = \frac{(\alpha_1(0)-\alpha_2(0))+\sqrt{(\alpha_1(0)-\alpha_2(0))^2+4}}{2} > 0.$$

Similarly, one has that $\alpha_2(0)+\lambda^- < 0$. An eigenvector (z_1, z_2) corresponding to λ^- satisfies

$$z_1-(\alpha_2(0)+\lambda^-)z_2 = 0,$$

or the slope of the eigenvector is negative. Hence the stable manifold of the origin lies exterior to $\mathbb{R}_+^2$. If $u(t) = (u_1(t), u_2(t))$ is a trajectory of (4.2) with the origin as an omega limit point then $u(t)$ cannot be on the stable manifold, and hence the omega limit set must contain a point of the stable manifold other than the origin (Butler–McGehee theorem), a contradiction. Thus the origin is not an omega limit point of any nontrivial trajectory with positive initial conditions. $\square$

LEMMA 4.2. *If the origin is an attractor for* (4.2) *(i.e., if* $\alpha_1(0)\alpha_2(0) > 1$ *and* $\alpha_1(0) + \alpha_2(0) > 0$*), then*

(a) *there is no rest point of the form* $(\hat{u}_1, \hat{u}_2)$ *with* $\hat{u}_i > 0$, $i = 1, 2$, *and*
(b) *the origin attracts all trajectories with initial conditions in* $\mathring{\mathbb{R}}_+^2 \cap \Gamma$, *where* $\mathring{\mathbb{R}}_+^2$ *denotes the interior of* $\mathbb{R}_+^2$.

Proof. A rest point $(\hat{u}_1, \hat{u}_2)$ with $\hat{u}_i > 0$ is a positive solution of

$$-2u_1 + u_2 + f_u(\tfrac{2}{3} - u_1)u_1 = 0,$$
$$u_1 - 2u_2 + f_u(\tfrac{1}{3} - u_2)u_2 = 0,$$

which can be rewritten as

$$(2 - f_u(\tfrac{2}{3} - u_1))u_1 - u_2 = 0,$$
$$-u_1 + (2 - f_u(\tfrac{1}{3} - u_2))u_2 = 0. \tag{4.3}$$

A nontrivial rest point can exist if and only if the determinant is zero; hence, $\alpha_1(\hat{u}_1)\alpha_2(\hat{u}_2) = 1$. However, $\alpha_i(z)$ is monotone increasing, so

$$\alpha_1(\hat{u}_1)\alpha_2(\hat{u}_2) > \alpha_1(0)\alpha_2(0) \geq 1.$$

Hence there is no rest point with positive coordinates if the origin is a (local) attractor.

Conversely, since trajectories are bounded, the positive quadrant is invariant, and there are no limit cycles in the invariant set $\mathbb{R}_+^2 \cap \Gamma$, so there must be a rest point of the form $(\hat{u}_1, \hat{u}_2) \in \mathring{\mathbb{R}}_+^2$ if the origin is not an attractor.

LEMMA 4.3. *If the origin is a repeller or a saddle point for* $\mathring{\mathbb{R}}_+^2 \cap \Gamma$, *then there is exactly one rest point of* (3.2) *in the interior of this set and it is a global attractor.*

Proof. The existence of a rest point with positive coordinates has been established in the paragraph preceding the statement of the lemma. Suppose there were two distinct rest points $(\hat{u}_1, \hat{u}_2)$ and $(\bar{u}_1, \bar{u}_2)$ with

$$\hat{u}_i > 0, \quad \tilde{u}_i > 0, \quad i = 1, 2.$$

Assume that the labeling is such that $\hat{u}_1 \le \tilde{u}_1$. Then it follows that

$$\hat{u}_2 = \hat{u}_1 \alpha_1(\hat{u}_1) \le \tilde{u}_1 \alpha_1(\tilde{u}_1) = \tilde{u}_2,$$

with equality only if $\hat{u}_1 = \tilde{u}_1$. Hence, if $(\hat{u}_1, \hat{u}_2) \ne (\tilde{u}_1, \tilde{u}_2)$ it follows that

$$1 = \alpha_1(\hat{u}_1)\alpha_2(\hat{u}_2) < \alpha_1(\tilde{u}_1)\alpha_2(\tilde{u}_2) = 1,$$

which is a contradiction. Thus there is only one equilibrium point in the interior of the positive quadrant. Since the only equilibrium point on the boundary – the origin – cannot be an omega limit point of trajectories with positive initial conditions, all trajectories must tend to the interior equilibrium.   □

REMARK. Similar lemmas apply to the equations for the v population,

$$v_1' = -2v_1 + v_2 + f_v(\tfrac{2}{3} - v_1)v_1,$$

$$v_2' = v_1 - 2v_2 + f_v(\tfrac{1}{3} - v_2)v_2,$$

with $\alpha_3(z) = 2 - f_v(\tfrac{2}{3} - z)$ and $\alpha_4(z) = 2 - f_v(\tfrac{1}{3} - z)$ replacing $\alpha_1(z)$ and $\alpha_2(z)$, respectively.

The conclusion of the preceding arguments is very simple. There are only two possibilities: The origin is an attractor and all trajectories tend to it; or the origin is not the omega limit point of any trajectory with positive initial conditions, and there exists a unique rest point with positive co-ordinates to which all trajectories tend. In biological terms, either the organism can survive in the gradostat in all vessels without competition or it cannot survive at all. We are interested in determining the parameter ranges wherein survival occurs.

5. Local Stability

We turn now to computing the local stability of the rest points of the full system. The arguments are based on standard linearization techniques, but the size of the variational matrix makes some of the computations difficult. The variational matrix for (2.4) takes the form

$$J = \begin{bmatrix} -\alpha_1 - \beta_1 & 1 & -\beta_1 & 0 \\ 1 & -\alpha_2 - \beta_2 & 0 & -\beta_2 \\ -\beta_3 & 0 & -\alpha_3 - \beta_3 & 1 \\ 0 & -\beta_4 & 1 & -\alpha_4 - \beta_4 \end{bmatrix}, \tag{5.1}$$

where the α_is and β_is are given by

$$\alpha_1 = 2 - f_u(\tfrac{2}{3} - u_1 - v_1),$$

$$\alpha_2 = 2 - f_u(\tfrac{1}{3} - u_2 - v_2),$$

$$\alpha_3 = 2 - f_v(\tfrac{2}{3} - u_1 - v_1),$$

$$\alpha_4 = 2 - f_v(\tfrac{1}{3} - u_2 - v_2),$$

$$\beta_1 = \frac{m_u a_u u_1}{(a_u + \tfrac{2}{3} - u_1 - v_1)^2},$$

$$\beta_2 = \frac{m_u a_u u_2}{(a_u + \tfrac{1}{3} - u_2 - v_2)^2},$$

$$\beta_3 = \frac{m_v a_v v_1}{(a_v + \tfrac{2}{3} - u_1 - v_1)^2},$$

$$\beta_4 = \frac{m_v a_v v_2}{(a_v + \tfrac{1}{3} - u_2 - v_2)^2}.$$

When evaluated at a hyperbolic rest point, J will determine the (local) asymptotic stability of that point. The computational problem is increased by the size of the matrix and the complexity of the entries. Fortunately, for some of the rest points there will be a large number of zero entries in J.

For a fixed set of parameters, let Ω denote the rest point set of the system (2.4) in Γ. There are four possible types of rest points, which we denote as follows:

$$E_0 = (0, 0, 0, 0);$$

$$E_1 = (\hat{u}_1, \hat{u}_2, 0, 0) \quad \text{with } \hat{u}_i > 0, \ i = 1, 2;$$

$$E_2 = (0, 0, \bar{v}_1, \bar{v}_2), \quad \text{with } \bar{v}_i > 0, \ i = 1, 2;$$

$$E_* = (u_1^*, u_2^*, v_1^*, v_2^*) \quad \text{with } u_i^* > 0 \text{ and } v_i^* > 0, \ i = 1, 2.$$

The rest point E_0 always exists. The results in Section 4 provide conditions for E_1 and E_2 to exist. The existence and stability of E_* is a major consideration.

The matrix J at E_0 has $\beta_i = 0$, $i = 1, 2, 3, 4$, so

$$J = \begin{bmatrix} J_u & 0 \\ 0 & J_v \end{bmatrix},$$

where J_u and J_v are the variational matrices for the two-dimensional systems considered in Section 4. Hence the origin is asymptotically stable if and only if both of the two-dimensional systems (the systems without competition) are stable. From Section 4, this is the case if and only if

$$\alpha_1(0)\alpha_2(0) > 1, \quad \alpha_1(0) > 0 \tag{5.2a}$$

and

$$\alpha_3(0)\alpha_4(0) > 1, \quad \alpha_3(0) > 0. \tag{5.2b}$$

If one of the conditions in (5.2) is reversed, then the origin is a repeller in the corresponding two-dimensional system (u_1–u_2 if $\alpha_1(0)\alpha_2(0) < 1$ or $\alpha_1(0) < 0$, and v_1–v_2 if $\alpha_3(0)\alpha_4(0) < 1$ or $\alpha_3(0) < 0$), and hence a nontrivial equilibrium point exists in the corresponding two-dimensional subset of the boundary.

The rest point E_1 corresponds to a rest state without the v competitor. When it is asymptotically stable, the v competitor will become extinct (will wash out of the system) for nearby initial conditions. If the stability is global (which will turn out to be the case when stability is local), then v becomes extinct for all positive initial conditions. At E_1, J takes the form

$$\begin{bmatrix} -\alpha_1 - \beta_1 & 1 & -\beta_1 & 0 \\ 1 & -\alpha_2 - \beta_2 & 0 & -\beta_2 \\ 0 & 0 & -\alpha_3 & 1 \\ 0 & 0 & 1 & -\alpha_4 \end{bmatrix},$$

where the α_is and β_is are evaluated at $(\hat{u}_1, \hat{u}_2, 0, 0)$. The zero block in the lower left corner makes the computation of eigenvalues easy, for (using $\alpha_1\alpha_2 = 1$) they satisfy

$$[\lambda^2 + (\alpha_1 + \alpha_2 + \beta_1 + \beta_2)\lambda + \alpha_2\beta_1 + \alpha_1\beta_2 + \beta_1\beta_2]$$
$$\times [\lambda^2 + (\alpha_3 + \alpha_4)\lambda + \alpha_3\alpha_4 - 1] = 0. \tag{5.3}$$

Since $\alpha_i > 0$ and $\beta_i > 0$, $i = 1, 2$, the eigenvalues

$$\lambda = \frac{-(\alpha_1 + \alpha_2 + \beta_1 + \beta_2) \pm \sqrt{(\alpha_1 + \alpha_2 + \beta_1 + \beta_2)^2 - 4(\beta_1\alpha_2 + \alpha_1\beta_2 + \beta_1\beta_2)}}{2}$$

from the first square bracket are real (the radical simplifies) and always have negative real parts. These eigenvalues correspond to eigenvectors in the subspace $(u_1, u_2, 0, 0)$, which is the stable manifold of $(\hat{u}_1, \hat{u}_2)$ viewed as a rest point of the two-dimensional system (4.2). This is expected, in view of the results in Section 4 on the stability of the two-dimensional system.

The eigenvalues corresponding to the second square bracket in (5.3) satisfy

$$\lambda = \frac{-(\alpha_3 + \alpha_4) \pm \sqrt{(\alpha_3 - \alpha_4)^2 + 4}}{2}.$$

The eigenvalues are real, and the sign depends on the values of $\alpha_3(\hat{u}_1)$ and $\alpha_4(\hat{u}_2)$. Clearly, if $\alpha_3(\hat{u}_1)\alpha_4(\hat{u}_2) > 1$ then the two eigenvalues are of the same sign. Then E_1 will be an attractor if $\alpha_3(\hat{u}_1) > 0$ and have a two-dimensional unstable manifold if $\alpha_3(\hat{u}_1) < 0$.

If $\alpha_3(\hat{u}_1)\alpha_4(\hat{u}_2) < 1$, there is one negative and one positive eigenvalue. Let λ^- denote the negative eigenvalue and let $z = (z_1, z_2, z_3, z_4)$ be the corresponding eigenvector. Since

$$\lambda^- + \alpha_3 = \frac{\alpha_3 - \alpha_4 - \sqrt{(\alpha_3 - \alpha_4)^2 + 4}}{2} < 0$$

and $(\lambda^- + \alpha_3(\hat{u}_1))z_3 = z_4$, z_3 and z_4 are of opposite signs and hence z must point out of the positive cone at $(\hat{u}_1, \hat{u}_2, 0, 0)$. In particular, the stable manifold of E_1 does not intersect the interior of the positive cone. This statement is trivially true in the other case, since the unstable manifold is two-dimensional and the stable manifold lies in the boundary. In this case no trajectory with initial conditions in the interior of Γ can tend to E_1. This argument is summarized in the following lemma.

LEMMA 5.1. *If $\alpha_3(\hat{u}_1)\alpha_4(\hat{u}_2) > 1$, then E_1 is an attractor if $\alpha_3(\hat{u}_1) > 0$ and has a two-dimensional unstable manifold if $\alpha_3(\hat{u}_1) < 0$. If $\alpha_3(\hat{u}_1)\alpha_4(\hat{u}_2) < 1$, or if $\alpha_3(\hat{u}_1)\alpha_4(\hat{u}_2) > 1$ and $\alpha_3(\hat{u}_1) < 0$, then E_1 is not the limit of any trajectory with initial conditions in the interior of Γ. Similar statements apply at E_2 using $\alpha_1(\tilde{v}_1)$ and $\alpha_2(\tilde{v}_2)$. When either E_1 or E_2 is unstable, the stable manifold of that rest point lies in the corresponding two-dimensional face containing E_1 or E_2.*

The difficult part remains – the stability of an interior rest point. Conditions for the existence of such a rest point will be established, on geometric grounds, later in the chapter.

LEMMA 5.2. *If $a_u \neq a_v$ or $m_u \neq m_v$ and if E_* exists, then E_* is locally asymptotically stable.*

Proof. A sufficient condition for J, evaluated at E_*, to have eigenvalues with negative real parts is that if the off-diagonal elements are replaced by their absolute values, then the determinants of the principal minors alternate in sign (Theorem A.11).

Let d_i, $i = 1, 2, 3, 4$, denote the determinants of the principal minors of

$$\begin{bmatrix} -\alpha_1 - \beta_1 & 1 & \beta_1 & 0 \\ 1 & -\alpha_2 - \beta_2 & 0 & \beta_2 \\ \beta_3 & 0 & -\alpha_3 - \beta_3 & 1 \\ 0 & \beta_4 & 1 & -\alpha_4 - \beta_4 \end{bmatrix}.$$

It must be shown that $d_1 < 0$, $d_2 > 0$, $d_3 < 0$, and $d_4 > 0$. It follows at once that $d_1 = -\alpha_1 - \beta_1 < 0$. Moreover,

$$
\begin{aligned}
d_2 &= \alpha_1\alpha_2 + \alpha_1\beta_2 + \alpha_2\beta_1 + \beta_1\beta_2 - 1 \\
&= \alpha_1\beta_2 + \alpha_2\beta_1 + \beta_1\beta_2 \\
&> 0,
\end{aligned}
$$

since $\alpha_1\alpha_2 = 1$ by Lemma 3.1. An easy computation shows that

$$
\begin{aligned}
d_3 &= -\alpha_1\beta_2\beta_3 - \alpha_1\alpha_3\beta_2 - \alpha_2\alpha_3\beta_1 - \alpha_3\beta_1\beta_2 \\
&< 0.
\end{aligned}
$$

The more difficult problem is that of computing d_4. The estimates in Lemma 3.1 are crucial here. The determinant d_4 can be expanded in elements of the last row to obtain

$$
d_4 = \beta_4 D_1 - D_2 - (\alpha_4 + \beta_4)d_3,
$$

where d_3 is as given previously. If each of the 3×3 determinants D_1 and D_2 are expanded (a tedious computation that we omit) and heavy use is made of the fact that $\alpha_1\alpha_2 = \alpha_3\alpha_4 = 1$ (Lemma 3.1), one can simplify this to

$$
d_4 = (\alpha_1\alpha_4 - 1)\beta_2\beta_3 + (\alpha_2\alpha_3 - 1)\beta_1\beta_4. \tag{5.4}
$$

The idea behind the following computations is to express all of the quantities as a function of the two parameters α_1 and α_4, and to show positivity of the resulting expression for the allowable parameter range.

First write

$$
\beta_1 = \frac{a_u}{m_u(S_1^*)^2}(\alpha_1 - 2)^2 u_1^*, \qquad \beta_2 = \frac{a_u}{m_u(S_2^*)^2}(\alpha_2 - 2)^2 u_2^*,
$$

$$
\beta_3 = \frac{a_v}{m_v(S_1^*)^2}(\alpha_3 - 2)^2 v_1^*, \qquad \beta_4 = \frac{a_v}{m_v(S_2^*)^2}(\alpha_4 - 2)^2 v_2^*,
$$

by multiplying and dividing by the square of the appropriate S. Then replace α_2, α_3, u_2^*, and v_1^* by the substitutions (obtained from Lemma 3.1 and the equations determining rest points)

$$
\alpha_2 = 1/\alpha_1, \qquad \alpha_3 = 1/\alpha_4;
$$

$$
u_2^* = \alpha_1 u_1^*, \qquad v_1^* = \alpha_4 v_2^*.
$$

The first term in (5.4) takes the form

$$
(\alpha_1\alpha_4 - 1)\beta_2\beta_3 = (\alpha_1\alpha_4 - 1)\frac{a_u a_v}{m_u m_v}\frac{1}{(S_1^* S_2^*)^2}\left(\frac{1}{\alpha_1} - 2\right)^2\left(\frac{1}{\alpha_4} - 2\right)^2\alpha_1\alpha_4 u_1^* v_2^*,
$$

while the second term is given by

$$(\alpha_2\alpha_3-1)\beta_1\beta_4 = \left(\frac{1}{\alpha_1\alpha_4}-1\right)\frac{a_u a_v}{m_u m_v}\frac{1}{(S_1^* S_2^*)^2}(\alpha_1-2)^2(\alpha_4-2)^2 u_1^* v_2^*.$$

Hence one can write

$$d_4 = \frac{a_u a_v}{m_u m_v}\frac{u_1^* v_2^*}{S_1^* S_2^*}R(\alpha_1,\alpha_4),$$

where

$$R(\alpha_1,\alpha_4) = (\alpha_1\alpha_4-1)\left(\frac{1}{\alpha_1}-2\right)^2\left(\frac{1}{\alpha_4}-2\right)^2\alpha_1\alpha_4$$

$$+\left(\frac{1}{\alpha_1\alpha_4}-1\right)(\alpha_1-2)^2(\alpha_4-2)^2.$$

It is necessary only to show that $R(\alpha_1,\alpha_4)$ is positive in the allowable range of α_1 and α_4 given by Lemma 3.1 (i.e., $\frac{1}{2}<\alpha_1<1$, $1<\alpha_4<2$). We may rewrite R as

$$R(\alpha_1,\alpha_4) = \frac{3(\alpha_1\alpha_4-1)^2}{\alpha_1\alpha_4}Q(\alpha_1,\alpha_4),$$

where $Q(\alpha_1,\alpha_4) = 5+5\alpha_1\alpha_4-4\alpha_1-4\alpha_4$. Because Q is an increasing function of α_1 in the allowable range of α_4, $Q>3-3\alpha_4/2$ which must then be positive.

If $\alpha_1\alpha_4 \neq 1$ this completes the proof. If equality holds then it follows that $\alpha_1 = \alpha_3$ and $\alpha_2 = \alpha_4$ or that $f_u(S_1^*) = f_v(S_1^*)$ and $f_u(S_2^*) = f_v(S_2^*)$. The functions f_u and f_v cross at the origin and at most one other point. Since $0 < S_2^* < S_1^*$, the two functions are exactly the same. This contradiction completes the proof of the lemma. $\square$

6. Order Properties

In the preceding sections, the possible rest points for the gradostat equations were determined and their stability analyzed. The problem that remains is to determine the global behavior of trajectories. In this regard, the theory of dynamical systems plays an important role. First of all, some information can be obtained from the general theorem on inequalities discussed in Appendix B. We illustrate this with an application to the gradostat equations.

Let $u(t) = (u_1(t), u_2(t), v_1(t), v_2(t))$ be the solution of (2.4) with omega limit point p, and let $x(t) = (x_1(t), x_2(t))$ be a solution of (4.2) with

$x_i(0) = u_i(0)$, $i = 1, 2$. Then, because of the monotonicity of f_u, the first two components of $u(t)$ satisfy

$$u_1'(t) < u_2(t) - (2 - f_u(\tfrac{2}{3} - u_1(t)))u_1(t).$$

$$u_2'(t) < u_1(t) - (2 - f_u(\tfrac{1}{3} - u_2(t)))u_2(t),$$

while the components of $x(t)$ satisfy an equation with the same right-hand side:

$$x_1'(t) = x_2(t) - (2 - f_u(\tfrac{2}{3} - x_1(t)))x_1(t),$$

$$x_2'(t) = x_1(t) - (2 - f_u(\tfrac{1}{3} - x_2(t)))x_2(t).$$

Hence, for all $t > 0$, Theorem B.1 (the basic comparison theorem) states that

$$u_1(t) < x_1(t) \quad \text{and} \quad u_2(t) < x_2(t).$$

However, if (5.2a) is satisfied then $\lim_{t \to \infty} x_i(t) = 0$ $(i = 1, 2)$, so the omega limit point p must be of the form $(0, 0, *, *)$; that is, the u population becomes extinct. Similarly, if (5.2b) holds then p is of the form $(*, *, 0, 0)$ and the v population is eliminated. Note that this is independent of initial conditions and hence is a global result. The component marked $*$ will be zero or positive depending on the other inequality.

If one of the inequalities in (5.2a) is reversed then

$$\lim_{t \to \infty} x_1(t) = \hat{u}_1 \quad \text{and} \quad \lim_{t \to \infty} x_2(t) = \hat{u}_2,$$

so $p_1 \le \hat{u}_1$ and $p_2 \le \hat{u}_2$. A similar argument shows that if one of the inequalities in (5.2b) is reversed, $p_3 \le \bar{v}_1$ and $p_4 \le \bar{v}_2$. These comments are summarized in the next lemma.

LEMMA 6.1. *Let $p = (p_1, p_2, p_3, p_4)$ be an omega limit point of any trajectory with initial condition in $\mathring{\Gamma}$. If E_1 exists, $p_1 \le \hat{u}_1$ and $p_2 \le \hat{u}_2$. If E_2 exists, $p_3 \le \bar{v}_1$ and $p_4 \le \bar{v}_2$. If E_1 does not exist, $p_1 = p_2 = 0$; if E_2 does not exist, $p_3 = p_4 = 0$.*

The system of interest, (2.4), is neither cooperative nor competitive with respect to the obvious order. The u_1-u_2 populations cooperate, as do the v_1-v_2 populations, but u and v compete. In ecological terms, the problem involves families of competing mutualists [S2]. Let $x = (u_1, u_2, v_1, v_2)$ and $y = (\bar{u}_1, \bar{u}_2, \bar{v}_1, \bar{v}_2)$. Define the partial order relation $x \le_K y$ by $u_i \le \bar{u}_i$, $v_i \ge \bar{v}_i$. We shall also use $x <_K y$ to mean $u_i < \bar{u}_i$, $v_i > \bar{v}_i$. (This corresponds to the notation $x \ll y$ used by some authors.) The system (2.4) is monotone (see Appendix C) with respect to this order (see also the discussion in Appendix B). Let $Df(x)$ be the variational matrix evaluated at an arbitrary point. A sufficient condition for strong monotonicity of a monotone system is given in Theorem C.1 – namely, that the matrix J be irreducible. From (5.1) this is easily seen to be the case in the interior of Γ and some of its boundary. The dynamical system generated by (2.4) is strongly monotone in the interior of Γ.

PROPOSITION 6.2. *The system* (2.4) *is uniformly persistent if and only if the rest points E_1 and E_2 are unstable.*

In the previous section it was shown that the rest points E_1 and E_2, when unstable, had no part of their stable manifolds in the positive cone (Lemma 5.1). This shows that (H) of Theorem D.2 holds. If the covering is taken to be the set of rest points, the flow on the boundary is acyclic (in terms of Appendix D), for the rest points in the faces $u_1 = u_2 = 0$ and the $v_1 = v_2 = 0$ attract all points in that face. An application of Theorem D.2 completes the proof: take $X = \mathbb{R}^4_+$ and $E = \{(u_1, u_2, v_1, v_2) \in \mathbb{R}^4_+ \,|\, u_i > 0$ for some i and $v_j > 0$ for some $j\}$.

Recall that hyperbolicity is a generic assumption in this chapter and that dissipativeness has been established. This has the following consequence.

PROPOSITION 6.3. *The rest point E_* exists if and only if E_1 and E_2 are unstable.*

Proof. If E_1 and E_2 are unstable, then dissipativeness and uniform persistence (previous proposition) yield the existence of an interior rest point E_* for $\pi(x, t)$ (Theorem D.3). If E_* exists then it is unique and has all eigenvalues negative (Lemma 5.2). Suppose that E_* exists and that E_1 is asymptotically stable. Then, since $E_* <_K E_1$ and both are asymptotically stable, Theorem E.1 contradicts the uniqueness of E_*. A similar argument applies if E_2 is asymptotically stable. Note that the computations leading up to Lemma 5.1 explicitly determine the signs of the eigenvalues for linearization about E_1 and E_2. $\square$

7. Global Behavior of Solutions

Enough information has now been collected to classify the behavior of all solutions of the gradostat equations. This classification will be given as a function of the set Ω of rest points. The existence and stability of the rest points has already been established. Indeed, the coordinates of the rest points can be given in terms of the system parameters – the a_u, a_v, m_u, m_v that are functions of the organism being cultured. The (local) stability of equilibria is determined by α and β evaluated at these points. The set Ω belongs to one of four categories, and for each category the global behavior is given.

THEOREM 7.1. *Suppose all of the rest points of (2.4) are nondegenerate.*

(a) *If $\Omega = \{E_0\}$, then E_0 is an attractor for all trajectories with initial conditions in $\mathring{\Gamma}$.*

(b) *If $\Omega = \{E_0, E_1\}$ or $\{E_0, E_2\}$, then the nontrivial rest point is an attractor for all trajectories with initial conditions in $\mathring{\Gamma}$.*

(c) *If $\Omega = \{E_0, E_1, E_2\}$, then exactly one of E_1 or E_2 is an attractor for all trajectories with initial conditions in $\mathring{\Gamma}$.*

(d) *If $\Omega = \{E_0, E_1, E_2, E_*\}$, then E_* is an attractor for all trajectories with initial conditions in $\mathring{\Gamma}$.*

Proof. Case (a) is covered by Lemmas 4.2 and 6.1. This case occurs if and only if the inequalities (5.2) are satisfied.

Case (b) occurs if and only if one of the inequalities in (5.2a) or (5.2b) is reversed and the other set of inequalities holds. Lemma 6.1 again yields that E_1 belongs to the omega limit set of any trajectory with positive initial conditions.

For case (c) to hold, one inequality in each part of (5.2) must be reversed and (by Proposition E.2) both E_1 and E_2 cannot be stable. Thus, one of E_1 and E_2 is a local attractor, say E_1. (Lemma 5.1 provides the explicit conditions for determining which is stable.) Similarly, if case (d) holds, both E_1 and E_2 must be unstable.

The next lemma will be useful in our proof. It will be convenient to let $[P_1, P_2]_K$ denote the order interval, that is

$$[P_1, P_2]_K = \{p \mid P_1 \leq_K p \leq_K P_2\}.$$

LEMMA 7.2. *Suppose that E_1 and E_2 belong to Ω and that E_1 is unstable. If hypothesis* (c) *holds, then every solution starting at a point $x_0 = (u_1(0), u_2(0), v_1(0), v_2(0))$ belonging to $\mathring{\Gamma} \cap [E_2, E_1]_K$ satisfies*

$$\lim_{t \to \infty} \pi(x_0, t) = E_2.$$

If hypothesis (d) *holds, then every solution starting at a point $x_0 \in \mathring{\Gamma} \cap [E_2, E_1]_K$ satisfies*

$$\lim_{t \to \infty} \pi(x_0, t) = E_*.$$

Proof. Let x denote a point (u_1, u_1, v_1, v_2), and let $F(x)$ denote the vector field on the right-hand side of (2.4). Let λ denote the largest eigenvalue of J, the Jacobian of F at $x = E_1$. (J is displayed just prior to (5.3).) Since E_1 is unstable, $\lambda > 0$. An eigenvector $\bar{x} = (u, v)$, $u = (u_1, u_2)$, $v = (v_1, v_2)$ corresponding to λ must satisfy

$$Au + Bv = \lambda u,$$

$$Cv = \lambda v,$$

where A, B, and C are (respectively) the three nonzero matrices in the upper left, upper right, and lower right corners of J, λ is an eigenvalue of C, and the eigenvalues of A are both negative. By direct calculation (or by an appeal to Theorem A.5), one can choose $v > 0$. Then u satisfies

$$u = -(A - \lambda I)^{-1} Bv.$$

Since $Bv < 0$ and $-(A - \lambda I)^{-1} > 0$, by Theorem A.12(i) (or by direct calculation) one concludes that $u < 0$ and $\bar{x} <_K 0$.

Consider a point $x_r = E_1 + r\bar{x}$ for $0 < r$ and r sufficiently small. A calculation yields that

$$F(x_r) = F(E_1 + r\bar{x})$$
$$= rJ\bar{x} + o(r),$$

where $o(r)$ represents a term satisfying

$$\lim_{r \to 0}(o(r)/r) = 0.$$

Thus

$$F(x_r) = r(\lambda \bar{x} + o(r)/r) <_K 0$$

for all sufficiently small $r > 0$, since $\lambda \bar{x} <_K 0$. By Theorem C.2 and monotonicity of (2.4),

$$E_2 <_K \pi(x_r, t) \leq_K x_r <_K E_1$$

and $\pi(x_r, t)$ converges monotonically to a rest point for all small positive r. In Theorem 7.1(c) it is clear that

$$\lim_{t \to \infty} \pi(x_r, t) = E_2,$$

since E_2 is the only rest point contained in $[E_2, x_r]_K$. In Theorem 7.1(d), the inequalities may be strengthened to read

$$E_* <_K \pi(x_r, t) \leq_K x_r <_K E_1,$$

since $E_* <_K x_r <_K E_1$ for small positive r. Consequently, in case (d),

$$\lim_{t \to \infty} \pi(x_r, t) = E_*.$$

The assertions of Lemma 7.2 will now follow by comparison arguments. We give the argument in case (d). Let x_0 be an arbitrary point of $\mathring{\Gamma} \cap [E_2, E_1]_K$. There exist points x_r as before, and x_s chosen similarly with respect to E_2, such that

$$E_2 <_K x_s <_K x_0 <_K x_r <_K E_1.$$

Since

$$\lim_{t \to \infty} \pi(x_r, t) = E_* \quad \text{and} \quad \lim_{t \to \infty} \pi(x_s, t) = E_*,$$

it follows that

$$\lim_{t \to \infty} \pi(x_0, t) = E_*. \qquad \square$$

Proof of Theorem 7.1 (cont.). Suppose that the hypothesis of (c) holds and that E_2 is locally asymptotically stable. (There is a similar proof if E_1 is locally asymptotically stable, and by Theorem E.1 both are not.) To establish the conclusion in case (c), we must show that E_2 is globally asymptotically stable. Since $\mathring{\Gamma} \cap [E_2, E_1]_K$ is positively invariant for (2.4), and belongs to the basin of attraction of E_2 as a consequence of Lemma 7.2, one need only show that

$$\lim_{t \to \infty} \pi(x, t) = E_2$$

for $x \in \mathring{\Gamma} \setminus [E_2, E_1]_K$. Assume this is false; that is, assume there is an $x \in \mathring{\Gamma} \setminus [E_2, E_1]_K$ such that the $\gamma^+(x)$ does not enter $[E_2, E_1]_K$.

By Lemma 6.1, every limit point p of $\pi(x, t)$ must satisfy $p \in [E_2, E_1]_K$. By Lemma 5.1, $\pi(x, t)$ cannot converge to E_1. Hence, one can find a limit point $P = (p_1, p_2, p_3, p_4)$ with $p_3 > 0$ or $p_4 > 0$ or both. By the invariance of limit sets, $\pi(p, t)$ must have third and fourth entries positive for t positive and hence belong to $[E_2, E_1]_K$. By Lemma 7.2,

$$\lim_{t \to \infty} \pi(p, t) = E_2.$$

Thus

$$\lim_{t \to \infty} \pi(x, t) = E_2,$$

since its limit set contains the asymptotically stable rest point E_2. This completes the proof of case (c).

In case (d), both E_1 and E_2 are unstable. An argument similar to Lemma 7.2 shows that

$$\lim_{t \to \infty} \pi(x_0, t) = E_*$$

for $x_0 \in \mathring{\Gamma} \cap [E_2, E_*]_K$. If $x_0 \in \mathring{\Gamma} \cap [E_2, E_1]_K$, choose $x_1 \in \mathring{\Gamma} \cap [E_2, E_*]_K$ and $x_2 \in \mathring{\Gamma} \cap [E_*, E_1]_K$ such that

$$x_1 \leq_K x_0 \leq_K x_2.$$

(This can be accomplished by choosing x_1 close enough to E_2 and x_2 close enough to E_1.) Strong monotonicity and the fact that $\lim_{t \to \infty} \pi(x_i, t) = E_*$, $i = 1, 2$, implies that $\lim_{t \to \infty} \pi(x_0, t) = E_*$. Therefore, the positively invariant set $\mathring{\Gamma} \cap [E_2, E_1]_K$ belongs to the basin of attraction of E_*.

Suppose that x_0 is a point such that

$$\pi(x_0, t) \notin \mathring{\Gamma} \cap [E_2, E_1]_K$$

for any $t > 0$. As before, to prove the theorem it is sufficient to show that a limit point p of $\gamma^+(x_0)$ belongs to $\mathring{\Gamma} \cap [E_2, E_1]_K$, since all such points are attracted to E_*. Since $\pi(x_0, t)$ cannot converge to either E_1 or E_2 and remains outside of $\mathring{\Gamma} \cap [E_2, E_1]_K$, there must be a limit point of $\gamma^+(x_0)$ distinct from E_1 and E_2 in $\mathring{\Gamma} \cap [E_2, E_1]_K$, with $p_i \geq 0$, $p_1 + p_2 > 0$, and $p_3 + p_4 > 0$. By the invariance of limit sets and strong monotonicity, $\pi(p, t) \in \mathring{\Gamma} \cap [E_2, E_1]_K$. This completes the proof of case (d), since $\lim_{t \to \infty} \pi(p, t) = E_*$. □

8. Numerical Example

The theoretical results in section 6 show the global asymptotic behavior of the gradostat. The outcome of the competition was always a steady state, but there were several different possibilities. In each case, however, computations were provided not only to locate the rest points but also to determine their stability. It remains to show that these computations can be carried out. In doing so, a single parameter will be changed to illustrate all three of the nontrivial cases. (The total washout case is not illustrated because it is uninteresting and clearly achievable.)

The outcomes are presented in Table 8.1. The parameters a_u and m_u are fixed, which in turn fixes the value of E_1 as shown. The a_v term is fixed but m_v is assigned three different values (as shown), which in turn yields three different sets of coordinates for E_2. Given the coordinates of these points, α can be computed. The terms α_1 and α_2 determine the stability of E_2, while α_3 and α_4 determine the stability of E_1. All three permutations are illustrated: E_1 stable and E_2 unstable; E_1 unstable and E_2 stable; both E_1 and E_2 unstable, in which case E_* exists and is stable. As a consequence of Theorem 6.1, stability means global stability (with respect to the interior of the cone).

The interesting case is coexistence. Figure 8.1 illustrates the time course for this choice of parameters, showing the asymptotic approach to a steady state with all limits clearly above zero.

Table 8.1. *Predicted outcomes for different values of the parameter m_v with m_u, a_u, a_v fixed*

	m_v		
	15	20.18	25
E_2	$(0, 0, 0.3092, 0.1348)$	$(0, 0, 0.3123, 0.2076)$	$(0, 0, 0.3794, 0.2434)$
α_1	0.3917	0.6918	0.8841
$\alpha_1\alpha_2$	0.4590	0.9972	1.4036
E_2	Unstable	Unstable	Stable
α_3	1.0006	0.6557	0.3344
$\alpha_3\alpha_4$	1.6463	0.9984	0.4711
E_1	Stable	Unstable	Unstable
E_*	Does not exist	$(0.2140, 0.1471, 0.0968, 0.0638)$	Does not exist
Outcome	u wins	u and v coexist	v wins

Note: $m_u = 5$, $a_u = 1$, $a_v = 5$; $E_1 = (0.3098, 0.2122, 0, 0)$.
Source: [JSTW], data Copyright 1987, *Journal of Mathematical Biology*. Reproduced by permission.

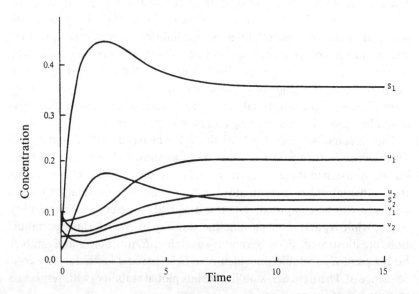

Figure 8.1. The coexistence case with the parameters from Table 8.1. (From [SW2], Copyright 1991, *Microbial Ecology*. Reproduced by permission.)

9. Discussion

The important conclusion from this chapter is that coexistence for two populations can occur in the gradostat for an open set in the parameter space. Since the purpose of the gradostat was to mirror behavior along a nutrient gradient, one can speculate that some of the coexistence observed in nature can be attributed to the existence of such gradients. The gradostat analyzed had only two vessels, but we anticipate that coexistence of two populations could occur in more complicated gradostats. A complete analysis of the asymptotic behavior was obtained in terms of the parameters of the system, and the computability of the conditions was also demonstrated. In the next chapter, the general gradostat will be studied and, unfortunately, some of this computability will be lost.

A surprising consequence of the analysis was that, when the interior rest point existed, it was unique and globally asymptotically stable.

Competition in a modified gradostat was considered in Smith and Tang [STa]. There the rate E between the vessels (called the *communication* rate) was allowed to differ from the rate D (the *dilution* rate) from the feed bottle and to the overflow vessel. This, of course, still maintains the assumption that the volume in each vessel is constant. It was shown that the outcome of competition can be sensitive to the ratio E/D in the following sense: As E/D is increased, first one competitor wins the competition, then coexistence occurs, and finally the second competitor wins. The analysis in the case $E \neq D$ is entirely similar to the case $E = D$ discussed in this chapter. In [STa], a number of operating diagrams were determined numerically. For fixed population parameters a_u, a_v, m_u, m_v, these operating diagrams depict regions in the E–D plane in which the various outcomes occur. One such diagram is shown as Figure 9.1 (see page 128).

128   *The Simple Gradostat*

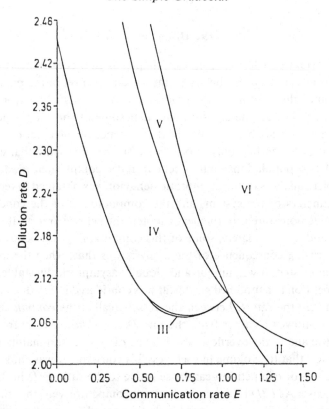

Figure 9.1. Operating diagram for two-species competition with a varying communication rate E different from the washout rate D. In region VI, $\Omega = \{E_0\}$;
in region V, $\Omega = \{E_0, E_2\}$; in region II, $\Omega = \{E_0, E_1\}$; in regions I and IV, $\Omega =
\{E_0, E_1, E_2\}$; in region III, $\Omega = \{E_0, E_1, E_2, E_*\}$. (From [STa], Copyright 1989,
Journal of Mathematical Biology. Reproduced by permission.)

6

The General Gradostat

1. Introduction

In the previous chapter the gradostat was introduced as a model of competition along a nutrient gradient. The case of two competitors and two vessels with Michaelis–Menten uptake functions was explored in considerable detail. In this chapter the restriction to two vessels and to Michaelis–Menten uptake will be removed, and a much more general version of the gradostat will be introduced. The results in the previous chapter were obtained by a mixture of dynamical systems techniques and specific computations that established the uniqueness and stability of the coexistence rest point. When the number of vessels is increased and the restriction to Michaelis–Menten uptake functions is relaxed, these computations are inconclusive. It turns out that unstable positive rest points are possible and that non-uniqueness of the coexistence rest point cannot be excluded. The main result of this chapter is that coexistence of two microbial populations in a gradostat is possible in the sense that the concentration of each population in each vessel approaches a positive equilibrium value. The main difference with the previous chapter is that we cannot exclude the possibility of more than one coexistence rest point.

Throughout this chapter we rely extensively on the results contained in Appendices A, B, and C. The presentation here follows closely that in [STW] (see also the review [SW2]).

The most straightforward generalization of the work of Chapter 5 would be simply to extend the number of vessels from two to an arbitrary number, say n. This is the original proposal of Lovitt and Wimpenny [LW1], and would take the form indicated in the schematic in Figure 1.1. Since we are considering only one nutrient, the input at the right-hand end is only medium without nutrient. Let S denote the nutrient concentration

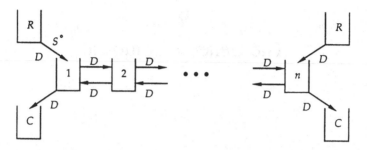

Figure 1.1. The standard *n*-vessel gradostat. The left vessel labeled R is a reservoir containing nutrient at concentration $S^{(0)}$, C is an overflow vessel, and D denotes the dilution rate. All vessels have the same volume.

and u and v the concentration of the two competitors. Then, using the subscript i to denote concentrations of S, u, and v in vessel i, the equations take the form

$$S_i' = (S_{i-1} - 2S_i + S_{i+1})D - \frac{u_i}{\gamma_u}f_u(S_i) - \frac{v_i}{\gamma_v}f_v(S_i),$$

$$u_i' = (u_{i-1} - 2u_i + u_{i+1})D + u_i f_u(S_i),$$

$$v_i' = (v_{i-1} - 2v_i + v_{i+1})D + v_i f_v(S_i),$$

$$i = 1, \ldots, n,$$

where

$$S_0 = S^{(0)}, \quad u_0 = v_0 = 0,$$

$$S_{n+1} = u_{n+1} = v_{n+1} = 0,$$

$$S_i(0) \geq 0, \ u_i(0) \geq 0, \ v_i(0) \geq 0,$$

and f_u and f_v satisfy the following:

(i) $f: \mathbb{R}^+ \to \mathbb{R}^+$ is continuously differentiable;
(ii) $f(0) = 0$ and $f'(S) > 0$ for $S > 0$.

The Michaelis–Menten function

$$f(S) = \frac{mS}{a+S}$$

is the prototypical example. In Chapter 2 we noted other functions with these properties that have appeared in the literature. The principal reason for allowing quite general monotone uptake functions in this case is that we are unable to obtain sharper results under the stronger hypotheses that the uptake functions are Michaelis–Menten.

The $S^{(0)}$ is the input concentration of nutrient (to the leftmost vessel), and D is the washout rate. These two parameters are under the control of the experimenter. The terms γ_u and γ_v are the yield coefficients. For convenience, one can scale substrate concentrations S_i by $S^{(0)}$, time by $1/D$ (making m_i nondimensional and $D = 1$), and microorganism concentrations by $y_u S^{(0)}$ and $y_v S^{(0)}$ to obtain the less cluttered system

$$S_i' = S_{i-1} - 2S_i + S_{i+1} - u_i f_u(S_i) - v_i f_v(S_i),$$

$$u_i' = u_{i-1} - 2u_i + u_{i+1} + u_i f_u(S_i),$$

$$v_i' = v_{i-1} - 2v_i + v_{i+1} + v_i f_v(S_i),$$ (1.1)

$$i = 1, ..., n,$$

where we use the same conventions as in the unscaled equations except that $S_0 = 1$. Hereafter, we refer to (1.1) as the "standard" n-vessel gradostat model.

Mathematically and experimentally there is no reason to connect the vessels linearly, to restrict the source to the left-hand vessel, or to keep the washout rates D equal so long as the volume of the fluid in each vessel is kept constant (see [S7]). We next describe a class of gradostat models which is sufficiently general to include all cases of biological interest and yet remain mathematically tractable.

Suppose that our gradostat consists of n vessels. Let E_{ij} be the constant (volumetric) flow rate from vessel j to i ($i \neq j$), with the convention that $E_{ii} = 0$ for $i = 1, ..., n$. Let V_i be the volume of fluid in the ith vessel, D_i the flow rate from a reservoir to vessel i ($D_i = 0$ if no such reservoir exists), $S_i^{(0)}$ the concentration of substrate in the reservoir feeding vessel i ($S_i^{(0)} = 0$ if $D_i = 0$), and C_i the flow rate from vessel i to an overflow vessel ($C_i = 0$ if no such vessel exists). The notation diag(β_i) is used to denote a diagonal matrix whose diagonal elements are given by β_i; E is the matrix of flow rates E_{ij}.

The rate of change of the vector $S(t) = (S_1(t), ..., S_n(t))$ at time t in a general gradostat, in the absence of any consumers, is given by

$$[\text{diag}(V_i)]S' = \bar{A}S + g,$$

where

$$\bar{A} = E - \text{diag}[C_i] - \text{diag}\left[\sum_{l=1}^{n} E_{li}\right],$$

$$g = (D_1 S_1^{(0)}, D_2 S_2^{(0)}, ..., D_n S_n^{(0)}).$$

Of course, the volume V_i of fluid in vessel i must be constant if this system is to describe a gradostat. This requires that

$$\sum_{j} E_{ij} + D_i = \sum_{l} E_{li} + C_i \qquad (1.2)$$

or that the volumetric flow rates in and out of any fixed vessel are the same.

It is convenient to multiply through by $[\operatorname{diag} V_i]^{-1}$ and obtain

$$S' = AS + e_0, \qquad (1.3)$$

where $e_0 = \operatorname{diag}(V_i^{-1})g$ and $A = \operatorname{diag}(V_i^{-1})\bar{A}$. We assume that at least one vessel receives substrate, since otherwise no microorganisms can survive. Mathematically this means that $S_i^{(0)} > 0$ for some i, so $e_0 \neq 0$. From the definition of A we have that $A_{ii} < 0$, since $E_{ii} = 0$ (excluding one trivial case of no input or output). In addition, A satisfies

$$A_{ij} \geq 0, \quad i \neq j, \qquad (1.4)$$

and

$$\sum_{j=1}^{n} A_{ij} = -V_i^{-1}D_i \leq 0 \qquad (1.5)$$

by virtue of (1.2). Our assumption that $S_i^{(0)} > 0$ for some i implies $D_i > 0$, and hence strict inequality holds in (1.5) for some i.

Our principal hypothesis is that the matrix A (or, equivalently, the matrix E) will be assumed to be irreducible (see Appendix A for a mathematical definition). This means that the set of vessels comprising the gradostat may not be partitioned into two disjoint non-empty subsets, I and J, such that no vessel in subset J receives input from any vessel in subset I. Note that the standard gradostat has this property but that the gradostat in Figure 1.2 does not. In that figure, the subset J consisting of the two vessels on the left receives no input from the vessel on the right, which comprises subset I. If that gradostat were of interest, one could

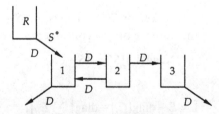

Figure 1.2. A three-vessel gradostat that does not satisfy the irreducibility hypothesis. The two vessels on the left receive no input from the vessel on the right. Arrows pointing down from the first and last vessel represent flow to overflow vessels not depicted. Notation is the same as in Figure 1.1.

simply treat the third vessel as an overflow vessel and effectively ignore it. The subset J could be viewed as the gradostat and the results to be described in this chapter could be applied to it. Once this subgradostat is understood, the input to the third vessel is known and it is then a simple matter to describe what happens in this last vessel. Thereby, the behavior of the entire gradostat can be worked out. More generally, if the matrix A does not have the property of being irreducible, then one can always partition the gradostat into irreducible subsets (subgradostats) that can be studied sequentially (see [BP]). In this sense, there is really no loss in generality in assuming irreducibility from the start. We also mention another way to view the hypothesis of irreducibility: for any pair of distinct vessels i and j, material from vessel i can travel to vessel j, though perhaps indirectly by first passing through intermediate vessels before entering vessel j.

While we focus on irreducible gradostats, reducible gradostats may be of biological interest as well. They could be used to model a system of mountain lakes situated at different elevations, where a lake at higher elevation feeds a lake at lower elevation.

Let $F_u = \text{diag}[f_u(S_1), f_u(S_2), ..., f_u(S_n)]$, and let F_v be defined analogously with subscript v replacing subscript u. Then, introducing consumption, the general model takes the form

$$S' = e_0 + AS - F_u(S)u - F_v(S)v,$$
$$u' = Au + F_u(S)u, \qquad (1.6)$$
$$v' = Av + F_v(S)v.$$

The standard model (1.1) is a special case of (1.6), where e_0 is the vector with first component equal to 1 and all others equal to 0 and where A is the matrix with -2 in the main diagonal entries, 1 in the superdiagonal and subdiagonal entries, and 0 elsewhere. Although the standard model is of primary interest, the general gradostat described by (1.6) can be treated with the same mathematics. In Figure 1.3, two irreducible gradostat configurations are described. See [LW1; LW2; S7] for other possibilities.

2. The Conservation Principle

In this section it will be shown that the conservation principle holds and allows the reduction of the $3n$-dimensional system (1.6) to a $2n$-dimensional system, eliminating the nutrient from consideration. In order to do this we will need to make use of those results of Appendix A dealing

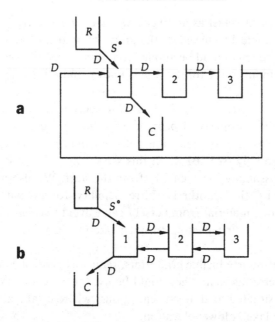

Figure 1.3. Two irreducible gradostats: **a** cyclic gradostat; **b** "dead-end" gradostat. Note that the inflow to each vessel balances the outflow.

with quasipositive matrices. Recall that a real matrix is quasipositive if its
off-diagonal entries are nonnegative. The matrix A in (1.6) has this property. (The reader may wish to briefly review Appendix A at this time.) If
M is a matrix, we denote by $\sigma(M)$ the set of eigenvalues of M and

$$s(M) = \max\{\Re(\lambda) : \lambda \in \sigma(M)\},$$

where $\Re(\lambda)$ denotes the real part of λ; $s(M)$ is called the *stability modulus*
of M. The next result is crucial for establishing the conservation principle. The properties of the matrix A in the hypotheses of this result have
already been noted in the previous section.

LEMMA 2.1. *Let $A = (a_{ij})$ be an irreducible matrix with nonnegative off-
diagonal entries. Suppose that*

$$\sum_{j=1}^{n} a_{ij} \leq 0$$

*for each i and that strict inequality holds for some i. Then $s(A) < 0$ and
$-A^{-1} > 0$.*

Proof. Since A has nonnegative off-diagonal entries and is irreducible, Theorem A.5 asserts that $s(A)$ is a simple eigenvalue of A, larger than the real parts of all other eigenvalues. The inequality hypothesis and the Gershgorin circle theorem (Theorem A.1) together imply that $s(A) \le 0$. If $s(A) < 0$, then the final assertion of the lemma follows from Theorem A.12. If $s(A) = 0$, then Theorem A.5 implies that there exists $x > 0$ such that $Ax = 0$. We can assume that $x_j \le 1$ for all j and that $x_i = 1$ for a non-empty subset I of indices. If J is the complementary set of indices then J is non-empty by our assumptions on the row sums of A. For $i \in I$ we have

$$0 = \sum_l a_{il} x_l = \sum_{l \in I} a_{il} + \sum_{l \in J} a_{il} x_l$$

$$= \sum_j a_{ij} + \sum_{l \in J} a_{il}(x_l - 1)$$

$$\le \sum_{l \in J} a_{il}(x_l - 1).$$

As $x_l < 1$ for $l \in J$ and $a_{il} \ge 0$, it must be that $a_{il} = 0$ for all $l \in J$; otherwise, the sum is negative. Thus $a_{ij} = 0$ for all $i \in I$ and $j \in J$. This contradicts the irreducibility of A and so proves the lemma. $\square$

The conservation principle is stated next. Like its analog, Lemma 2.2 of Chapter 5, it states that the total nutrient in each vessel, consisting of both pure nutrient and that making up the biomass of microorganisms, approaches a constant value (which depends on how the vessels making up the gradostat are configured) exponentially fast. As in the previous chapter, the conservation principle is crucial for our analysis because it allows the reduction of (1.6) to a lower-dimensional dynamical system.

LEMMA 2.2. *Solutions of* (1.6) *with nonnegative initial data exist and are nonnegative and bounded for* $t \ge 0$; *moreover,*

$$\lim_{t \to \infty} S(t) + u(t) + v(t) = z,$$

where $z > 0$ *is the unique solution of*

$$Az + e_0 = 0.$$

Proof. Solutions of (1.6) remain nonnegative by Proposition B.7. Let $w(t) = S(t) + u(t) + v(t)$. Then

$$w' = Aw + e_0, \qquad w(0) = S(0) + u(0) + v(0) \ge 0.$$

By Lemma 2.1, the equation $Az + e_0 = 0$ has a unique solution $z = (-A)^{-1}e_0$, which is positive because $e_0 \geq 0$ does not vanish and $(-A)^{-1} > 0$. Putting $y = w - z$ in the linear system, we find that $y' = Ay$. Since $s(A) < 0$, $\lim_{t \to \infty} y(t) = 0$, completing the proof.   □

It follows from (1.3) that $S = z$ is the steady-state distribution of nutrient in the gradostat when no species of microorganisms are present ($u = v = 0$). In the case of the standard gradostat of Figure 1.1, modeled by (2.1), one can easily calculate that $z_i = S^{(0)}[1 - i/(n+1)]$, $1 \leq i \leq n$. The concentration of nutrient declines linearly from the value $S^{(0)}$ in the leftmost vessel to the value $S^{(0)}/(n+1)$ in the rightmost vessel, just as one might expect by analogy with a diffusion process. Thus, a nutrient gradient is established. This observation explains the term "gradostat" coined by Wimpenny and Lovitt to describe this continuous culture device.

Lemma 2.2 says that on the omega limit set, solutions of the system (1.6) satisfy

$$u' = [A + F_u(z - u - v)]u, \quad u(0) = u_0 \geq 0,$$
$$v' = [A + F_v(z - u - v)]v, \quad v(0) = v_0 \geq 0$$

(2.1)

on

$$\Gamma = \{(u, v) \in \mathbb{R}^{2n}_+ : u + v \leq z\}.$$

Note that Γ is positively invariant for (2.1). In fact, solutions remain nonnegative (by Proposition B.7) and, if $u_i + v_i = z_i$,

$$(u_i + v_i)' = [A(u + v)]_i \leq (Az)_i = -(e_0)_i \leq 0.$$

From this inequality we conclude that the vector field points into Γ on the hyperplane $u_i + v_i = z_i$.

We will use several notations for a solution of (2.1). The notation

$$(u(t, u_0, v_0), v(t, u_0, v_0))$$

will be used to indicate the solution of (2.1) satisfying

$$(u(0, u_0, v_0), v(0, u_0, v_0)) = (u_0, v_0).$$

The initial condition (u_0, v_0) may be dropped from the notation when no confusion can occur over which initial condition is being considered. We will also find it convenient to let $x = (u_0, v_0)$ and write $\pi(x, t)$ for the solution of (2.1), that is,

$$\pi(x, t) = (u(t, u_0, v_0), v(t, u_0, v_0)) = \pi(u_0, v_0, t).$$

When $v_0 = 0$ in (2.1) then v vanishes identically, so we write $u(t, u_0)$ for the solution of the first of equations (2.1). A similar notation is used for v if $u_0 = 0$.

The next result states that a solution of (2.1) - starting at $t = 0$ from an initial distribution of each microbial population among the vessels of the gradostat - which has the property that each population is present in some (not necessarily the same) vessel, thereafter has the property that both populations are present in every vessel. A population must occupy all of the vessels or none of the vessels. This property is an important consequence of the irreducibility assumption on the matrix A. Obviously, it can fail for the gradostat of Figure 1.2.

LEMMA 2.3. *If $(u(t), v(t))$ satisfies* (2.1) *and $u_i(0) > 0$ (resp. $v_i(0) > 0$) for some i, then $u(t) > 0$ (resp. $v(t) > 0$) for all $t > 0$.*

Proof. The linear equation $u'(t) = A(t)u(t)$ is satisfied by $u(t)$, where $A(t) = A + F_u(z - u(t) - v(t))$ is quasipositive and irreducible. If $u_i(0) > 0$ for some i, then $u(t) > 0$ for $t > 0$ follows from the positivity of the fundamental matrix (see Theorem B.3). $\qquad\Box$

3. Growth without Competition

When there is only one population, say u, one equation in (2.1) drops out. The growth of population u is then governed by

$$u' = [A + F_u(z - u)]u, \quad u(0) = u_0. \tag{3.1}$$

The initial data u_0 must belong to the order interval $[0, z] = \{u : 0 \le u \le z\}$. It is easy to see that $[0, z]$ is positively invariant. Solutions of (3.1) exist for all $t \ge 0$ and are nonnegative and bounded, by Lemma 2.2 and the fact that $(u(t), 0)$ is a solution of (2.1). The next result is the analog of Lemmas 4.2 and 4.3 of Chapter 5. It describes the global behavior of solutions of (3.1). The proof, which is similar to the one given in [Ta], makes use of the fact that (3.1) generates a strongly monotone dynamical system in the interior of $[0, z]$ (see Appendix C). The reader may be well advised to skip the proof, which is quite technical, on first reading. Following the proof, a simple argument is given which makes the result seem plausible.

THEOREM 3.1. *If $s(A + F_u(z)) \le 0$, then $\lim_{t \to \infty} u(t, u_0) = 0$ for every $u_0 \in [0, z]$. If $s(A + F_u(z)) > 0$, then there exists a unique equilibrium $\hat{u}$, $\hat{u} > 0$ and $\lim_{t \to \infty} u(t, u_0) = \hat{u}$, for every nontrivial u_0 in $[0, z]$.*

Proof. Suppose that $s = s(A + F_u(z)) < 0$. Then

$$u' \le [A + F_u(z)]u,$$

so $u(t) \le y(t)$ where $y(0) = u(0)$ and y satisfies $y' = [A + F_u(z)]y$ (by Theorem B.1). Since $y(t) \to 0$ as $t \to \infty$, the same holds for $u(t)$. The case where $s = 0$ is more difficult; we refer the reader to [S7]. Now suppose that $s > 0$ and let v be a corresponding eigenvector for $A + F_u(z)$ such that $v > 0$. Such an eigenvector exists by Theorem A.5. Put $G(u) = [A + F_u(z - u)]u$, the right-hand side of (3.1). Then $G(rv) = rG'(0)v + o(r)$, where $G'(0)$ is the Jacobian of G at $u = 0$ and $o(r)/r \to 0$ as $r \to 0$. It follows that $G(rv) > 0$ for all small positive r. By Theorem C.2(a), we conclude that the solution $u(t)$ of (3.1) satisfying $u(0) = rv$ converges to an equilibrium $e \ge rv$ belonging to $[0, z]$. Therefore, if $s > 0$ then there exists a positive rest point of (3.1). Lemma 3.3 implies that any nontrivial rest point in $[0, z]$ must be positive. We must show that this rest point is both unique and a global attractor for $[0, z] \setminus \{0\}$.

Suppose that the uniqueness of the positive rest point has been established and denote it by $\hat{u}$. If $u(t)$ is any solution of (3.1) with $u_i(0) > 0$ for some i, then $u(t) > 0$ for $t > 0$ by Lemma 3.3, so we can assume without loss of generality that $u(0) > 0$. The dynamical system generated by (3.1) is strongly monotone in the interior of $[0, z]$. Let J be the portion of the line joining 0 to $u(0)$ which lies in the interior of $[0, z]$. By Theorem C.5, the solutions starting at all (but at most finitely many) points of J converge to a rest point. In fact, these convergent solutions must converge to $\hat{u}$, since the solutions starting at rv converge to $\hat{u}$ for all small $r > 0$ and we may compare any solution starting on J with one of these. The solution $u(t)$ must converge to $\hat{u}$, since we can find points u_1, u_2 on J satisfying $u_1 < u(0) < u_2$ and such that the solutions $u_i(t)$ starting at u_i converge to $\hat{u}$ and $u_1(t) \le u(t) \le u_2(t)$ holds for all $t \ge 0$. We have established the convergence of nontrivial solutions of (3.1) to $\hat{u}$, provided this rest point is unique. We remark that if A is symmetric, as in (1.1), then (3.1) is a gradient system [Ta]; in this case, convergence to equilibrium could be established by appealing to the LaSalle corollary of Chapter 2.

Choose $p > 0$ sufficiently large so that, for each i, the function $u_i \to pu_i + f_u(z_i - u_i)u_i$ has a positive derivative on $[0, z_i]$. It is easy to see that nontrivial rest points of (3.1) are positive fixed points of the map $T: [0, z] \to \mathbb{R}^n_+$ defined by

$$Tu = [-(A - p)^{-1}][F_u(z - u) + pI]u,$$

where I is the identity matrix. Note that the matrix in the first bracket on the right side is positive for the same reasons that $[-A]^{-1}$ is. The map T is monotone in the sense that if $x \le y$ then $Tx \le Ty$. Furthermore, for each $u > 0$ and for each $r \in (0, 1)$, there exists $q > 0$ (depending on u, r) such that $T(ru) \ge (1+q)rT(u)$. Now if u_1 and u_2 are distinct positive fixed points of T, then one of the relations $u_1 \le u_2$ or $u_2 \le u_1$ must fail to hold. Suppose that $u_2 \le u_1$ does not hold. Then we can find $r \in (0, 1)$ with the property that $ru_2 \le u_1$ and r is the largest positive number with this property. But then we have a contradiction to the maximality of r from

$$(1+q)ru_2 = (1+q)rTu_2 \le T(ru_2) \le Tu_1 = u_1.$$

This establishes the uniqueness of the nontrivial rest point. $\qquad \square$

In words, Theorem 3.1 states that a microorganism can either survive in the gradostat, eventually reaching a positive equilibrium concentration in each vessel, or it cannot survive in the gradostat and it is washed out of every vessel. One number, the stability modulus of the matrix $A + F_u(z)$, determines which of the two outcomes holds. The terms A and z contain all the information about the physical apparatus: the flow rates between vessels, the volumes of fluid in each vessel, and the nutrient inputs to each vessel. The function F_u contains the relevant information about the microorganism. For example, if f_u is Michaelis–Menten then the maximum growth rate m_u and the half-saturation constant a_u are the key biological data. Considering the freedom allowed in the construction of the general gradostat, it is remarkable that, according to the model, microorganism survival is determined by the sign of a single number, $s = s(A + F_u(z))$. If $s < 0$ then washout occurs but if $s > 0$ then survival is assured.

In fact, this result is not difficult to anticipate. Imagine introducing an infinitesimally small concentration of a microbial population into a gradostat that has been operating for a considerable time without any microbes present. According to Lemma 2.2 and (1.3), the nutrient concentration will have had time to equilibrate to the level z_i in vessel i. The microbial population will see this level of nutrient in each vessel and therefore its concentration $u(t)$ will approximately obey the linear equation $u' = [A + F_u(z)]u$, where we have neglected to subtract u from z in the argument of F_u because we are assuming that u is negligible compared to z. Thus the microbial population will grow if $s > 0$ and decay if $s < 0$.

In the corresponding single-population equation for v, we denote the positive equilibrium, which exists if and only if $s(A + F_v(z)) > 0$, by $\bar{v}$. For the full system (2.1), both $(\hat{u}, 0)$ and $(0, \bar{v})$ will be equilibria.

4. Competition

We turn now to the question of competition. In the previous chapter, we established a classification of the dynamic behavior based on the set of rest points. Unfortunately, our computations – which established the stability of any interior rest point and thereby led to the conclusion that such an equilibrium is unique in the case of two vessels and Michaelis–Menten uptake functions – are extremely difficult for n vessels and general uptake functions [HSo], so the results in this case are not as simple as in that chapter. In the present context we attempt to classify the dynamics in terms of both the set Ω of rest points and the sign of the stability modulus of certain key matrices. The theory of monotone dynamics is then used to resolve global questions. The principal result is Theorem 4.4.

There are three obvious candidates for equilibria:

$$E_0 = (0,0), \quad E_1 = (\hat{u},0), \quad E_2 = (0,\bar{v}),$$

where $\hat{u}$ and $\bar{v}$ were defined in Section 3. Throughout this section we let Ω denote the set of equilibria of (2.1). Clearly, Ω depends on the parameters of (2.1). The next result summarizes what we already know about Ω.

LEMMA 4.1. *The following hold:*

(a) E_0 *always exists;*
(b) E_1 *exists if and only if* $s(A + F_u(z)) > 0$;
(c) E_2 *exists if and only if* $s(A + F_v(z)) > 0$;
(d) *if* $E_* = (\bar{u}, \bar{v})$ *is an equilibrium distinct from* E_0, E_1, E_2, *then* $E_* > 0$.

Proof. Part (a) is clear, and (b) and (c) follow from Theorem 3.1. Part (d) follows from Lemma 2.3. □

The stability of the rest points is determined by the linearization at these points. The variational matrix for (2.1) takes the form

$$J = \begin{bmatrix} A + F_u(z-u-v) - D_u & -D_u \\ -D_v & A + F_v(z-u-v) - D_v \end{bmatrix}, \quad (4.1)$$

where

$$D_u = \mathrm{diag}(u_1 f_u'(z_1 - u_1 - v_1), \ldots, u_n f_u'(z_n - u_n - v_n))$$

and

$$D_v = \mathrm{diag}(v_1 f_v'(z_1 - u_1 - v_1), \ldots, v_n f_v'(z_n - u_n - v_n)).$$

At E_0, both D_u and D_v vanish so that J has only the two diagonal blocks $A+F_u(z)$ and $A+F_v(z)$. Clearly, the eigenvalues of J consist of the union of the eigenvalues of the two blocks. Consequently, E_0 is stable in the linear approximation (all eigenvalues have negative real parts) if and only if $s(A+F_u(z)) < 0$ and $s(A+F_v(z)) < 0$. It is unstable if either of these is positive, in which case Ω contains either E_1 or E_2.

We now turn to the question of the stability of E_1. The main results are stated formally, since they are crucial for our later analysis. The basic idea is simple. Imagine that a microbial population is allowed to grow in the gradostat in the absence of competition. This population will approach the equilibrium concentration $\hat{u}_i$ in vessel i corresponding to the rest point E_1. Now add an infinitesimally small concentration of the competing population and ask whether or not it can survive. If the vector of concentrations of the competing population is $v(t)$, then $v(t)$ approximately satisfies the linear system $v' = [A+F_v(z-\hat{u})]v$, since v is assumed to be negligibly small. Consequently, the competing population v survives or decays as $s(A+F_v(z-\hat{u}))$ is positive or negative. This is all made precise in what follows.

LEMMA 4.2. *If E_1 exists then $s = s(A+F_v(z-\hat{u}))$ is an eigenvalue of the variational matrix corresponding to E_1, which is asymptotically stable in the linear approximation if and only if $s < 0$. If $s > 0$, then the variational matrix at E_1 has a corresponding eigenvector w satisfying $w <_K 0$.*

Proof. At E_1 the matrix (4.1) takes the form

$$J = \begin{bmatrix} A+F_u(z-\hat{u})-D_u & -D_u \\ 0 & A+F_v(z-\hat{u}) \end{bmatrix},$$

where submatrices are $n \times n$. The existence of E_1 implies that

$$[A+F_u(z-\hat{u})]x = 0$$

has the solution $x = \hat{u} > 0$. This fact and Theorem A.5 imply that

$$s(A+F_u(z-\hat{u})) = 0$$

and that $\hat{u}$ is the corresponding eigenvector. Since $D_u \geq 0$ is not the zero matrix, $s(A+F_u(z-\hat{u})-D_u) < s(A+F_u(z-\hat{u})) = 0$ by Theorem A.5. The structure of the matrix J implies that the eigenvalues of J are just the eigenvalues of the two blocks on the diagonal. Since the upper block is stable, the stability of E_1 is determined by s. If it is negative then E_1 is stable in the linear approximation, and if it is positive then E_1 is unstable in the linear approximation.

If $s > 0$, a corresponding eigenvector $w = (w_1, w_2)$ satisfies

$$[A + F_u(z - \hat{u}) - D_u]w_1 - D_u w_2 = sw_1,$$

$$[A + F_v(z - \hat{u})]w_2 = sw_2.$$

Thus we may take $w_2 > 0$, by Theorem A.5. Once w_2 is fixed, w_1 is determined by

$$w_1 = [A + F_u(z - \hat{u}) - D_u - sI]^{-1}D_u w_2 < 0,$$

where we have used $-[A + F_u(z - \hat{u}) - D_u - sI]^{-1} > 0$. This holds by first appealing to Theorem A.5 to conclude that $s(A + F_u(z - \hat{u}) - D_u - sI) = s(A + F_u(z - \hat{u}) - D_u) - s \le s(A + F_u(z - \hat{u})) - s = -s$; then apply Theorem A.12. Thus, $w <_K 0$. $\qquad\qquad\square$

A similar result holds concerning the stability of E_2, which is asymptotically stable if $s = s(A + F_u(z - \bar{v})) < 0$ and unstable if $s > 0$.

An important observation concerning (2.1) is that it generates a strongly monotone dynamical system in the interior of Γ. This observation is immediate from (4.1) and Theorem C.1. Let $x = (u, v)$ and $y = (\bar{u}, \bar{v})$ be two points of $\mathbb{R}^{2n}$. We write $x \le_K y$ in case $u \le \bar{u}$ and $\bar{v} \le v$; we write $x <_K y$ if $u < \bar{u}$ and $\bar{v} < v$. If $x \le_K y$ we let $[x, y]_K = \{z : x \le_K z \le_K y\}$.

If x and y are distinct points belonging to the interior of Γ and satisfying $x \le_K y$, then $\pi(x, t) <_K \pi(y, t)$ for $t > 0$. An even stronger monotonicity property is needed. If x and y are distinct points belonging to Γ, $x \le_K y$, and either $x > 0$ or $y > 0$, then $\pi(x, t) <_K \pi(y, t)$ holds for $t > 0$.

The monotonicity properties of (2.1) provide additional information about Ω which is cataloged in the following result. The most important of these is the assertion that a positive rest point cannot exist unless each competitor can survive alone in the gradostat, that is, unless both E_1 and E_2 exist. Later we will see that coexistence of the two populations is possible, but only if there exists a positive rest point. Therefore, coexistence is only possible when each competitor can survive alone in the gradostat.

THEOREM 4.3. *Let Ω denote the set of rest points of (2.1). Then:*

(a) *if there exists an equilibrium $E_* \in \Omega$ satisfying $E_* > 0$, then Ω contains E_1 and E_2;*

(b) *if $\Omega = \{E_0\}$, then E_0 is a global attractor for (2.1);*

(c) *if* $\Omega = \{E_0, E_1\}$ *(resp.* $(\Omega = \{E_0, E_2\})$, *then the nontrivial rest point attracts all orbits of* (2.1) *with initial conditions* $(u_0, v_0) \in \Gamma$ *satisfying* $u_0 \neq 0$ *(resp.* $v_0 \neq 0$).

Proof. If $(u(t), v(t))$ is a solution of (2.1) then $u' \leq [A + F_u(z - u)]u$, so (by Theorem B.1) $u(t) \leq \bar{u}(t)$, where $\bar{u}$ is a solution of (3.1) satisfying $\bar{u}(0) = u(0)$. If E_1 does not exist then $\bar{u}(t) \to 0$ as $t \to \infty$, by Theorem 3.1. It follows that $u(t) \to 0$ as $t \to \infty$ if E_1 does not exist. A similar argument shows that $v(t) \to 0$ as $t \to \infty$ if E_2 does not exist. These observations prove (a) and (b).

If $\Omega = \{E_0, E_1\}$ then the omega limit set of every solution of (2.1) belongs to $\{(u, v) \in \Gamma : v = 0\}$, by the arguments of the previous paragraph. The equilibrium E_1 is asymptotically stable because $s(A + F_v(z - \hat{u})) < s(A + F_v(z)) \leq 0$, where we have used Theorem A.5 and Lemma 4.1(c). Therefore, if E_1 belongs to the omega limit set then it must *be* the omega limit set. As the omega limit set is invariant, we conclude (using Theorem 3.1) that the limit set is E_1 provided that it contains any point $(u, 0)$ with $u \neq 0$. Suppose that $x(t) = (u(t), v(t))$ is a solution satisfying $u(0) \neq 0$ and $v(0) \neq 0$. Then $x(t) > 0$ for $t > 0$. If $x(t) \to 0$ as $t \to \infty$ then, by monotonicity, so do all solutions $\bar{x}(t)$ with initial conditions $\bar{x}(0)$ satisfying $\bar{x}(0) \leq_K (u(1), v(1))$. Therefore E_0 attracts an open set of initial values, and it follows (see [Hi3]) that $s(J_0) \leq 0$ where J_0 is the Jacobian (4.1) evaluated at E_0. But this contradicts Lemma 4.1(b) and so proves that the limit set of $x(0)$ must contain a point $(u, 0)$ with $u \neq 0$. $\qquad\square$

We can now state the main result of this chapter. It provides sufficient conditions for the coexistence of the two populations in the gradostat and ensures that Ω contains a positive rest point E_*. In fact, it guarantees that the two populations are uniformly persistent in the sense of Appendix D. Briefly, Theorem 4.4 states that coexistence holds if each population can successfully invade its competitor's rest point.

THEOREM 4.4. *Suppose that* E_1 *and* E_2 *exist and*

$$s(A + F_u(z - \bar{v})) > 0, \qquad s(A + F_v(z - \hat{u})) > 0.$$

Then there exist rest points (*which may coincide*)

$$E_* = (u_*, v_*) > 0 \quad and \quad E_{**} = (u_{**}, v_{**}) > 0$$

of (2.1) *belonging to* Γ *and satisfying*

$$E_2 <_K E_{**} \leq_K E_* <_K E_1. \qquad (4.2)$$

The rest point E_* attracts all solutions $(u(t), v(t))$ with

$$E_* \leq_K (u(0), v(0)) \leq_K E_1$$

such that $v(0) \neq 0$; E_{**} attracts all solutions with

$$E_2 \leq_K (u(0), v(0)) \leq_K E_{**}$$

such that $u(0) \neq 0$. The set

$$O = [E_{**}, E_*]_K \cap \Gamma$$

attracts all orbits corresponding to initial data $(u_0, v_0) \in \Gamma$ satisfying $u_0 \neq 0$ and $v_0 \neq 0$. If $E_* = E_{**}$ then E_* attracts all orbits as before. If Ω has no accumulation points in Γ, then there exists a subset of Γ whose complement has zero Lebesgue measure consisting of points (u_0, v_0) for which $\pi(u_0, v_0, t)$ approaches a rest point in O as $t \to \infty$. Both E_* and E_{**} have the property that the stability modulus of the Jacobian of (2.1) at these points is not positive.

Figure 4.1 describes the theorem schematically.

Proof. Consider a solution of (2.1) starting from a point $x_r = E_1 + rw$, where $w <_K 0$ is the eigenvector corresponding to $s > 0$ in Lemma 4.2. Let $\mathcal{F}(x)$ denote the vector field defined by (2.1), where $x = (u, v)$. Let J denote the Jacobian of $\mathcal{F}$ at E_1. Then

$$\mathcal{F}(x_r) = \mathcal{F}(E_1) + rJw + o(r) = rsw + o(r) <_K 0$$

for small positive r, since $o(r)/r \to 0$ as $r \to 0$. Choose r_0 so small that the displayed inequality holds for $0 < r \leq r_0$. It follows from Theorem C.2(a) that the solution starting at x_r, $\pi(x_r, t)$, converges monotonically to an equilibrium E_r of (2.1) satisfying $E_2 \leq_K E_r \leq_K \pi(x_r, t) <_K x_r <_K E_1$ for $t \geq 0$. That $E_r = E_{r_0} = E_*$ is independent of r for $r_0 \geq r > 0$ will be established as follows. If $0 < r_1 < r_2 \leq r_0$ then $x_{r_2} <_K x_{r_1} <_K E_1$, so applying $\pi(\cdot, t)$ and monotonicity and letting $t \to \infty$ we find that $E_{r_2} \leq_K E_{r_1}$. Since $E_r <_K x_r = E_1 + rw$, we can choose $h > 0$ such that $r + h \leq r_0$, and $E_r \leq_K E_1 + (r+h)w = x_{r+h}$, and such that h is maximal with these properties. Then $E_r = \pi(E_r, t) \leq_K \pi(x_{r+h}, t)$, which implies that $E_r \leq_K E_{r+h}$. Since $E_{r+h} \leq_K E_r$, we conclude that $E_{r+h} = E_r$. Consequently, $E_r = E_{r+h} <_K E_1 + (r+h)w$, and the maximality of h implies that $r + h = r_0$. Thus $E_r = E_{r_0}$, and since r was an arbitrary element of $0 < r \leq r_0$, the equality holds for all such r.

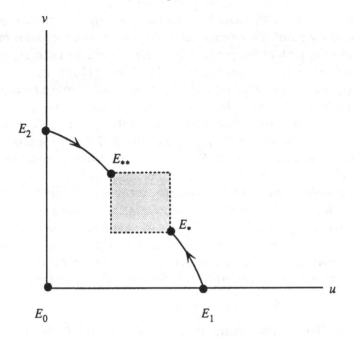

Figure 4.1. The attracting hypercube O is shaded. Each axis represents a copy of $\mathbb{R}^n_+$. The monotone trajectory emanating from near E_1 and converging to E_*, and a similar one emanating from near E_2 and converging to E_{**}, are described in the proof of Theorem 4.4.

If $(u_0, v_0) \in [E_*, E_1]_K$ and $v_0 \neq 0$ then $E_* <_K \pi(u_0, v_0, t) <_K E_1$ for $t > 0$, by strong monotonicity and Lemma 2.3. For all sufficiently small positive r, $\pi(u_0, v_0, 1) <_K x_r$ and monotonicity implies that

$$\lim_{t \to \infty} \pi(u_0, v_0, t) = E_*.$$

The key facts that allowed us to construct monotone converging orbits emanating from the ray through E_1 in the direction w were that $s(J) > 0$ and that the corresponding eigenvector could be chosen to satisfy $w <_K 0$ or $0 <_K w$. It follows that a similar construction can be carried out from E_2, where now the eigenvector w can be chosen to satisfy $0 <_K w$. This construction will yield the rest point E_{**} in the same way that E_* was obtained. It is easy to see that (4.2) must hold. Furthermore, if $(u_0, v_0) \in [E_2, E_{**}]$ and $u_0 \neq 0$ then $\pi(u_0, v_0, t)$ converges to E_{**} as $t \to \infty$.

If $s(J_*) > 0$, where J_* is the Jacobian of (2.1) at E_*, then (5.1) and Theorem A.6 give a corresponding eigenvector $0 <_K w_*$. But then a monotone increasing (with respect to $<_K$) orbit starting at a point of the ray through

E_* in the direction w_* could be constructed exactly as in the first paragraph of the proof. This would clearly contradict the conclusions of the second paragraph of the proof. Thus $s(J_*) \le 0$ and a similar argument shows that $s(J_{**}) \le 0$, where J_{**} is the Jacobian of (2.1) at E_{**}.

If $(u_0, v_0) \in \Gamma$ satisfies $E_2 <_K (u_0, v_0) <_K E_1$, then choose a point x_r on the ray through E_1 in the direction w such that $\pi(u_0, v_0, 1) <_K x_r$. Since $\pi(x_r, t) \to E_*$ as $t \to \infty$, monotonicity implies that $(u, v) \le_K E_*$ for every point (u, v) of the positive limit set of the orbit of (2.1) starting at (u_0, v_0). A similar argument establishes $E_{**} \le_K (u, v)$. Therefore, the limit set corresponding to the orbit through (u_0, v_0) is contained in O.

We complete the proof by showing that the omega limit set Λ of a point $(u(0), v(0)) \in \Gamma$ satisfying $u(0) \ne 0$ and $v(0) \ne 0$ is contained in O. We first show that $E_2 \le_K x \le_K E_1$ for each $x \in \Lambda$. By Lemma 2.3, $(u(t), v(t)) > 0$ for $t > 0$. The solution $(\bar{u}(t), \bar{v}(t))$ of (2.1) with $(\bar{u}(0), \bar{v}(0)) = (u(0), 0)$ must satisfy $(u(t), v(t)) <_K (\bar{u}(t), \bar{v}(t)) = (\bar{u}(t), 0)$ for $t > 0$ by strong monotonicity, since $(u(0), v(0)) \le_K (u(0), 0)$. Similarly we have

$$(\hat{u}(t), \hat{v}(t)) <_K (u(t), v(t)) \quad \text{for } t > 0,$$

where $(\hat{u}(t), \hat{v}(t))$ is the solution of (2.1) satisfying $(\hat{u}(0), \hat{v}(0)) = (0, v(0))$. Letting $t \to \infty$ in the inequalities $(\hat{u}(t), \hat{v}(t)) = (0, \hat{v}(t)) <_K (u(t), v(t)) <_K (\bar{u}(t), 0)$ and observing that $(0, \hat{v}(t)) \to E_2$ and $(\bar{u}(t), 0) \to E_1$ as $t \to \infty$, we see that every limit point of the orbit starting from $(u(0), v(0))$ must belong to $\Gamma \cap [E_2, E_1]_K$.

If $x \in \Lambda$ then $x = \pi(y, 1)$ for some $y \in \Lambda$, since Λ is invariant. This and Lemma 2.3 imply that if $x = (u, v) \in \Lambda$ then either $u = 0$ ($v = 0$) or $u > 0$ ($v > 0$). Furthermore, if $u > 0$ and $v > 0$ then $E_2 <_K x <_K E_1$. In this case, as x is a limit point of $(u(t), v(t))$, it follows that $E_2 <_K (u(t_0), v(t_0)) <_K E_1$ for some $t_0 > 0$. By arguments in a preceding paragraph we may then conclude that $\Lambda \subset O$. Therefore, this containment holds whenever Λ contains a point $x = (u, v)$ with $u > 0$ and $v > 0$. Suppose that Λ contains no such point. Then every point $x = (u, v) \in \Lambda$ satisfies either $u = 0$ or $v = 0$ but not both, since E_0 does not belong to Λ. Furthermore, Λ contains the entire orbit through x together with the alpha and omega limit sets of this orbit. Consider the case where $x = (u, 0)$. The omega limit set of the orbit through x is clearly E_1, and the alpha limit set of any entire orbit that belongs to $[0, z]$ is E_0 unless $x = E_1$. Since $E_0 \notin \Lambda$, we conclude that E_1 is the only point of the form $(u, 0)$ in Λ. A symmetric argument shows that E_2 is the only point of Λ of the type $(0, v)$. Since Λ is connected, it follows that $\Lambda \subset O$, $\Lambda = E_1$, or $\Lambda = E_2$.

Suppose that $\Lambda = E_1$. Then $u(t) + v(t) \to \hat{u}$ as $t \to \infty$. By continuity of $s(\cdot)$, we can find $\epsilon > 0$ such that

$$s = s(A + F_v(z(1-\epsilon) - \hat{u})) > 0.$$

Let $\bar{v} > 0$ be the eigenvector corresponding to s. Then there exists $t_0 > 0$ such that $z - u(t) - v(t) \geq z(1-\epsilon) - \hat{u}$ for all $t \geq t_0$. Consequently,

$$v'(t) \geq [A + F_v(z(1-\epsilon) - \hat{u})] v(t)$$

for $t \geq t_0$. As $v(t_0) > 0$, we may choose $\delta > 0$ such that $v(t_0) \geq \delta \bar{v}$. Then, by Theorem B.1,

$$v(t) \geq \delta \bar{v} e^{s(t-t_0)}, \quad t \geq t_0.$$

This contradiction to the boundedness of $v(t)$ shows that $\Lambda = E_1$ cannot hold. A similar argument shows that $\Lambda = E_2$ cannot hold. Therefore, $\Lambda \subset O$ as asserted.

Our assertion – that almost all initial conditions (u_0, v_0), in the sense of Lebesgue measure, belong to orbits converging to a rest point in O if the set Ω has no accumulation points – follows from Theorem C.8. $\quad\square$

The hypotheses of Theorem 4.4 are stable to perturbation, in the sense that if they hold for particular values of the parameters and uptake functions then they continue to hold for all nearby values of the parameters and for nearby uptake functions. By a "nearby" uptake function we mean an uptake function with the properties that (a) it satisfies requirements (i) and (ii) of Section 1, and (b) it and its derivative are uniformly close to the given uptake function and its derivative on the closed interval $[0, 1]$. The reason for this stability is that simple eigenvalues depend continuously on the entries of a matrix.

In [STW] it is shown that the assumption that Ω has no points of accumulation can be dropped from the hypotheses of Theorem 4.4 with no change in the conclusions. One can also replace the assertion that almost all initial data, in the sense of Lebesgue measure, belong to orbits converging to a rest point in O by the assertion that this holds for an open and dense set of initial data.

We remark that (generically) E_* and E_{**} are asymptotically stable, since it is expected that $s(J_*) \neq 0$ where J_* is the Jacobian of (2.1) at E_*. In this case, when $E_* \neq E_{**}$, by Theorem E.1 there must exist another positive rest point belonging to $[E_{**}, E_*]_K$ for which the corresponding Jacobian matrix is not stable.

In [STW], a global bifurcation theorem is used to show the existence of a connected global branch of ordered pairs (m_u, E_*) connecting (m_u^*, E_1) to (m_u^{**}, E_2), where the bifurcation parameter is the maximum growth rate m_u of population u and where m_u^*, m_u^{**} are critical values of this parameter.

The idea in the proof of Theorem 4.4 can be used to prove the follow-ing result, which describes some of the possible behavior of solutions of (2.1) when the two key stability moduli in Theorem 4.4 have different signs.

THEOREM 4.5. *Suppose that E_1 and E_2 exist, and that*

$$s_1 \equiv s(A + F_v(z - \hat{u})) \neq 0,$$

$$s_2 \equiv s(A + F_u(z - \bar{v})) \neq 0.$$

(a) *If $s_1 > 0$ and $s_2 < 0$, then either:*

(i) *E_2 attracts all solutions $(u(t), v(t))$ with $v(0) \neq 0$; or*

(ii) *there exists a positive rest point E_*, $E_2 <_K E_* <_K E_1$, that attracts all solutions satisfying $E_* \leq_K (u(0), v(0)) \leq_K E_1$ for which $v(0) \neq 0$, and $s(J_*) \leq 0$ where J_* is the Jacobian (4.1) at E_*. If $s(J_*) < 0$ then there exists another positive rest point $E_\heartsuit$ such that $E_2 <_K E_\heartsuit <_K E_*$ and $s(J_\heartsuit) \geq 0$, where $J_\heartsuit$ is the Jacobian (4.1) at $E_\heartsuit$.*

(b) *If $s_2 > 0$ and $s_1 < 0$, then either:*

(i) *E_1 attracts all solutions $(u(t), v(t))$ with $u(0) \neq 0$; or*

(ii) *there exists a positive rest point E_{**}, $E_2 <_K E_{**} <_K E_1$, that attracts all solutions satisfying $E_2 \leq_K (u(0), v(0)) \leq_K E_{**}$ for which $u(0) \neq 0$, and $s(J_{**}) \leq 0$ where J_{**} is the Jacobian (4.1) at E_{**}. If $s(J_{**}) < 0$ then there exists another positive rest point $E_\spadesuit$ such that $E_{**} <_K E_\spadesuit <_K E_1$ and $s(J_\spadesuit) \geq 0$, where $J_\spadesuit$ is the Jacobian (4.1) at $E_\spadesuit$.*

(c) *If $s_1 < 0$ and $s_2 < 0$ then there exists a positive rest point $E_\spadesuit$ such that $s(J_\spadesuit) \geq 0$, where $J_\spadesuit$ is the Jacobian (4.1) at $E_\spadesuit$, $E_2 <_K E_\spadesuit <_K E_1$. If $s(J_\spadesuit) > 0$ then there exist rest points $E_\natural$ and $E_\flat$ satisfying*

$$E_2 \leq_K E_\flat <_K E_\spadesuit <_K E_\natural \leq_K E_1,$$

where equality may hold in either the first or last inequalities. The rest point $E_\flat$ attracts all solutions satisfying $E_\flat \leq_K (u(0), v(0)) \leq_K E_\spadesuit$ except $E_\spadesuit$; $E_\natural$ attracts all solutions satisfying $E_\spadesuit \leq_K (u(0), v(0)) \leq_K E_\natural$ except $E_\spadesuit$. Both $s(J_\flat) \leq 0$ and $s(J_\natural) \leq 0$, where $J_\flat$ and $J_\natural$ are the Jacobians (4.1) at $E_\flat$ and $E_\natural$, respectively.

Figure 4.2 describes some of the possibilities schematically.

Proof. If $s_1 > 0$ then the rest point E_*, obtained in the proof of Theorem 4.4, exists and is either positive or coincides with E_2. The rest point E_*

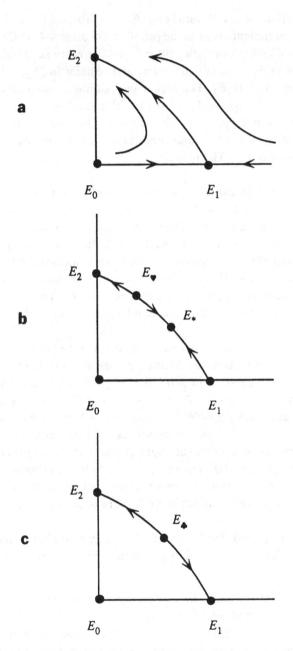

Figure 4.2. **a** Case (a)(i) of Theorem 4.5, where E_2 is the global attractor. **b** One of several possible scenarios in case (a)(ii); there could be more rest points. **c** A possible scenario in case (c); this case cannot be eliminated, but it is believed to be unlikely for biologically reasonable uptake functions.

attracts all solutions of (2.1) satisfying $E_* \leq_K (u(0), v(0)) \leq_K E_1$ and $v(0) \neq 0$, by the argument given in the proof of Theorem 4.4. If $E_* = E_2$ then E_2 attracts all solutions with $v(0) \neq 0$ because every such solution either converges to E_2 or eventually enters and remains in $[E_2, E_1]_K$, in which case it converges to E_2. The argument is similar to the one in the proof of Theorem 4.4 showing that O attracts all solutions. The assertion $s(J_*) \leq 0$ follows as in the proof of Theorem 4.4. This establishes all the assertions of case (a) except for those concerning the rest point $E_\heartsuit$, which follow from Theorem E.1. The proof of case (b) is analogous to that of case (a).

Consider case (c). The existence of the rest point $E_\spadesuit$ with $s(J_\spadesuit) \geq 0$ is proved in Proposition E.2. If $s(J_\spadesuit)$ is positive then the main construction of Theorem 4.4 can be applied to obtain a monotonically converging solution starting at $E_\spadesuit + rw$, where $r > 0$ and $0 <_K w$ is the principal eigenvector of $J_\spadesuit$. The limit of this solution is $E_\sharp$, which may coincide with E_1. Another monotonically converging solution starts at $E_\spadesuit - rw$, where $r > 0$ and this solution converges to $E_\natural$, which may coincide with E_2. The remaining assertions in this case follow from now standard arguments. $\square$

It is important to stress that, in the case (considered in the previous chapter) of two vessels and Michaelis–Menten uptake functions, there can be at most one positive rest point and it is asymptotically stable when it exists. In that case, $E_* = E_{**}$ in Theorem 4.4, and only cases (a)(i) and (b)(i) of Theorem 4.5 are possible. Furthermore, each of these possibilities has been observed. Because we cannot exclude the possibility that positive rest points are not unique or might be unstable, in the generality considered in the present chapter, we cannot exclude the other alternatives contained in the theorems. However, as of this writing, the authors are aware of no example for which there is more than one positive rest point.

In a recent paper [HSo], Hofbauer and So obtain a number of results bearing on the uniqueness and stability question. For general two-vessel gradostats they show:

(1) If f_u, f_v satisfy (i) and (ii) of Section 1, f_u'/f_v' is monotone, $f_u \not\equiv f_v$ on any interval $[0, c]$, and $-f_u(\bar{S}) \neq s(A)$ where $\bar{S}$ is the unique positive solution of $f_u(S) = f_v(S)$ and $s(A)$ is the stability modulus of A, then there exists at most one positive equilibrium. It is globally attracting with respect to positive initial data if it exists.

(2) If f_u, f_v satisfy (i) and (ii) and there exist S_i ($i = 1, 2$) satisfying $0 < S_2 < S_1$, $f_u(S_i) = f_v(S_i)$, and $f_u'(S_1)/f_v'(S_1) < 1 < f_u'(S_2)/f_v'(S_2)$, then

there exists a matrix A satisfying the hypotheses of Lemma 2.1 and an input vector e_0 such that (2.1) has an unstable positive rest point.

Note that, for the first result, the equation $f_u(S) = f_v(S)$ has at most one positive solution, and that the additional hypotheses imply uniqueness and global stability. Distinct Michaelis–Menten functions automatically satisfy all but the last hypothesis, which holds for almost all A. The second result shows that if the aforementioned equation can have two distinct solutions, yet $f_u \neq f_v$, then an unstable positive rest point exists for some two-vessel gradostat.

Hofbauer and So give two examples of three-vessel gradostats with Michaelis–Menten uptake functions where an unstable positive rest point exists. The data for the case where A is tridiagonal are: $f_u = S/(1+S)$, $f_v(S) = 3S/(2+6S)$, $a_{11} = -103/286$, $a_{12} = 35/143$, $a_{13} = 0$, $a_{21} = 1,049,864,998/300,040,001$, $a_{22} = -1,499,919,999/300,040,001$, $a_{23} = 1$, $a_{31} = 0$, $a_{32} = 3/32$, $a_{33} = -9/32$, $e_0 = (e_1, e_2, e_3)$, $e_1 = 262,703/2860$,

$$e_2 = 18,302,665,105,500/300,040,001,$$

$e_3 = 510,249/160$. The rest point $E_* = (u, v)$ (where $u = (u_1, u_2, u_3)$ and similarly for v) given by $u_1 = 10$, $u_2 = 11$, $u_3 = 9$, $v_1 = v_2 = v_3 = 22,000$ is unstable. One concludes that unstable positive rest points are possible for the class of gradostat models considered in this chapter.

5. The Standard Gradostat

It is possible to obtain interesting information on how each microbial population is distributed among the n vessels of the standard gradostat of Figure 1.1 at equilibrium if the number of vessels is large. The approach is to pass to a continuum limit. This section is devoted to an informal presentation of the results (following [S9]).

We are primarily interested in the steady-state values of u_i and v_i, so we set $u' = 0$ and $v' = 0$ in (2.1) where $z_i = 1 - i/(n+1)$ (see the paragraph following Lemma 2.2). We obtain

$$0 = (u_{i-1} - 2u_i + u_{i+1}) + u_i f_u(1 - i/(n+1) - u_i - v_i),$$

$$0 = (v_{i-1} - 2v_i + v_{i+1}) + v_i f_v(1 - i/(n+1) - u_i - v_i), \quad (5.1)$$

$$u_0 = v_0 = u_{n+1} = v_{n+1} = 0,$$

where the index i runs from 1 to n. If the vessel i is imagined to be located at $x_i = i/(n+1)$, with n large so that $\epsilon = 1/(n+1)$ is small, and using the approximation

$$u_{xx}(x_i) \approx \epsilon^{-2}(u_{i-1} - 2u_i + u_{i-1}),$$

we are led to considering the singularly perturbed boundary value problem

$$0 = \epsilon^2 u_{xx} + u f_u (1 - x - u - v),$$

$$0 = \epsilon^2 v_{xx} + v f_v (1 - x - u - v)$$

$$(5.2)$$

on $0 < x < 1$ with boundary conditions

$$u(0) = v(0) = u(1) = v(1) = 0.$$

The boundary conditions follow naturally from the conventions $u_0 = u_{n+1} = 0$, and similarly for v, which say that there are no microorganisms in the two reservoirs. They are justified by the agreement between the numerically computed rest points of (5.1) and the solutions of (5.2) obtained by using singular perturbation theory (as in [S9]).

Consider first the single-population equilibrium for u. Setting $v = 0$ in (5.2) leads to the boundary value problem

$$0 = \epsilon^2 u_{xx} + u f_u (1 - x - u),$$

$$u(0) = u(1) = 0.$$

$$(5.3)$$

A positive solution u must be concave on $0 < x < 1$, since $u_{xx} < 0$. The function

$$u = 1 - x \qquad (5.4)$$

satisfies the differential equation (5.3) and the boundary conditions at $x = 1$. Therefore, one guesses that (5.4) is an accurate approximation except for a boundary layer near $x = 0$. Standard singular perturbation techniques lead to the approximation

$$u \approx U(x/\epsilon) - x, \qquad (5.5)$$

where $U(X)$ is the solution of

$$0 = U'' + U f_u (1 - U),$$

$$U(0) = 0, \quad U(+\infty) = 1.$$

$$(5.6)$$

The condition $U(0) = 0$ ensures that the right side of (5.5) vanishes at $x = 0$. The condition $U(+\infty) = 1$ implies that $U(x/\epsilon) \approx 1$, so that (5.4) is approximated outside a neighborhood of $x = 0$ of size order ϵ. The system (5.6) is derived by passing to the variable $X = x/\epsilon$ in (5.3) and dropping all but the lowest-order terms in ϵ. Figure 5.1 shows the phase plane associated with (5.6), and Figure 5.2 gives a qualitative sketch of the approximation (5.5).

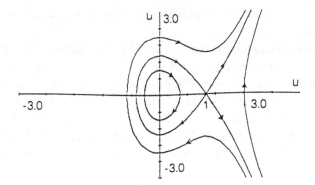

Figure 5.1. The phase portrait of (5.6). The orbit corresponding to the solution satisfying $U(0) = 0$, $U(\infty) = 1$, is the upper orbit asymptotic to the saddle point $(1, 0)$. (From [S9], Copyright 1991, *Journal of Mathematical Biology*. Reproduced by permission.)

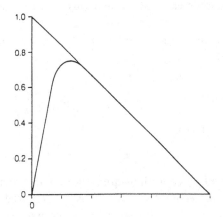

Figure 5.2. A qualitative sketch of the approximation (5.5). (From [S9], Copyright 1991, *Journal of Mathematical Biology*. Reprinted by permission.)

With u given approximately by (5.5) for small ϵ, the nutrient is given by

$$S \approx 1 - U(x/\epsilon).$$

Thus the nutrient is present in significant quantity only in an order-ϵ neighborhood of $x = 0$, and is virtually absent outside this region. Clearly, most of the growth must take place in the boundary layer near $x = 0$. Diffusion then distributes the population away from the boundary, which acts as a source, in a monotone decreasing manner. Basically, all the biology occurs in the boundary layer. The analogy for the standard gradostat with many vessels is that all growth and consumption occur in the first few vessels.

A coexistence equilibrium corresponds to a solution (u, v) of (5.2), where u and v are positive in some region and satisfy the boundary conditions. By the results of the previous section, we expect that such a solution exists provided each single-population equilibrium is unstable to invasion by its rival. The stability of these equilibria can be considered (see [S9]), but we do not require the formalities for our brief treatment here. Both u and v must be positive and concave on the interval $0 < x < 1$. There is a family of solutions of (5.2) given by

$$u = r(1-x), \qquad v = (1-r)(1-x) \tag{5.7}$$

(where r is a parameter satisfying $0 < r < 1$), which satisfies the boundary conditions at $x = 1$. Of course, the boundary conditions at $x = 0$ are not satisfied, and this suggests that (5.7) is valid outside a small neighborhood of $x = 0$ for some value of r which must be determined. In order to determine the solution in the boundary layer near $x = 0$, we set $X = x/\epsilon$, $u(x) = U(X)$, and $v(x) = V(X)$ in (5.2), keeping only the lowest-order terms in ϵ. This gives

$$0 = U'' + Uf_u(1 - U - V),$$
$$0 = V'' + Vf_v(1 - U - V), \tag{5.8}$$

where U and V must satisfy

$$U(0) = 0, \quad V(0) = 0, \quad U(+\infty) = r, \quad V(+\infty) = 1 - r. \tag{5.9}$$

The first two boundary conditions are necessary in order for u and v to vanish at $x = 0$. The second two conditions are required in order for the solution to match up with (5.7) near $x = 0$. Of course, we also expect that

$$U > 0, \quad V > 0, \quad U' > 0, \quad V' > 0, \quad U + V < 1 \tag{5.10}$$

hold for all $X > 0$.

The system (5.8) is four-dimensional for (U, U', V, V') if equations for U' and V' are included. In addition to the trivial rest point, there is a line of rest points given by $P_s = (s, 0, 1-s, 0)$ for $0 \le s \le 1$. In mathematical terms, we seek a solution in the stable manifold of P_r, for some value of r satisfying $0 < r < 1$, which remains in the region (5.10) and at $X = 0$ satisfies $U(0) = 0$ and $U'(0) > 0$, and similarly for V. It turns out that the set of eigenvalues of the Jacobian of the vector field on the right side of (5.8) at the rest point P_s contains exactly one positive number (another is negative and there are two zero eigenvalues). The corresponding eigenvector can be chosen to point into the region (5.10). This implies that the stable manifold of P_s consists of two orbits, one of which lies, at least in part, in the region (5.10). In [S9], sufficient conditions are given for the existence of a value of r satisfying $0 < r < 1$ for which there is a solution of (5.8) satisfying (5.9) and (5.10). These sufficient conditions, which will not be given here because they are of a technical nature and shed no light on the biology, are at least consistent with the hypothesis that both single-population equilibria are unstable to invasion by the rival population. Unfortunately, the proof only gives the existence of r and provides no information about its value.

The corresponding coexistence equilibrium solution is thereby approximated by

$$u \approx U(x/\epsilon) - rx, \qquad v \approx V(x/\epsilon) - (1-r)x, \qquad (5.11)$$

where (U, V) is the solution described previously. Note that the form of (5.11) guarantees that (5.7) holds away from an order-ϵ neighborhood of $x = 0$, since $U(x/\epsilon) \approx r$ and $V(x/\epsilon) \approx 1 - r$ when ϵ is small.

The biological content of this analysis is revealed by considering also the nutrient, which is given approximately by

$$S \approx 1 - U(x/\epsilon) - V(x/\epsilon).$$

Away from the boundary layer near $x = 0$ there is a negligible amount of nutrient, and population densities are dominated by simple diffusion from the source which, in effect, is contained in the boundary layer. This is where most growth and consumption of nutrient occurs. The implications for the n-vessel standard gradostat are that if n is large then all of the interesting biology occurs in the first few vessels. It might reasonably be concluded that, at least for the standard gradostat, there is little point in considering gradostats with a large number of vessels.

It is particularly unfortunate that our analysis sheds no light on the value of r, since (5.11) implies that r determines which population is dominant in the gradostat.

In Figure 5.3 and Figure 5.4 (taken from [S9]), the numerically computed steady states of (5.1) for a five-vessel gradostat are displayed and compared with the appropriate approximate solutions (5.5) and (5.11). The agreement is quite good even for the second vessel. The value of r was estimated using the ratios $r/(1-r) = u_5/v_5$, which resulted in $r = 0.243$. As a check, that portion of the stable manifold of (5.8) corresponding to

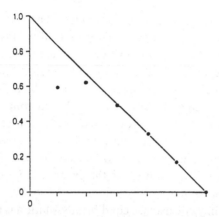

Figure 5.3. The five points represent computed values of the steady-state concentrations of a single species with $m = 5$ and $a = 1$ in a five-vessel gradostat. The last three points are very close to the solid line. (From [S9], Copyright 1991, *Journal of Mathematical Biology*. Reprinted by permission.)

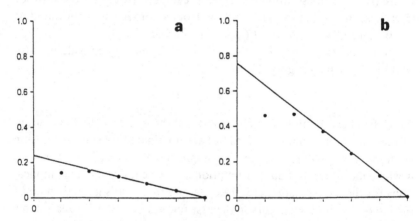

Figure 5.4. **a** The five points depicted represent computed values of the concentrations of population U in each of the five vessels for the coexistence rest point. For U, $m = 5$ and $a = 1$. The points are close to the line $\alpha(1 - x)$, where $\alpha = U_5/V_5$, the ratio of computed values of U and V in vessel 5. **b** The corresponding concentrations for the competitor V, with $m = 22.2$ and $a = 5.08$; compare with the line $(1 - \alpha)(1 - x)$. (From [S9], Copyright 1991, *Journal of Mathematical Biology*. Reprinted by permission.)

$(r, 0, 1 - r, 0)$, with r as before, was computed using PHASE PLANE [Er]. Its projection in the U–V plane passed through the origin to within screen resolution. The estimated value of r was used in Figure 5.4 to give a qualitative (not computed) picture of the approximation (5.11).

6. Discussion

In the previous chapter, a fairly complete analysis was given of the two-vessel gradostat with Michaelis–Menten uptake functions. In the present chapter we have seen that much, but not all, of this analysis carries over to very general gradostats consisting of an arbitrary number of vessels connected together in a wide variety of ways and using arbitrary monotone uptake functions. The main result of this chapter, as in the previous one, is the prediction that coexistence of two microbial populations in the gradostat is possible provided that each population can successfully invade its rival's single-population equilibrium. Coexistence occurs in the form of convergence to a positive equilibrium – which may not be unique, although no example of non-uniqueness is known. By extrapolation, the model suggests that coexistence is possible in a natural environment supporting a nutrient gradient.

Two key results of the previous chapter – the uniqueness of a positive rest point and the local asymptotic stability of a positive rest point – are left unsettled in the present chapter. As already mentioned, the second result no longer holds in the generality considered in the present chapter. In [HSo], examples are given in which an unstable positive rest point exists. While we know of no case where non-uniqueness of the positive rest point has been shown, the fact that positive rest points may be unstable suggests the possibility that uniqueness too may fail in the generality considered in this chapter. It is not clear at the time of this writing whether unstable positive rest points can occur in biologically interesting cases. The examples of Hofbauer and So are not conclusive in this sense, since apparently there are no counterexamples to the conjecture that the results of the previous chapter carry over to the standard n-vessel gradostat with monotone uptake functions whose graphs intersect in at most one positive value of S.

We know that coexistence of two microbial populations in the gradostat is possible. It is natural to ask how many microbial populations can coexist in an n-vessel gradostat. In [JST] it is shown that the number is not greater than n. This is done by first showing that, generically, there does not exist a coexistence rest point, and second by showing that there are solutions corresponding to the presence of all populations which converge to rest points on the boundary – that is, to rest points with certain populations absent. Numerical simulations in [BWu; CB] suggest that three competitors cannot coexist in a three-vessel gradostat but can in a four-vessel gradostat. Very little has been rigorously established concerning competition between more than two competitors in a gradostat. For more than two competitors, the equations no longer give rise to a monotone dynamical system.

7

The Chemostat with Periodic
Washout Rate

1. Introduction

The results of Chapters 1 and 2 demonstrate that coexistence cannot occur in the chemostat with constant values of the operating parameters (the washout rate D and the input nutrient concentration S^0). Roughly speaking, in a temporally homogeneous (constant operating parameters) and spatially homogeneous (well-stirred chemostat) environment, the model predicts competitive exclusion. Of course, real environments are far from being homogeneous, either in space or in time. In addition to the day-night variability in the environment, there are both seasonal effects and random effects caused by the variable climate. In this chapter, the simple chemostat model of competition will be modified to create a periodically varying environment, and it will be shown that coexistence of the two competitors can occur.

There are two basic ways to modify the chemostat model to create a periodic environment – make the reservoir nutrient concentration S^0 vary periodically in time, or make the washout rate D vary periodically. The first modification was studied by Hale and Somolinos [HaS], Hsu [Hsu2], Smith [S1], and Stephanopoulos, Fredrickson, and Aris [SFA]. This study is natural from the ecological point of view, as nutrient levels in many ecosystems might be expected to vary with the day-night cycle or with the season. The second modification was studied by Butler, Hsu, and Waltman [BHW2] and in [SFA]. It is more natural from the point of view of the experimenter working with a chemostat, since controlling the pump speed (so as to vary the washout rate) is simpler than controlling the nutrient concentration. The chemostat is sometimes viewed as a simple model of a wastewater treatment plant, especially by chemical and industrial engineers [SFA], and in this role also it is more sensible to vary

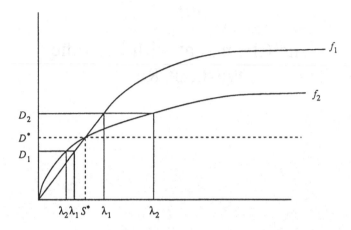

Figure 1.1. If $D = D_2$, then $\lambda_1 < \lambda_2$ and population 1 has a competitive advantage over population 2. If $D = D_1$, then $\lambda_2 < \lambda_1$ and the reverse holds. If $D = D(t)$ oscillates periodically between D_1 and D_2, then each population enjoys the competitive advantage during a part of the cycle.

the washout rate. We will vary the washout rate in this chapter. The mathematics, however, is the same regardless of which operating parameter is varied, and the end results are similar – coexistence is possible.

   The intuition supporting the expectation that coexistence should be possible in a chemostat with oscillatory washout rate D is fairly simple. It should be possible to vary D in such a way that first one competitor, then the other competitor, has the competitive advantage. More precisely, suppose the growth curves, or functional responses f_1 and f_2, of the two competitors are monotone and intersect in exactly one point: $f_1(S^*) = f_2(S^*) = D^*$. Let f_1 be smaller than f_2 for positive S smaller than S^* and larger than f_2 for S larger than S^*; see Figure 1.1. Then, in a chemostat with constant values of the operating parameters, competitor x_1 is favored when D is larger than D^* since λ_1 is smaller than λ_2, and competitor x_2 is favored when D less than D^* since in that case λ_2 is the smaller. It seems plausible to expect that if D is allowed to vary with time in such a way that alternately D is less than D^* and D is larger than D^*, then coexistence of the two competitors is possible. As simple as this reasoning sounds, it turns out to be surprisingly difficult to prove rigorously. The reason is that the corresponding equations are non-autonomous (i.e. time-dependent), and one can no longer explicitly compute the analogs of the steady states as was done for the autonomous equations of Chapter 1.

The focus in this chapter will be on the possibility of coexistence of two competitors competing for a single nutrient in a chemostat with an oscillatory washout rate. Therefore, an exhaustive study of sufficient conditions for competitive exclusion to hold, as was carried out in Chapter 1, will not be made here. The reference [BHW2] may be consulted for results of this type.

The differential equations of the model take the following form:

$$S' = (S^0 - S)D(t) - \gamma_1^{-1}x_1 f_1(S) - \gamma_2^{-1}x_2 f_2(S),$$

$$x_i' = x_i(f_i(S) - D(t)), \quad i = 1, 2,$$

$$x_i(0) > 0, \quad S(0) > 0.$$

The main change in this system compared to (5.1) of Chapter 1 is that the constant washout rate D of (5.1) is replaced by a positive, continuous, periodic function $D(t)$:

$$D(t + \omega) = D(t)$$

for some positive ω. A typical example is

$$D(t) = D_0 + d \sin(2\pi t/\omega).$$

We assume that f_i satisfies

(i) $f_i: \mathbb{R}^+ \to \mathbb{R}^+$ is continuously differentiable;
(ii) $f_i(0) = 0$, $f_i'(S) > 0$.

In other words, the functional response f_i is monotone increasing. Note that it need not be bounded.

In dealing with periodic functions the following notation is convenient. If g is a continuous ω-periodic function then $\langle g \rangle$ will denote the mean value of g:

$$\langle g \rangle = \omega^{-1} \int_0^\omega g(s)\, ds.$$

The system displayed previously will be scaled by measuring S in units of S^0 and x_i in units of $\gamma_i S^0$. If time is measured in units of $\langle D \rangle^{-1}$, the system takes the form:

$$S' = (1 - S)D(t) - x_1 f_1(S) - x_2 f_2(S),$$

$$x_i' = x_i(f_i(S) - D(t)), \quad i = 1, 2. \tag{1.1}$$

We have relabeled $\langle D \rangle$ as 1 in the equations (1.1). The $f_i(S)$ in (1.1) are actually $\langle D \rangle^{-1} f_i(S^0 S)$, but again we have relabeled. Furthermore, since time was rescaled, $D(t)$ in (1.1) is actually $\langle D \rangle^{-1} D(t/\langle D \rangle)$ in terms of the unscaled system. This affects both the period and the mean value of D. The former becomes $\langle D \rangle \omega$, which we relabel ω, and the latter becomes unity: $\langle D \rangle = 1$.

2. Periodic Differential Equations

Equations (1.1) are periodic in the time variable. As this is the first encounter with such equations, it is appropriate to review the basic strategy for dealing with them. For the purposes of this review, consider the general periodic system

$$x' = f(t, x), \tag{2.1}$$

where f and $\partial f/\partial x$, its derivative with respect to x, are continuous on $\mathbb{R} \times \mathbb{R}^n$ (or on $\mathbb{R} \times \Omega$ with Ω a suitable subset of $\mathbb{R}^n$) and, for some $\omega > 0$,

$$f(t+\omega, x) = f(t, x)$$

holds for all (t, x). The ω-periodic (hereafter simply "periodic") solutions $x(t+\omega) = x(t)$ play the same role for (2.1) as rest points do for autonomous systems.

Just as the stability of rest points can be determined by linearization, the stability of a periodic solution $x(t)$ can often be determined by linearizing (2.1) about $x(t)$. The linearization, or variational equation, corresponding to $x(t)$ is

$$z' = \frac{\partial f}{\partial x}(t, x(t))z. \tag{2.2}$$

This system is periodic and therefore the Floquet theory described in Section 4, Chapter 3, applies. Let $\Phi(t)$ be the fundamental matrix solution of (2.2). The Floquet multipliers of (2.2) are the eigenvalues of $\Phi(\omega)$; if μ is a Floquet multiplier and $\mu = e^{\omega \lambda}$ then λ is called a *Floquet exponent*. Only the real part of a Floquet exponent is uniquely defined.

A fundamental difference between (2.2) and (4.6) of Chapter 3 is that 1 is not necessarily a multiplier for (2.2) as it is for (4.6). Therefore, the analog of Theorem 4.2 of Chapter 3 is different.

THEOREM 2.1. *If $|\mu| < 1$ for all multipliers of (2.2) then $x(t)$ is a uniformly asymptotically stable periodic solution of (2.1). If $|\mu| > 1$ for some multiplier μ of (2.2) then $x(t)$ is unstable.*

Recall [H2] that $x(t)$ is *uniformly asymptotically stable* (hereafter simply "asymptotically stable") provided that (i) for every $\epsilon > 0$ there exists $\delta > 0$ such that, if $|x(t_0) - y(t_0)| < \delta$ for some $t_0 \geq 0$ and some solution $y(t)$, then $|x(t) - y(t)| < \epsilon$ for all $t \geq t_0$; and (ii) there exists $b > 0$ such that, if $|x(t_0) - y(t_0)| < b$ for some $t_0 \geq 0$, then $\lim_{t \to \infty} |x(t) - y(t)| = 0$ and this convergence is uniform in t_0.

In terms of Floquet exponents, the condition for stability is $\Re(\lambda) < 0$ for all exponents and the condition for instability is that $\Re(\lambda) > 0$ for some exponent λ. Here, $\Re(\lambda)$ denotes the real part of λ.

A powerful conceptual tool for the analysis of (2.1) is the *Poincaré map*. Let $x(t, x_0)$ denote the solution of (2.1) satisfying $x(0) = x_0$. The Poincaré map is defined by

$$Px_0 = x(\omega, x_0);$$

P simply advances a point one period along the trajectory through the point. The basic properties of P follow from standard theorems for ordinary differential equations: P is continuous and one-to-one, and $x(t, x_0)$ is a periodic solution of (2.1) with period ω if and only if $Px_0 = x_0$. The orbit of a point x_0 under P is the set $O^+(x_0) = \{x_0, Px_0, P^2 x_0, \ldots\}$, where P^n denotes the n-fold composition of P with itself. Note that $P^n x_0 = x(n\omega, x_0)$, so that the orbit of x_0 under P is just a sampling of the solution of (2.1) through x_0 at integral multiples of ω. The map P captures all of the dynamical features of (2.1). For example, if $\lim_{n \to \infty} P^n x_0 = \bar{x}$ then, by the continuity of P, $P\bar{x} = \bar{x}$. Therefore $x(t, \bar{x})$ is a periodic solution and $\lim_{t \to \infty} |x(t, x_0) - x(t, \bar{x})| = 0$; that is, $x(t, x_0)$ tends asymptotically to the periodic solution $x(t, \bar{x})$.

As f has a continuous partial derivative with respect to x, P is continuously differentiable and its Jacobian derivative is given by

$$P'(x_1) = \frac{\partial x}{\partial x_0}(\omega, x_1) = \Phi(\omega),$$

where $\Phi(t)$ is the fundamental matrix for (2.2) with $x(t) = x(t, x_1)$ (see [H2]). If $Px_1 = x_1$ so that $x(t, x_1)$ is a periodic solution, then the eigenvalues of the derivative of P at x_1 are precisely the Floquet multipliers.

The behavior of orbits of P near a fixed point $\bar{x}$ can be described in the case where $\bar{x}$ is a *hyperbolic fixed point,* that is, when no eigenvalue (multiplier) of the Jacobian of P at $\bar{x}$ has modulus equal to 1. In this case there exist (local) stable and unstable manifolds $M^+(\bar{x})$ and $M^-(\bar{x})$ (respectively) containing the point $\bar{x}$ which are tangent to the stable (resp. unstable) subspace of the Jacobian of P at $\bar{x}$. (The stable (unstable) subspace is that space spanned by the real parts of the generalized eigenvectors corresponding to eigenvalues with modulus less (greater) than 1.) The stable and unstable manifolds have the dimensions of the stable and unstable subspaces, respectively. The manifold $M^+(\bar{x})$ contains all those points near $\bar{x}$ whose orbit under P converges to $\bar{x}$, and $M^-(\bar{x})$ contains all those points near $\bar{x}$ whose orbit under P^{-1} converges to $\bar{x}$. Points near $\bar{x}$ not belonging to the stable manifold have the property that their orbit cannot remain in a neighborhood of $\bar{x}$.

The (local) unstable manifold $M^-(\bar{x})$ can be extended to a (global) object, which we will also call the unstable manifold (although its geometry can be quite complicated; in particular, it need not be a submanifold), simply by taking the union of the images of $M^-(\bar{x})$ under P^n, $n = 0, 1, 2, \ldots$. Then the (global) unstable manifold can be characterized as $\{x \mid P^{-n}x \to \bar{x}$ as $n \to \infty\}$. A similar construction gives the (global) stable manifold. Hereafter, we will not distinguish between the local and global objects.

In concluding this brief review, we remark that the principal motivation for working with the Poincaré map as opposed to the solutions of (2.1), neither of which is explicitly computable in general, is that the Poincaré map has its domain in Euclidean n-space whereas solutions of (2.1) must be viewed in Euclidean $(n+1)$-space – that is, as $(t, x(t))$. The advantage of working in a lower-dimensional space is the key. We hope the utility of the Poincaré map will be evident in the remaining sections of this chapter.

3. The Conservation Principle

The analysis of equations (1.1) begins by establishing that the conservation principle holds exactly as in the previous chapters and so allows for the reduction of (1.1) to a two-dimensional system. Letting $\Sigma = S + x_1 + x_2 - 1$ and adding the equations (1.1) yields the periodic linear equation

$$\Sigma' = -D(t)\Sigma. \tag{3.1}$$

Therefore, as $\langle D \rangle = 1$, we have

$$\Sigma(t) = q(t)e^{-t},$$

where

$$q(t) = \Sigma(0)\exp\left(-\int_0^t [D(s) - 1]\,ds\right)$$

is a periodic function. Consequently,

$$\lim_{t \to \infty} \Sigma(t) = 0.$$

In other words, independently of initial conditions, solutions of (1.1) asymptotically approach the plane, $S + x_1 + x_2 = 1$, at an exponential rate. The asymptotic behavior of (1.1) is therefore determined by the two-dimensional system obtained from (1.1) by deleting the equation for S and replacing S by $1 - x_1 - x_2$, just as in the previous chapters. This yields

$$x_i' = x_i[f_i(1 - x_1 - x_2) - D(t)],$$

$$i = 1, 2.$$

(3.2)

Biologically relevant initial data for (3.2) belong to

$$\Omega = \{(x_1, x_2) \in \mathbb{R}_+^2 : x_1 + x_2 \le 1\},$$

which is positively invariant for (3.2).

In the remainder of this chapter, our attention will be restricted to the system (3.2). The results of Appendix F can be shown to hold for discrete dynamical systems (e.g., generated by the Poincaré map for the system of equations for (Σ, x_1, x_2)); this makes it possible to deduce the asymptotic behavior of (1.1) from that of (3.2).

We begin by obtaining a sufficient condition for the washout of a competitor from the chemostat which is independent of the presence or absence of an adversary. The following result is "sharp" in the sense that if it fails to hold then the competitor can survive in the absence of competition in the chemostat.

PROPOSITION 3.1. *If $f_i(1) \le 1$ then $\lim_{t \to \infty} x_i(t) = 0$.*

Proof. If $x_i(0) = 0$ then $x_i(t) = 0$ and there is nothing to prove, so we assume that $0 < x(0) \le 1$. It follows that $x_i' < x_i(f_i(1) - D(t))$, which implies that

$$x_i(t + \omega) < x_i(t) \exp[(f_i(1) - 1)\omega] \le x_i(t).$$

We conclude that

$$x_i((n+1)\omega) < x_i(n\omega) \quad \text{for } n = 1, 2, \ldots,$$

so $\lim_{n \to \infty} x_i(n\omega) = l \ge 0$ exists. If $l > 0$ then it is easy to see that

$$\liminf_{t \to \infty} x_i(t) = \epsilon > 0.$$

Then, for all large t, one has $1 - x_1(t) - x_2(t) \le 1 - (\epsilon/2)$, and for such t it follows that

$$x_i'(t) \le x_i(f_i(1 - (\epsilon/2)) - D(t)).$$

Therefore, for all large s and all $t \ge 0$,

$$x_i(t+s) \le x_i(s)\exp\left[-\int_s^{t+s} [f_i(1 - (\epsilon/2)) - D(r)]\,dr\right].$$

Since $f_i(1 - (\epsilon/2)) - 1 < 0$, the exponential term tends to zero as $t \to \infty$, and we have reached a contradiction to our assumption that $l > 0$. Consequently, $\lim_{n \to \infty} x_i(n\omega) = 0$, which implies the assertion of the proposition. $\Box$

It is easy to see that the numbers $(f_i(1) - 1)$, $i = 1, 2$, are the Floquet exponents corresponding to the identically zero periodic solution of (3.2). Consequently, the solution is asymptotically stable when both exponents are negative and unstable when one of the exponents is positive. Proposition 3.1 says more than this; it states that competitor x_i is washed out of the chemostat if $f_i(1) \le 1$, but this outcome has nothing to do with competition since it occurs even in the absence of the other competitor. As our main interest is in the effects of competition, we assume hereafter that

$$f_i(1) > 1, \quad i = 1, 2. \tag{3.3}$$

When (3.3) holds for the ith competitor, that competitor can survive in the chemostat in the absence of competition and with its concentration oscillating in response to the periodically varying washout rate. This is the content of the next result.

PROPOSITION 3.2. *There exist unique, positive, periodic functions* $\xi(t)$ *and* $\eta(t)$ *such that* $(\xi(t), 0)$ *and* $(0, \eta(t))$ *are solutions of* (3.2). *If* $(x_1(t), 0)$ *is a solution of* (3.2) *satisfying* $x_1(0) > 0$, *then*

$$\lim_{t \to \infty} |x_1(t) - \xi(t)| = 0.$$

If $(0, x_2(t))$ *is a solution of* (3.2) *satisfying* $x_2(0) > 0$, *then*

$$\lim_{t \to \infty} |x_2(t) - \eta(t)| = 0.$$

Proof. The assertion essentially concerns the scalar equation

$$x' = x[f(1-x) - D(t)],$$

where we have omitted the subscripts. It is assumed that $f(1) > 1$. Let P denote the Poincaré map $P: [0, 1] \to [0, 1]$ given by $Px_0 = x(\omega, x_0)$. Then P is continuously differentiable, one-to-one, and satisfies $P0 = 0$

and $P1 < 1$. The derivative $P'(x_0) = (\partial x / \partial x_0)(\omega, x_0) = v(\omega)$, where $v(t)$ is the solution of

$$v' = v[f(1 - x(t, x_0)) - D(t) - x(t, x_0) f'(1 - x(t, x_0))]$$

satisfying $v(0) = 1$. The equation can be solved, giving

$$P'(x_0) = \exp[\omega \langle f(1 - x(t, x_0)) - D(t) - x(t, x_0) f'(1 - x(t, x_0)) \rangle].$$

It is apparent that $P'(x_0) > 0$ for all x_0, and that

$$P'(0) = \exp[\omega(f(1) - 1)] > 1.$$

Recalling that $P1 < 1$, it follows that $Px - x$ changes sign in $(0, 1)$ so there exists at least one fixed point of P in that interval. If $\bar{x}$ is a fixed point of P, it then follows from the differential equation that

$$\langle f(1 - x(t, \bar{x})) - D(t) \rangle = 0.$$

Consequently,

$$P'(\bar{x}) = \exp[-\omega \langle x(t, \bar{x}) f'(1 - x(t, \bar{x})) \rangle] < 1.$$

It is now easy to see that P can have only one fixed point, $\bar{x}$, in $(0, 1)$ and, moreover, that $\bar{x}$ is an asymptotically stable fixed point of P.

If $x_0 \neq \bar{x}$ and belongs to $(0, 1]$ then either $Px_0 > x_0$ or $Px_0 < x_0$. Consider the latter case, as the argument in the former case is simpler. Since P is strictly increasing, it follows that $\{P^n x_0\}_n$ is a strictly decreasing sequence. Therefore, it converges to a point y in $[0, 1]$ which is a fixed point of P. As $Px > x$ for all small x, $y \neq 0$ and so $y = \bar{x}$. We have shown that every orbit of P starting at a point in $(0, 1]$ converges to $\bar{x}$. Consequently, if $\psi(t) = x(t, \bar{x})$ then ψ is periodic and $\lim_{t \to \infty} |x(t) - \psi(t)| = 0$ for every positive solution $x(t)$. $\qquad \Box$

It will be convenient to use the familiar notation E_1 and E_2 for the single-competitor periodic solutions whose existence is asserted in Proposition 3.2:

$$E_1(t) = (\xi(t), 0) \quad \text{and} \quad E_2(t) = (0, \eta(t)).$$

The outcome of competition between two competitors depends on the stability properties of the single-competitor periodic solutions E_1 and E_2. It turns out that the stability of these solutions is determined, in each case, by a single Floquet exponent in a biologically intuitive way. Suppose that a chemostat is charged at $t = 0$ with only the competitor x_1. According to Proposition 3.2, the concentration x_1 very rapidly approaches the level

$\xi(t)$. Now imagine that a small concentration of competitor x_2 is added to the chemostat, where x_1 is essentially at concentration $\xi(t)$. Will x_2 increase or decrease in concentration? To find out, it is reasonable to linearize the second of equations (3.2) about $x_2 = 0$, setting $x_1 = \xi(t)$, since x_2 is assumed to be very small. This results in the equation

$$x_2' = x_2[f_2(1-\xi(t))-D(t)],$$

which can be solved in terms of ξ as

$$x_2(t+\omega) = x_2(t)\exp[\omega(\langle f_2(1-\xi)\rangle - 1)].$$

Setting $\lambda_{12} = \langle f_2(1-\xi)\rangle - 1$, one sees that x_2 increases when it is at very low concentration if $\lambda_{12} > 0$ and decreases if $\lambda_{12} < 0$.

Similar arguments apply to the solution E_2. Our next result makes these heuristic arguments precise.

PROPOSITION 3.3. *Floquet exponents of the periodic solution E_1 are given by*

$$\lambda_{11} = -\langle \xi f_1'(1-\xi)\rangle < 0 \quad and \quad \lambda_{12} = \langle f_2(1-\xi)\rangle - 1.$$

Consequently, E_1 is asymptotically stable (unstable) if $\lambda_{12} < 0$ ($\lambda_{12} > 0$). An analogous statement holds for the periodic solution E_2.

Proof. The variational equation corresponding to E_1 is

$$z' = \begin{pmatrix} [f_1(1-\xi)-D-\xi f_1'(1-\xi)] & -f_1'(1-\xi)\xi \\ 0 & f_2(1-\xi)-D \end{pmatrix} z, \qquad (3.4)$$

where we have suppressed the t dependence of the coefficients. A computation yields the fundamental matrix $\Phi(t)$:

$$\Phi(t) = \begin{pmatrix} \exp[\int_0^t a_{11}(s)\,ds] & u(t) \\ 0 & \exp[\int_0^t a_{22}(s)\,ds] \end{pmatrix},$$

where (a_{ij}) denotes the coefficients in (3.4) and where $u(t)$ is given by

$$u(t) = \int_0^t \exp\left[\int_r^t a_{11}(s)\,ds\right] a_{12}(r)\exp\left[\int_0^r a_{22}(s)\,ds\right] dr.$$

Note for future reference that $u(t) < 0$ for $t > 0$ since $a_{12}(r) < 0$. Evaluating Φ at $t = \omega$, we obtain the multipliers $\exp[\int_0^\omega a_{ii}(s)\,ds]$, $i = 1, 2$. It follows immediately that λ_{11} and λ_{12} are Floquet exponents. The remaining assertions follow from the discussion of Section 2. $\square$

Denote by λ_{21} and λ_{22} the Floquet exponents of the periodic solution E_2:

$$\lambda_{22} = -\langle \eta f_2'(1-\eta) \rangle \quad \text{and} \quad \lambda_{21} = \langle f_1(1-\eta) \rangle - 1.$$

The solution E_2 is asymptotically stable if $\lambda_{21} < 0$ and unstable if $\lambda_{21} > 0$.

Further analysis of the system (3.2) requires understanding the dynamics generated by the Poincaré map in the interior of Ω. For general periodic systems, such an understanding is beyond current knowledge. Fortunately, system (3.2) – in addition to being two-dimensional – has the property that it is competitive. A beautiful theory for such systems has recently been constructed. The next section is devoted to the principal result of this theory.

4. Periodic Competitive Planar Systems

The study of mathematical models of competition has led to the discovery of some very beautiful mathematics. This mathematics, often referred to as monotone dynamical systems theory, was largely developed by M. W. Hirsch [Hi1; Hi3], although others have made substantial contributions as well. In this section we describe a result that was first obtained in a now classical paper of DeMottoni and Schiaffino [DS] for the special case of periodic Lotka–Volterra systems. Later, it was recognized by Hale and Somolinos [HaS] and Smith [S4; S5] that the arguments in [DS] hold for general competitive and cooperative planar periodic systems. The result says that every bounded solution of such a system converges to a periodic solution that has the same period as the differential equation.

We proceed to state and prove this result. Consider the system

$$\begin{aligned} x_1' &= f_1(t, x_1, x_2), \\ x_2' &= f_2(t, x_1, x_2), \end{aligned} \tag{4.1}$$

where $f_i(t+\omega, x_1, x_2) = f_i(t, x_1, x_2)$ and

$$\frac{\partial f_i}{\partial x_j}(t, x_1, x_2) \le 0, \quad i \ne j. \tag{4.2}$$

The requirement (4.2) means that (4.1) is a *competitive* system – an increase in x_2 has a negative effect on the growth rate of x_1 and vice versa. The system is said to be a *cooperative* system if the reverse inequalities hold in (4.2). Our interest here will be in the competitive case since (3.2) satisfies (4.2) in Ω, as is easily checked. However, the cooperative case is also of interest, so the main result will be stated so as to include it as well. Some definitions are required before stating the result.

If $x, y \in \mathbb{R}^2$ then we write $x \leq y$ whenever $x_i \leq y_i$ holds for $i = 1, 2$. We write $x \leq_K y$ whenever $x_1 \leq y_1$ and $x_2 \geq y_2$. If one imagines the x_1 axis pointing east and the x_2 axis pointing north, then $x \leq y$ means that y lies to the northeast of x and $x \leq_K y$ means that y lies to the southeast of x.

The key facts about competitive and cooperative planar systems are summarized next in terms of the Poincaré map P for (4.1).

LEMMA 4.1. *The following inequalities hold for* (4.1) *satisfying* (4.2):

 (i) *if* $x \leq_K y$, *then* $Px \leq_K Py$;
(ii) *if* $Px \leq Py$, *then* $x \leq y$.

If (4.1) *is cooperative, then analogous implications hold with the two partial orders interchanged in (i) and (ii).*

Proof. We consider only the competitive case. Case (i) follows immediately from Theorem B.4 (see Appendix B). (ii) asserts that if $x(\omega, x) \leq x(\omega, y)$ then $x \leq y$. To see this, let $u(t) = x(\omega - t, x)$ and $v(t) = x(\omega - t, y)$, and observe that u and v are solutions of

$$x_1' = -f_1(t, x_1, x_2), \qquad x_2' = -f_2(t, x_1, x_2).$$

This system is cooperative and so, as $u(0) \leq v(0)$, we conclude from Corollary B.2 that $u(\omega) \leq v(\omega)$. This proves (ii). □

The main result can now be stated.

THEOREM 4.2. *If* $O^+(x_0)$, *the orbit of* x_0 *under* P, *is bounded then*

$$\lim_{n \to \infty} P^n x_0 = \bar{x} = P\bar{x}.$$

Consequently, if $x(t) = x(t, x_0)$ *and* $p(t) = x(t, \bar{x})$, *then* p *is periodic and*

$$\lim_{t \to \infty} |x(t) - p(t)| = 0.$$

Proof. Given two points $x, y \in \mathbb{R}^2$, one or more of the four relations $x \leq y$, $y \leq x$, $x \leq_K y$, or $y \leq_K x$ must hold. Now, if $P^{n_0} x_0 \leq_K P^{n_0+1} x_0$ (or the reverse inequality) holds for some $n_0 \geq 0$, then Lemma 4.1 implies that $P^n x_0 \leq_K P^{n+1} x_0$ (or the reverse inequality) holds for all $n \geq n_0$. Therefore, $\{P^n x_0\}_n$ converges to some $\bar{x}$, since the sequence is monotone and bounded. The proof is complete in this case, so we assume that there does not exist such an n_0 as just described. In particular, it follows that x_0 is not a fixed point of P. Then it follows that for each n we must have either that $P^{n+1}x_0 \leq P^n x_0$ or that the reverse inequality holds. Suppose for definiteness that $x_0 \leq Px_0$, the other case being similar. We claim that $P^n x_0 \leq P^{n+1}x_0$ for all n. If not, there exists n_0 such that

$$x_0 \leq Px_0 \leq P^2 x_0 \leq \cdots \leq P^{n_0-1}x_0 \leq P^{n_0}x_0$$

but $P^{n_0}x_0 \geq P^{n_0+1}x_0$. Clearly, $n_0 \geq 1$ since $x_0 \leq Px_0$. Applying Lemma 4.1(ii) to the displayed inequality yields $P^{n_0-1}x_0 \geq P^{n_0}x_0$ and therefore $P^{n_0-1}x_0 = P^{n_0}x_0$. As P is one-to-one, x_0 must be a fixed point in contradiction to our assumption. This proves the claim and implies that the sequence $\{P^n x_0\}$ converges to some $\bar{x}$. The theorem now follows from the facts described in Section 2. $\qquad\qquad\qquad\qquad\qquad\qquad\qquad\square$

We remark that if (4.1) is autonomous and is either competitive or cooperative then we are free to choose ω and a corresponding Poincaré map P_ω. Theorem 4.2 implies that every bounded solution of (4.1) is asymptotic to an ω-periodic solution. Since ω is arbitrary, it follows that every bounded solution converges to a rest point.

As an immediate corollary of Theorem 4.2, we have the following result for (3.2).

COROLLARY 4.3. *Let $x(t, x_0)$ be a solution of (3.2) corresponding to $x_0 \in \Omega$. Then there exists a periodic solution $p(t)$ of (3.2) such that*

$$\lim_{t \to \infty} |p(t) - x(t, x_0)| = 0.$$

5. Coexistence

In the previous section we saw that every solution of (3.2) converges to a periodic solution. The goal of this section is to provide sufficient conditions for the existence of a positive periodic solution of (3.2) possessing strong stability properties. Hereafter, a solution (or vector) $x(t) = (x_1(t), x_2(t))$ will be called *positive* provided both components are positive. A positive stable periodic solution of (3.2) corresponds to the coexistence of both competitors in the chemostat. In this case, each competitor's concentration oscillates between a positive minimum and maximum value. It is important to point out that coexistence of two competitors can take place only if there exists a positive periodic solution of (3.2). If coexistence means that the concentrations of each of the two populations must remain above some fixed positive concentration for all future time, then the main result of the previous section shows that the corresponding solution of (3.2) must approach a positive periodic solution.

This fact can be exploited to give simple necessary conditions for coexistence (or sufficient conditions for competitive exclusion). Suppose that $x(t)$ is a positive periodic solution of (3.2) and let $S(t) = 1 - x_1(t) - x_2(t)$. It is apparent that $0 < S(t) < 1$. From (3.2) we have

$$x_i(t + \omega) = x_i(t) = x_i(t) \exp(\omega[\langle f_i(S(t)) \rangle - 1])$$

and, consequently

$$\langle f_i(S(t)) \rangle - 1 = 0, \quad i = 1, 2.$$

It follows immediately from these two equalities that if $S(t)$ is nonconstant and the two growth functions f_i ($i = 1, 2$) are not identical on any open subset of $0 < S < 1$, then $f_1(S(t)) - f_2(S(t))$ must change sign, from positive to negative and back again, as t increases. Therefore, first one and then the other competitor must have the competitive advantage. In particular, coexistence can only occur if the graphs of the two growth functions intersect at some value S^* of S, where $0 < S^* < 1$. The reader will recall that this is precisely the intuition which suggested that coexistence should be possible.

Our goal then is to show that positive periodic solutions exist under suitable hypotheses. It turns out that it is simpler to state our results in terms of the discrete dynamical system generated by the Poincaré map, $P: \Omega \to \Omega$, rather than in terms of solutions of (3.2); (3.3) is assumed to hold throughout this section.

The following notation will be used in this section:

$$E_0 = (0, 0),$$

$$E_1 = E_1(0) = (\xi(0), 0),$$

$$E_2 = E_2(0) = (0, \eta(0)).$$

Recall that in the previous section E_i denoted the periodic solutions obtained in Proposition 3.2, whereas now we reserve this same notation for the corresponding fixed points of the Poincaré map. In order to avoid confusion we will write $E_i(t)$ from now on whenever we wish to refer to the solutions.

Solutions of (3.2) possess a stronger order-preserving property than was used in Section 4, one that will be needed for the present section. The reader may wish to review Appendix B at this time. For the statement of our main result here, the following definitions are used. For vectors x and y in $\mathbb{R}^2$, $x <_K y$ means that $x_1 < y_1$ and $x_2 > y_2$. Recall from Section 3 that $x \leq_K y$ means that the corresponding weak inequalities hold – for example, the inequalities, $E_2 <_K E_1$ and $E_2 \leq_K E_0 \leq_K E_1$ hold.

One of the main results of this section is the following.

THEOREM 5.1. *If* $\lambda_{12} > 0$ *then there exists a fixed point* E_* *of P, possibly coinciding with* E_2, *with the following properties:*

(i) $E_2 \leq_K E_* <_K E_1$.
(ii) *For all* $x_0 \in \Omega$ *satisfying* $E_* \leq_K x_0 <_K E_1$,

$$\lim_{n \to \infty} P^n x_0 = E_*. \tag{5.1}$$

(iii) *If* $E_* = E_2$ *then* (5.1) *holds for all positive* x_0 *and, furthermore,* $\lambda_{21} \leq 0$. *If* $E_* \neq E_2$ *then* E_* *is positive and* $E_2 <_K E_* <_K E_1$.
(iv) *The Floquet multipliers (eigenvalues of* $P'(E_*)$) μ_1, μ_2 *corresponding to* E_* *satisfy* $0 < \mu_1 < \mu_2 \leq 1$.

An analogous statement holds if $\lambda_{21} > 0$.

As the statement of the theorem is a bit technical, we offer Figure 5.1 as a geometrical description of the result. The somewhat lengthy proof is postponed until the end of this section. Essentially, Theorem 5.1 says that if λ_{12} is positive (meaning that E_1 is unstable) then there exists a fixed point E_* of P, corresponding to a periodic solution of (3.2), which has strong stability properties. The domain of attraction of E_*, denoted by

$$W^+(E_*) = \{x_0 \in \Omega : P^n x_0 \to E_*\},$$

is substantial. In fact it includes the set $\{x_0 \in \Omega : E_* \leq_K x_0 <_K E_1\}$ if $E_* \neq E_2$, and it coincides with $\{x_0 \in \Omega : x_0 \text{ is positive}\}$ when $E_* = E_2$. In either case, $W^+(E_*)$ contains an open set. Furthermore, the Floquet multipliers of E_* have modulus less than or equal to 1. Generically, we expect E_* to be hyperbolic, in which case it is an attractor.

Let us be specific about the final assertion of Theorem 5.1. If we assume that λ_{21} rather than λ_{12} is positive then there exists a fixed point E_{**} of P satisfying:

(i) $E_2 <_K E_{**} \leq_K E_1$.
(ii) $\lim_{n \to \infty} P^n x_0 = E_{**}$ for all $x_0 \in \Omega$ satisfying $E_2 <_K x_0 \leq_K E_{**}$.
(iii) If $E_{**} = E_1$, then λ_{12} is nonpositive and the orbit of every positive point of Ω converges to E_{**}; if $E_{**} \neq E_1$ then E_{**} is positive and $E_2 <_K E_{**} <_K E_1$.
(iv) The Floquet multipliers of E_{**} have property (iv) of Theorem 5.1.

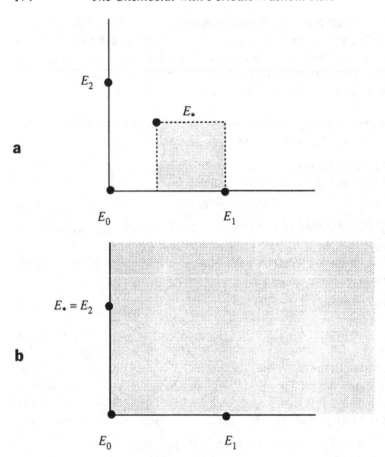

Figure 5.1. The shaded areas denote $W^+(E_*)$, the region of attraction of E_*. **a** E_* is positive and represents a positive periodic solution (coexistence). **b** $E_* = E_2$ which therefore attracts all positive initial data.

Obviously, the cases where $\lambda_{12} > 0$ and $E_* = E_2$, or where $\lambda_{21} > 0$ and $E_{**} = E_1$, correspond to competitive exclusion and therefore are of less interest. On the other hand, if both λ_{12} and λ_{21} are positive then Theorem 5.1(iii) implies that both E_* and E_{**} are positive. In fact, in this case every solution starting with positive initial values converges to a positive periodic solution by the following result, which is the main result of this chapter. Its proof is given following the proof of Theorem 5.1.

COROLLARY 5.2. *Suppose that both λ_{12} and λ_{21} are positive. Then there exist positive fixed points E_* and E_{**}, possibly identical, satisfying:*

(i) $E_2 <_K E_{**} \leq_K E_* <_K E_1$;

(ii) $\lim_{n \to \infty} P^n x_0 = E_{**}$ if $E_2 <_K x_0 \leq_K E_{**}$;

(iii) $\lim_{n \to \infty} P^n x_0 = E_*$ if $E_* \leq_K x_0 <_K E_1$;

(iv) *if x_0 is positive then there exists a fixed point $\bar{x}$ of P satisfying $E_{**} \leq_K$
     $\bar{x} \leq_K E_*$ such that $\lim_{n \to \infty} P^n x_0 = \bar{x}$;*

(v) *the multipliers of both E_{**} and E_* satisfy Theorem 5.1(iv); and*

(vi) *if $E_{**} = E_*$ then $\lim_{n \to \infty} P^n x_0 = E_*$ for all positive $x_0 \in \Omega$.*

Corollary 5.2 can be paraphrased as follows. Provided both λ_{12} and λ_{21} are positive, so that both E_1 and E_2 are unstable, the orbit of every positive point of Ω is attracted to a positive fixed point of P belonging to the "box"

$$B = \{x \in \Omega : E_{**} \leq_K x \leq_K E_*\}.$$

Equivalently, every solution of (3.2) corresponding to positive initial data is asymptotic to a positive periodic solution. The two populations coexist in the chemostat. Figure 5.2 illustrates the result.

It is important to point out that the box B is a fixed positive distance from the two coordinate axes. Because every orbit of P corresponding to a positive starting point approaches B, it follows that there exists a positive number δ, independent of the positive initial value, such that each coordinate of the corresponding solution of (3.2) exceeds δ for all sufficiently large t. This fixed distance δ provides a cushion against the extinction of either population. Later, we will say that the system (3.2) is *uniformly persistent* when this situation holds. (See Appendix D.)

Another important observation concerning Corollary 5.2 is that if the hypotheses are satisfied for a particular set of growth functions f_i and parameters, then the hypotheses continue to hold if these functions and parameters are perturbed by a small amount. Such a property is clearly important, because parameters and functions are never known precisely. The stability of the conditions to perturbations follows from the well-known continuity of simple eigenvalues of matrices to changes in their entries.

It must be stressed that the hypotheses of Corollary 5.2 give sufficient, but not necessary, conditions for the existence of a positive periodic solution possessing strong stability properties. Furthermore, since the single-population periodic solutions $E_1(t)$ and $E_2(t)$ are not explicitly computable, as the corresponding rest points were in Chapter 1, it does not seem possible to obtain explicit formulas for λ_{12} and λ_{21}. However, these crucial Floquet exponents can be easily approximated numerically. One must

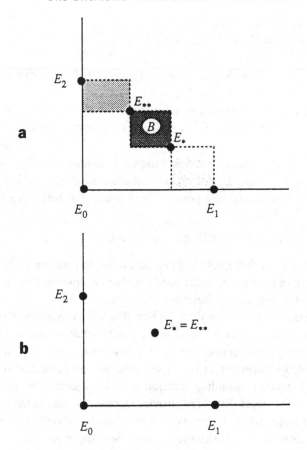

Figure 5.2. **a** The general case $E_* \neq E_{**}$; all positive starting points iterate to a fixed point in the box B. Points in the uppermost box are attracted to E_{**}, while those in the lowermost box are attracted to E_*. **b** The case $E_* = E_{**}$. Here, E_* attracts all positive starting points.

simply obtain accurate numerical approximations of the $E_i(t)$ so that the mean values in the expressions for λ_{21} and λ_{12} (see Proposition 3.3) can be estimated.

In our numerical simulations of (3.2) using PHASE PLANE [Er], we chose Monod-type functional responses $f_i(S) = m_i S/(a_i + S)$ and a sinusoidal dilution rate $D(t) = 1 + d \sin(t)$. An apparently periodic solution was found using the parameter values $m_1 = 1.155$, $a_1 = 0.05$, $m_2 = 3$, $a_2 = 0.65$, and $d = 0.1$ with initial values $x_1 = 0.4283661$ and $x_2 = 0.2878206$ at $t = 0$. See Figure 5.3. We also computed λ_{12} and λ_{21} by computing $E_1(t)$ and $E_2(t)$ and then integrating to find

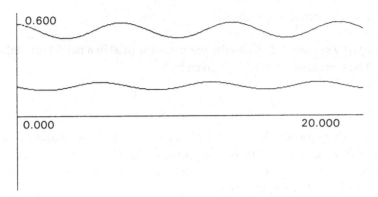

0.600

0.000   20.000

Figure 5.3. A positive, periodic, numerically computed solution representing coexistence of the two populations. Parameter values are given in the text.

$$T^{-1}\int_0^T f_2(1-\xi)\,ds \quad \text{and} \quad T^{-1}\int_0^T f_1(1-\eta)\,ds$$

for large values of T. These computations yielded $\xi(0) = 0.7346227$, $\eta(0) = 0.6966805$, $\lambda_{12} = 0.0021145$, and $\lambda_{21} = 0.000472$.

We now return to the proof of Theorem 5.1. It is convenient to begin with a preliminary result. The system (3.2) possesses stronger monotonicity properties than we have exploited so far, and these properties will be needed in the proof of Theorem 5.1. They are stated next.

LEMMA 5.3. *Let* $x_0 \in \Omega$ *be positive. Then*

$$P'(x_0) = \begin{pmatrix} a & b \\ c & d \end{pmatrix}$$

satisfies:

(i) $a > 0$, $d > 0$, $c < 0$, $b < 0$;

(ii) $P'(x_0)$ *has eigenvalues* μ_i *satisfying* $0 < \mu_1 < \mu_2$;

(iii) *there exists an eigenvector* u_2 *corresponding to* μ_2 *and satisfying* $0 <_K u_2$;

(iv) *if* x *and* y *are distinct points of* Ω, *at least one of which is positive, and if* $x \leq_K y$, *then* $Px <_K Py$.

Proof. Assertion (i) follows from Theorem B.6 since, for (3.2), strict inequality holds in (4.2). (ii) is a consequence of Theorem A.6, or can easily be established by observing that $ad - bc > 0$ by (4.3) of Chapter 3. (iii) follows from Theorem A.6, and (iv) is a consequence of Theorem B.6. □

Now we can prove Theorem 5.1.

Proof of Theorem 5.1. Consider the behavior of P in a neighborhood of E_1. The derivative of P at E_1 is given by

$$P'(E_1) = \Phi(\omega) = \begin{pmatrix} e^{\omega\lambda_{11}} & u(\omega) \\ 0 & e^{\omega\lambda_{12}} \end{pmatrix}$$

(see the discussion following (3.4)), where $\Phi(t)$ is the fundamental matrix of (3.4). Clearly, $(1, 0)^T$ is an eigenvector for $P'(E_1)$ corresponding to the eigenvalue $e^{\omega\lambda_{11}} < 1$. An eigenvector corresponding to the eigenvalue $e^{\omega\lambda_{12}} > 1$ is easily computed to be

$$v = (-1, [e^{\omega\lambda_{11}} - e^{\omega\lambda_{12}}]/u(\omega)).$$

Note that $v <_K 0$ since $u(\omega) < 0$.

Consider the action of P on the points $w(\epsilon) = E_1 + \epsilon v$, where $\epsilon > 0$ is small. We have

$$
\begin{aligned}
Pw(\epsilon) - w(\epsilon) &= E_1 + \epsilon P'(E_1)v + o(\epsilon) - E_1 - \epsilon v \\
&= \epsilon[(e^{\omega\lambda_{12}} - 1)v + o(\epsilon)/\epsilon] \\
&<_K 0
\end{aligned}
$$

for small ϵ, since $o(\epsilon)/\epsilon \to 0$ as $\epsilon \to 0$. It follows that for some $\epsilon_0 > 0$ and all $0 < \epsilon \le \epsilon_0$, $Pw(\epsilon) <_K w(\epsilon)$. Assume that ϵ_0 is so small that $E_2 <_K w(\epsilon) <_K E_1$ for $0 < \epsilon \le \epsilon_0$. by Lemma 5.3(iv) we conclude that $E_2 <_K P^{n+1}w(\epsilon) <_K P^n w(\epsilon) <_K E_1$ for $n = 1, 2, \ldots$. Obviously, $P^n w(\epsilon) \to E_\epsilon$ as $n \to \infty$, where E_ϵ is a fixed point of P satisfying $E_2 \le_K E_\epsilon <_K E_1$.

If $0 < \epsilon < \epsilon'$ then $E_2 <_K w(\epsilon') <_K w(\epsilon) <_K E_1$, so $E_2 \le_K E_{\epsilon'} \le_K E_\epsilon <_K E_1$. Put $E_* = \lim_{\epsilon \to 0^+} E_\epsilon$. Then $PE_* = E_*$ by continuity and $E_2 \le_K E_\epsilon \le_K E_* \le_K E_1$. In fact, $E_* <_K E_1$ and $E_* = E_\epsilon$ for all ϵ, $0 < \epsilon \le \epsilon_0$. To see this, fix ϵ_1 with $0 < \epsilon_1 \le \epsilon_0$. We claim that $E_{\epsilon_1} <_K w(\epsilon_0)$. If not, then, as $E_{\epsilon_1} <_K w(\epsilon_1)$, there exists ϵ_2 $(\epsilon_1 < \epsilon_2 \le \epsilon_0)$ such that $E_{\epsilon_1} \le_K w(\epsilon_2)$ but $E_{\epsilon_1} <_K w(\epsilon_2)$ does not hold. Applying P to the former inequality and using Lemma 4.1 gives $E_{\epsilon_1} \le_K Pw(\epsilon_2) <_K w(\epsilon_2)$, contradicting that $E_{\epsilon_1} <_K w(\epsilon_2)$ does not hold and so proving the assertion. Therefore $E_{\epsilon_1} <_K w(\epsilon_0)$ and, since ϵ_1 was arbitrary, $E_\epsilon <_K w(\epsilon_0)$ for all ϵ, $0 < \epsilon \le \epsilon_0$. Consequently, $E_* \le_K w(\epsilon_0)$. Combining inequalities, for $0 < \epsilon \le \epsilon_0$ we have $E_{\epsilon_0} \le_K E_\epsilon \le_K E_* \le_K w(\epsilon_0)$; hence, by Lemma 4.1, $E_{\epsilon_0} \le_K E_\epsilon \le_K E_* \le_K P^n w(\epsilon_0)$. Letting n go to infinity, we find that $E_{\epsilon_0} \le_K E_\epsilon \le_K E_* \le_K E_{\epsilon_0}$. This proves that $E_\epsilon = E_*$ for $0 < \epsilon \le \epsilon_0$, and (i) follows.

If $x_0 \in \Omega$ and $E_* \le_K x_0 <_K E_1$, then $x_0 <_K w(\epsilon)$ for all small ϵ. Therefore $E_* \le_K P^n x_0 <_K P^n w(\epsilon)$ and, as $P^n w(\epsilon) \to E_*$, (ii) follows immediately.

If $E_* = E_2$ then $P^n x_0 \to E_2$ for all x_0 with $E_2 \leq_K x_0 <_K E_1$. This is obviously incompatible with a positive value of λ_{21} since an argument parallel to that just given for E_1 shows, when applied to E_2, that (5.1) cannot hold for an x_0 satisfying $E_2 <_K x_0 <_K E_1$. Indeed, the stable manifold of E_2 is contained in the x_2 axis if λ_{21} is positive. This proves that $\lambda_{21} \leq 0$ if $E_* = E_2$.

Consider the first assertion of (iii). If $x_0 \in \Omega$ is positive then we can find $w = (0, w_2)$ and $v = (v_1, 0)$ in Ω such that $w <_K x_0 <_K v$. Applying Lemma 4.1, we have $P^n w \leq_K P^n x_0 \leq_K P^n v$. As $P^n w \to E_2$ and $P^n v \to E_1$, and since $P^n x_0 \to \bar{x}$ and $P\bar{x} = \bar{x}$, by Theorem 4.2 we have $E_2 \leq_K \bar{x} \leq_K E_1$. Since λ_{12} is positive, the stable manifold of E_1 belongs to the x_1 axis and so $\bar{x} \neq E_1$. Therefore, by the fundamental theorem of calculus,

$$E_1 - \bar{x} = P E_1 - P\bar{x} = \int_0^1 P'(t E_1 + (1-t)\bar{x})(E_1 - \bar{x}) \, dt.$$

Applying Lemma 5.3(i), we conclude that $0 <_K E_1 - \bar{x}$ or (equivalently) $\bar{x} <_K E_1$. But then $\bar{x} = E_2$, since $\bar{x} = P^n \bar{x} \to E_2$ as $n \to \infty$ by (ii).

If $E_* \neq E_2$ then, since $0 \leq_K E_* - E_2$, an argument identical to that given previously shows that $0 <_K E_* - E_2$, so $E_2 <_K E_*$ and it follows that E_* is positive. This proves (iii).

Finally, consider (iv). If $E_* = E_2$ then (iv) follows from λ_{21} being nonpositive, as noted in (iii), and Proposition 3.3. Therefore, assume that $E_* \neq E_2$ and so, by (iii), $E_2 <_K E_* <_K E_1$. The eigenvalues μ_1, μ_2 of $P'(E_*)$ satisfy $0 < \mu_1 < \mu_2$ by Lemma 5.3. Suppose that $\mu_2 > 1$. By Lemma 5.3, there is a corresponding eigenvector u_2 satisfying $0 <_K u_2$. By considering the action of P on points $w(\epsilon) = E_* + \epsilon u_2$ for small positive ϵ, exactly as in the beginning of the proof, we find that $E_* <_K w(\epsilon) <_K Pw(\epsilon) <_K E_1$. As before, this implies that if $E_* <_K x_0 <_K E_1$ and x_0 is near E_* then $P^n x_0$ converges to a fixed point $\bar{x}$ of P satisfying $E_* <_K \bar{x}$. But, on the contrary, $P^n x_0 \to E_*$ as $n \to \infty$ by (ii), so we have contradicted the assumption that $\mu_2 > 1$. This establishes (iv). $\square$

Proof of Corollary 5.2. Since λ_{12} is positive, Theorem 5.1 implies the existence of a fixed point E_* of P satisfying assertions (i)–(iv) of the theorem. In particular, as λ_{21} is also positive, Theorem 5.1(iii) implies that E_* is positive and $E_2 <_K E_* <_K E_1$. Similarly, since λ_{21} is positive there exists a positive fixed point E_{**} of P satisfying $E_2 <_K E_{**} <_K E_1$. By (ii) of the theorem, we can find $x_0 \in \Omega$ satisfying $E_{**} <_K x_0 <_K E_1$ sufficiently near E_1 such that (5.1) holds. It follows by monotonicity of P that $E_{**} \leq_K E_*$. Therefore, (i) of the corollary holds. Assertions (ii) and (iii) are immediate from (ii) of the theorem. Assertion (v) of the corollary follows from (iv) of the theorem.

It remains to show that (iv) holds, since (vi) follows from (iv). Let $x_0 \in \Omega$ be positive. Arguing as in the proof of the theorem, it can be shown that $\lim_{n \to \infty} P^n x_0 = \bar{x}$, where $\bar{x}$ is a fixed point of P satisfying $E_2 \leq_K \bar{x} \leq_K E_1$. Since the stable manifold of E_1 is the x_1 axis, $\bar{x} \neq E_1$. Similarly, $\bar{x} \neq E_2$ and $\bar{x} \neq E_0$. Consequently, as in the proof of the theorem, $E_2 <_K \bar{x} <_K E_1$ and $\bar{x}$ is positive. Choose x_i, $i = 1, 2$, such that $x_1 \leq_K \bar{x} \leq_K x_2$ and $E_2 <_K x_1 <_K E_{**} \leq_K E_* <_K x_2 <_K E_1$. By monotonicity of P, $P^n x_1 \leq_K \bar{x} \leq_K P^n x_2$ and, letting $n \to \infty$, we find that $E_{**} \leq_K \bar{x} \leq_K E_*$. This completes the proof.   □

6. Discussion

It was observed in Section 1 that coexistence of two competitors competing for a single growth-limiting nutrient in a chemostat with a periodically varying washout rate should be possible, provided that the graphs of the functional responses of the two competitors intersect for some positive value of S and that the washout rate varies in such a way that alternately one, then the other, competitor has the advantage. In Section 5 it was shown that coexistence of the two competitors is only possible if there exists a positive periodic solution of (3.2). This allowed us to show rigorously that the intuition described previously is, in fact, necessary for coexistence to occur.

The very beautiful theory of the dynamics generated by planar, competitive maps, developed in [DS], was used to show that every solution of (3.2) asymptotically approaches a periodic solution. Building on this result, the main result of this chapter, Corollary 5.2, gives a sufficient condition for the coexistence of two competitors (equivalently: for the existence of a positive periodic solution of (3.2)). This condition is that each of the two single-competitor periodic solutions of (3.2) be unstable to invasion by its rival. Either numerical computations or perturbation techniques [S1] are apparently required to verify the conditions of Corollary 5.2. The numerical calculations required are straightforward: one need simply compute each single-competitor periodic solution (the solution of a scalar periodic equation) and then compute the average value along this solution of one of the growth functions. This gives the key Floquet exponent, and one only needs to know the sign of each of these – if both are positive, then coexistence is assured.

Our limited numerical study, described in Section 5, detected a case where the hypotheses of Corollary 5.2 were satisfied. We were able to find what appears to be a positive periodic solution for parameter values such that Corollary 5.2 applies, since both λ_{21} and λ_{12} are positive. Therefore, coexistence holds for these parameter values. Furthermore, as remarked in that section, coexistence continues to hold for all sufficiently nearby values of the parameters.

Very little information exists on the robustness of the parameter region for which coexistence occurs. Studying the case where the nutrient concentration, rather than the washout rate, is varied periodically, Hsu [Hsu2] obtains very interesting information about the parameter region corresponding to coexistence. Perturbation methods are used in [S1] to explore this region in the case studied by Hsu. See also [SFA] for other numerical work in both cases.

It is worth noting that positive periodic solutions of (3.2), though possibly unstable, can be shown to exist by using the bifurcation techniques described in Chapter 3, Section 6. We have chosen not to pursue this route here because we are unable to carry out the necessary calculations to determine the stability of the bifurcating periodic solutions. See [BHW2; Cu1], where this approach is used.

More is known about the dynamics generated by P than has been discussed in this chapter. In [DS] and [HaS] it is shown that there is a curve C joining E_1 and E_2 which is the graph of a strictly decreasing continuous function. This curve C forms the boundary of the unstable manifold of E_0 and every fixed point, except E_0 must lie on C. Therefore, every orbit of P except E_0 is attracted to a fixed point on C. If each fixed point of P is assumed to be hyperbolic, then there are finitely many fixed points. Moving along the curve C, the fixed points alternate between saddle points and attractors. In particular, if the hypotheses of Corollary 5.2 hold then there are an odd number of positive fixed points on C, at least one of which is an attractor. See [S5] for more details.

8

Variable-Yield Models

1. Introduction

In the classical model of the chemostat, discussed in Chapter 1, it is assumed (following Monod [Mo1; Mo2]) that the nutrient uptake rate is proportional to the reproductive rate. The constant of proportionality, which converts units of nutrient to units of organism, is called the *yield constant*. As a consequence of the assumed constant value of the yield, the classical model is sometimes referred to as the "constant-yield" model.

In phytoplankton ecology, it has long been known that the yield is not constant and that it can vary depending on the growth rate [D1]. This led to the formulation of the variable-yield model, also called the variable-internal-stores model [G1] and the Caperon-Droop model [CN1]. This model effectively decouples specific growth rate from external nutrient concentration by introducing an intracellular store of nutrient. The specific growth rate is hypothesized to depend on a quantity, called the *cell quota,* which may be viewed as the average amount of stored nutrient in each cell of the particular organism in the chemostat. The cell quota increases with nutrient uptake and decreases with cell division, which acts to spread the total stored nutrient over more cells. The uptake rate is assumed to depend on the ambient nutrient concentration and, perhaps, the cell quota. In fact, it is reasonable to assume that when the latter is at a high value then uptake will be at a lower level, for a given nutrient concentration, than would be the case if the quota were at a low level.

The purpose of this chapter is to give a complete global analysis of the variable-yield model. Essentially, we confirm that the variable-yield models make the same predictions – concerning the growth of a single population, and concerning the outcome of competition between two

182

microbial populations – as do the simpler constant-yield models discussed in Chapter 1. Our presentation follows [SW3].

The chapter proceeds as follows. In the next section the variable-yield model of single-population growth is derived and analyzed. In Section 3, the competition model is formulated and its equilibrium solutions identified. The conservation principle is introduced in Section 4 in order to reduce the dimension of the system of equations by one; local stability properties of the equilibrium solutions are also determined. The global behavior of solutions of the reduced system is treated in Section 5, and the global behavior of solutions of the original competitive system is discussed in Section 6. The chapter concludes with a discussion of the main results.

2. The Single-Population Growth Model

In this section, the variable-yield model of growth of a single population in the chemostat is derived and analyzed. Let $S(t)$ denote the free nutrient in the chemostat at time t (the quantity we have used all along), and let $x(t)$ denote the concentration of microorganism at time t. It is not reasonable to model the actual process of storing nutrient in a single cell, so a new variable, the cell quota $Q(t)$, is introduced. The variable $Q(t)$ is the average amount of stored nutrient per cell at time t; $x(t)Q(t)$ is the total amount of stored nutrient at time t. Note that this $Q(t)$ can also be thought of as the average biomass of the cells. The uptake of free nutrient, ρ, will be assumed to be a function of S and Q, the idea being that if there is a considerable stored nutrient then the uptake rate will be smaller. The equation for free nutrient is then

$$S' = (S^{(0)} - S)D - x\rho(S, Q),$$

where ρ is a function to be specified. The organism will reproduce according to the level of the cell quotient, so the growth equation becomes

$$x' = x(\mu(Q) - D),$$

where $\mu(Q)$ is to be specified. Note that in this formulation $x(t)$ corresponds to the number of cells (or a concentration) and $x(t)Q(t)$ is the total "biomass" in the vessel. It remains to determine the equation for Q.

Throughout this book we have made use of the conservation principle, which is the defining property of a chemostat. In the previous discussions this principle followed from the form of the equations, but in this formulation we assume the principle and use it to construct the equation for Q.

The principle simply states that if everything is expressed in nutrient equivalents then the sum of the variables should behave as a chemostat without consumption. Nutrient is neither created nor lost, but rather is merely converted from a free to a "stored" state. For the situation just described, the sum at time t of free and stored nutrient, called Σ, should satisfy

$$\Sigma' = (S^{(0)} - \Sigma)D.$$

In previous chapters we have usually made use of the consequence of this – namely, that

$$\lim_{t \to \infty}[\Sigma(t) - S^{(0)}] = 0.$$

In the present case,

$$\Sigma(t) = S(t) + x(t)Q(t)$$

so

$$\begin{aligned}
\Sigma' &= S' + x'Q + xQ' \\
&= S^{(0)}D - SD - x\rho(S, Q) + x\mu(Q)Q - xQD + xQ' \\
&= (S^{(0)} - S - xQ)D.
\end{aligned}$$

From the last equality it follows that

$$x(t)[Q'(t) - \rho(S(t), Q(t)) + \mu(Q(t))Q(t)] = 0.$$

Assuming that $x(t)$ remains positive, one has the following equation for the cell quota:

$$Q' = \rho(S, Q) - Q\mu(Q).$$

Thus, the model takes the form

$$\begin{aligned}
x' &= x(\mu(Q) - D), \\
Q' &= \rho(S, Q) - \mu(Q)Q, \qquad\qquad (2.1) \\
S' &= (S^{(0)} - S)D - x\rho(S, Q).
\end{aligned}$$

Because of the division by $x(t)$ in the derivation of the equation for Q, some comments are in order. First of all, this division means that for $x = 0$ there is no biologically meaningful equation for Q. However, the equation for Q makes mathematical sense even if x is zero. In the mathematical analysis, the case where $x = 0$ and $S = S^{(0)}$ corresponds to a biological steady state of the system, with no organism and with free nutrient at the input level. To mathematically achieve a rest point for (2.1), there must also be a value of Q that satisfies $\rho(S^{(0)}, Q) - \mu(Q)Q = 0$. Although this may seem strange, and biologically unmeaningful, the resulting linear analysis is benign (and valid). (The reader may be more familiar with this anomaly as it occurs in the analysis of linear oscillators in polar coordinate form, when the equation for the polar angle makes mathematical sense even if the polar radius is zero.)

The functions $\mu(Q)$ and $\rho(S, Q)$ are, respectively, the per-capita growth rate and the per-capita uptake rate. To motivate appropriate hypotheses for these functions, we consider some examples from the literature. The following form for the growth rate is attributed to Droop [D1; D2; CN1; CN2]:

$$\mu = \mu_{max} \frac{(Q - Q_{min})_+}{K + (Q - Q_{min})_+},$$

where Q_{min} is the minimum cell quota necessary to allow any cell division. The term $(Q - Q_{min})_+$ is the positive part of $Q - Q_{min}$ and therefore vanishes when the quantity is negative. Motivated by this example, we assume that μ is defined, continuous, and nondecreasing, and that there exists $P \geq 0$ such that:

$$\mu(Q) \geq 0,$$

$$\mu'(Q) > 0 \text{ and continuous for } Q \geq P, \tag{2.2}$$

$$\mu(P) = 0.$$

The growth rate increases with cell quota. The following form for the uptake rate appears in [G2], where Q has the range $Q_{min} \leq Q \leq Q_{max}$:

$$\rho(S, Q) = \rho_{max}(Q) \frac{S}{K + S},$$

$$\rho_{max}(Q) = \rho_{max}^{high} - (\rho_{max}^{high} - \rho_{max}^{low}) \frac{Q - Q_{min}}{Q_{max} - Q_{min}}.$$

In other words, ρ has the Monod form in S but the saturation value of the Monod function, ρ_{max}, decreases with cell quota Q. Cunningham and Nisbet [CN1; CN2] take ρ_{max} to be constant. Therefore, we assume that ρ is continuously differentiable in (S, Q) for $S \geq 0$ and $Q \geq P$ and satisfies

$$\rho(0, Q) = 0,$$

$$\frac{\partial \rho}{\partial S} > 0, \tag{2.3}$$

$$\frac{\partial \rho}{\partial Q} \leq 0.$$

In particular, $\rho(S, Q) > 0$ when $S > 0$. Equation (2.2) requires that the uptake rate vanish in the absence of nutrient, increase with increasing nutrient, and decrease as the cell quota increases.

Observe that (2.2) and (2.3) imply that $Q' \geq 0$ if $Q = P$, and that therefore the interval of Q values, $[P, \infty)$, is positively invariant under the dynamics of (2.1).

It will be convenient to scale the variables appearing in (2.1) as follows:

$$\bar{t} = Dt,$$

$$\bar{S} = S/S^{(0)},$$

$$\bar{Q} = Q/Q^*,$$

$$\bar{x} = xQ^*/S^{(0)}.$$

The term Q^* is an arbitrarily chosen value of the variable Q. If we also define

$$\bar{\mu}(\bar{Q}) \equiv D^{-1}\mu(Q^*\bar{Q}),$$

$$\bar{\rho}(\bar{S}, \bar{Q}) \equiv (DQ^*)^{-1}\rho(S^0\bar{S}, Q^*\bar{Q}),$$

then (2.1) becomes

$$x' = x(\mu(Q) - 1),$$

$$Q' = \rho(S, Q) - \mu(Q)Q, \qquad (2.4)$$

$$S' = 1 - S - x\rho(S, Q),$$

where for convenience we have omitted the bars over the variables.

To determine the equilibrium points of (2.4), we use the following consequence of (2.2) and (2.3): For a fixed value of S, $\rho(S, Q) - \mu(Q)Q$ is strictly decreasing in Q for $Q \geq P$. Also note that $Q\mu(Q)$ increases without bound as Q increases. Equation (2.4) has at most two equilibrium solutions. One of these, which we label E_0, corresponds to the absence of the microorganism. It is given by

$$E_0 = (x, Q, S) = (0, Q^0, 1)$$

and it always exists. Here Q^0 is the unique solution of $\rho(1, Q) - Q\mu(Q) = 0$. The other possible equilibrium, labeled E_1, corresponds to the presence of the population:

$$E_1 = (\hat{x}, \hat{Q}, \hat{S}),$$

where

$$\mu(\hat{Q}) = 1,$$

$$\rho(\hat{S}, \hat{Q}) = \hat{Q},$$

$$\hat{x} = (1 - \hat{S})/\hat{Q}.$$

Examination of these formulas reveals that E_1 exists, in the sense that $\hat{x} > 0$ and $\hat{Q} \geq P$, if and only if

$$\mu(Q) = 1 \text{ has a solution } Q = \hat{Q} \qquad (2.5a)$$

and

$$\rho(1, \hat{Q}) > \hat{Q}. \qquad (2.5b)$$

By the monotonicity assumptions mentioned previously, E_1 is unique if it exists.

The conservation principle allows the reduction of (2.4) to a planar system. Let

$$T = S + Qx,$$

where T consists of unbound free nutrient plus stored nutrient. An easy computation shows that T satisfies

$$T' = 1 - T.$$

Therefore, all solutions of (2.4) asymptotically approach the surface

$$S + Qx = 1 \qquad (2.6)$$

as $t \to \infty$; that is, $T(t) \to 1$ as $t \to \infty$. Consequently, as a first step in the analysis of (2.4), we consider the restriction of (2.4) to the exponentially attracting invariant subset given by (2.6). Dropping S from (2.4), we obtain the system

$$x' = x(\mu(Q) - 1),$$
$$Q' = \rho(1 - Qx, Q) - Q\mu(Q) \qquad (2.7)$$

in

$$Ł = \{(x, Q) \in \mathbb{R}_+^2 : xQ \leq 1, Q \geq P\},$$

where $Ł$ is positively invariant for (2.7).

The equilibria for (2.7) are obtained from those of (2.4) by deleting the S coordinate and replacing $\hat{S}$ by $1 - \hat{Q}\hat{x}$. To conserve notation we use the same letters, E_0 and E_1, to denote the equilibria of (2.7). This should not result in confusion so long as the equation is clear from the context.

A straightforward calculation shows that if E_1 exists as an equilibrium of (2.7), then it is locally asymptotically stable. In this case, E_0 is a saddle point. The first result of this section describes the asymptotic behavior of (2.7).

THEOREM 2.1. *If E_1 does not exist, then every solution of (2.7) satisfies*

$$\lim_{t \to \infty}(x(t), Q(t)) = E_0.$$

If E_1 exists, then every solution of (2.7) satisfying $x(0) > 0$ satisfies

$$\lim_{t \to \infty}(x(t), Q(t)) = E_1.$$

Proof. Solutions of (2.7) are bounded in Ł. Indeed, $Q' < 0$ for all large Q independent of x and $x' < 0$ when Q is near P – that is, when x is large. The result is now a standard application of the Poincaré–Bendixson theorem using the Dulac criterion (discussed in Chapter 1) to eliminate nontrivial periodic orbits and cycles of steady states in Ł. Indeed, since

$$\frac{\partial x'}{\partial x} + \frac{\partial Q'}{\partial Q} = \mu(Q) - 1 + \frac{\partial \rho}{\partial Q} - x\frac{\partial \rho}{\partial S} - \mu(Q) - Q\mu'$$

$$= -1 + \frac{\partial \rho}{\partial Q} - x\frac{\partial \rho}{\partial S} - Q\mu' < 0,$$

there can be no periodic orbits in Ł. $\qquad\square$

Using the results from Appendix F, we obtain the following result for (2.4).

THEOREM 2.2. *If $E_0 = (0, Q_0, S^0)$ is the only steady state of (2.4) and $\mu(Q_0) \neq 1$, then E_0 attracts all solutions of (2.4). If E_0 and $E_1 = (\hat{x}, \hat{Q}, \hat{S})$ exist as steady states, then E_1 attracts all solutions of (2.4) for which $x(0) > 0$.*

As usual, the proof of Theorem 2.2 involves consideration of the equations

$$x' = x(\mu(Q) - 1),$$

$$Q' = \rho(1 - Z - Qx, Q) - \mu(Q)Q,$$

$$Z' = -Z,$$

where $Z = 1 - T$. The condition $\mu(Q_0) \neq 1$ ensures that E_0 is hyperbolic.

3. The Competition Model

Consider two populations, with densities x_1 and x_2, competing for a single nutrient of concentration S in the chemostat. Competition occurs in the sense that each population consumes nutrient and so makes it unavailable for the competitior. The average amount of stored nutrient per individual of population x_1 is denoted by Q_1, and for population x_2 by Q_2. Following the derivation of Section 2, we have the following equations:

$$x_1' = x_1(\mu_1(Q_1) - D),$$

$$Q_1' = \rho_1(S, Q_1) - \mu_1(Q_1)Q_1,$$

$$x_2' = x_2(\mu_2(Q_2) - D), \tag{3.1}$$

$$Q_2' = \rho_2(S, Q_2) - \mu_2(Q_2)Q_2,$$

$$S' = D(S^0 - S) - x_1\rho_1(S, Q_1) - x_2\rho_2(S, Q_2).$$

The functions $\mu_i(Q_i)$ and $\rho_i(S, Q_i)$ are, respectively, the per-capita growth rate and the per-capita uptake rate of population x_i. We assume that μ_i is defined and continuously differentiable for $Q_i \geq P_i$, where $P_i \geq 0$, and satisfies:

$$\mu_i(Q_i) \geq 0,$$

$$\mu_i'(Q_i) > 0, \tag{3.2}$$

$$\mu_i(P_i) = 0.$$

We assume that ρ_i is continuously differentiable in (S, Q_i) for $S \geq 0$ and $Q_i \geq P_i$, and that ρ_i satisfies

$$\rho_i(0, Q_i) = 0,$$

$$\frac{\partial \rho_i}{\partial S} > 0, \tag{3.3}$$

$$\frac{\partial \rho_i}{\partial Q_i} \leq 0.$$

In particular, $\rho_i(S, Q_i) > 0$ when $S > 0$.

Observe that (3.2) and (3.3) imply that $Q_i' \geq 0$ if $Q_i = P_i$, and that therefore the interval of Q_i values, $[P_i, \infty)$, is positively invariant under the dynamics of (3.1). Biologically relevant initial values for (3.1) are

$$x_i(0) > 0, \quad Q_i(0) \geq P_i, \quad S(0) \geq 0.$$

We will repeatedly use the following consequence of (3.2) and (3.3): For a fixed value of S, $\rho_i(S, Q_i) - \mu_i(Q_i)Q_i$ is strictly decreasing in Q_i for $Q_i \geq P_i$. Also note that $Q_i\mu_i(Q_i)$ increases without bound as Q_i increases.

Continuing as in Section 2, we scale the variables appearing in (3.1) as follows:

$$\bar{t} = Dt,$$

$$\bar{S} = S/S^0,$$

$$\bar{Q}_i = Q_i/Q_i^*,$$

$$\bar{x}_i = x_iQ_i^*/S^0.$$

The Q_i^* are arbitrarily chosen representative values of the variables Q_i. If we define

$$\bar{\mu}_i(\bar{Q}_i) \equiv D^{-1}\mu_i(Q_i^*\bar{Q}_i),$$

$$\bar{\rho}_i(\bar{S}, \bar{Q}_i) \equiv (DQ_i^*)^{-1}\rho_i(S^0\bar{S}, Q_i^*\bar{Q}_i),$$

then (3.1) becomes

$$x_1' = x_1(\mu_1(Q_1) - 1),$$

$$Q_1' = \rho_1(S, Q_1) - \mu_1(Q_1)Q_1,$$

$$x_2' = x_2(\mu_2(Q_2) - 1),$$
$$\quad (3.4)$$

$$Q_2' = \rho_2(S, Q_2) - \mu_2(Q_2)Q_2,$$

$$S' = 1 - S - x_1\rho_1(S, Q_1) - x_2\rho_2(S, Q_2),$$

where we have omitted the bars over the variables for convenience and because, hereafter, we treat only (3.4). The hypotheses (3.2) and (3.3) are carried over without a change in notation. In particular, the Q_i range over the interval $Q_i \geq P_i$.

Generically, (3.4) has at most three steady-state solutions. One of these, which we label E_0, corresponds to the absence of both competitors. It is given by

$$E_0 = (x_1, Q_1, x_2, Q_2, S) = (0, Q_1^0, 0, Q_2^0, 1)$$

and it always exists. Here, Q_i^0 is the unique solution of

$$\rho_i(1, Q_i) - Q_i\mu_i(Q_i) = 0.$$

The two other possible steady states, labeled E_1 and E_2, correspond to the presence of one population and the absence of the other. For example,

$$E_1 = (\hat{x}_1, \hat{Q}_1, 0, \hat{Q}_2, \hat{S}),$$

where

$$\mu_1(\hat{Q}_1) = 1,$$

$$\rho_1(\hat{S}, \hat{Q}_1) = \hat{Q}_1,$$

$$\hat{x}_1 = (1 - \hat{S})/\hat{Q}_1,$$
$$\quad (3.5)$$

$$\rho_2(\hat{S}, \hat{Q}_2) - \hat{Q}_2\mu_2(\hat{Q}_2) = 0.$$

Examination of (3.5) reveals that E_1 exists – in the sense that all components are nonnegative, $Q_i \geq P_i$, and x_1 is positive – if and only if

$$\mu_1(Q_1) = 1 \text{ has a solution } Q_1 = \hat{Q}_1 \tag{3.6a}$$

and

$$\rho_1(1, \hat{Q}_1) > \hat{Q}_1. \tag{3.6b}$$

Conditions (3.6) state that the population x_1 can achieve a steady-state population provided that: (a) the dilution rate is not too large; and (b) the reservoir contains sufficient nutrient, that is, $1 > \hat{S}$.

An analogous steady state in which only population x_2 is present is given by

$$E_2 = (0, \tilde{Q}_1, \tilde{x}_2, \tilde{Q}_2, \tilde{S}),$$

where

$$\mu_2(\tilde{Q}_2) = 1,$$

$$\rho_2(\tilde{S}, \tilde{Q}_2) = D\tilde{Q}_2,$$

$$\tilde{x}_2 = (1 - \tilde{S})/\tilde{Q}_2, \tag{3.7}$$

$$\rho_1(\tilde{S}, \tilde{Q}_1) - \tilde{Q}_1 \mu_1(\tilde{Q}_1) = 0.$$

The steady state E_2 exists if and only if

$$\mu_2(Q_2) = 1 \text{ has a solution } Q_2 = \tilde{Q}_2 \tag{3.8a}$$

and

$$\rho_2(1, \tilde{Q}_2) > \tilde{Q}_2. \tag{3.8b}$$

It is possible, but highly unlikely, that there exist steady states with both x_1 and x_2 present. This can happen if and only if both (3.6) and (3.8) are satisfied and

$$\tilde{S} = \hat{S}, \tag{3.9}$$

where $\hat{S}$ and $\tilde{S}$ are as defined in (3.5) and (3.7), respectively. In this case, there is a line segment of steady states of (3.4) joining E_1 to E_2. Since (3.9) is highly unlikely, this case will be ignored. That is, we assume that $\tilde{S} \neq \hat{S}$ when both are defined.

It will be assumed hereafter that if both (3.6) and (3.8) hold then

$$\tilde{S} < \hat{S}. \tag{3.10}$$

Inequality (3.10) can always be achieved by simply renumbering the two populations, if necessary, since we assume that (3.9) does not hold. In biological terms, we label as x_2 the competitor able to grow at the least amount of nutrient.

4. The Conservation Principle

In this section, following a now familiar argument, the conservation principle is used to reduce the dimension of system (3.4) by eliminating the equation for the nutrient. The local stability properties of the equilibria of the reduced system are also investigated.

Let

$$T = S + Q_1 x_1 + Q_2 x_2,$$

where T consists of unbound free nutrient plus stored nutrient and satisfies

$$T' = 1 - T. \tag{4.1}$$

Therefore, all solutions of (3.4) asymptotically approach the surface

$$S + Q_1 x_1 + Q_2 x_2 = 1 \tag{4.2}$$

as $t \to \infty$; that is, $T(t) \to 1$ as $t \to \infty$. Consequently, as a first step in the analysis of (3.4), we consider its restriction to the exponentially attracting invariant subset given by (4.2). Dropping S from (3.4), we obtain the system

$$
\begin{aligned}
x_1' &= x_1(\mu_1(Q_1) - 1), \\
Q_1' &= \rho_1(1 - Q_1 x_1 - Q_2 x_2, Q_1) - \mu_1(Q_1)Q_1, \\
x_2' &= x_2(\mu_2(Q_2) - 1), \\
Q_2' &= \rho_2(1 - Q_1 x_1 - Q_2 x_2, Q_2) - \mu_2(Q_2)Q_2.
\end{aligned}
\tag{4.3}
$$

The biologically relevant domain for (4.3) is

$$\Omega = \{(x_1, Q_1, x_2, Q_2) \in \mathbb{R}_+^4 : Q_1 x_1 + Q_2 x_2 \le 1, \ Q_i \ge P_i\}.$$

It is immediate from the form of (4.3) that Ω is a positively invariant set. We will refer to (4.3) as the "reduced system."

The equilibria of (4.3) are obtained from those of (3.4) by deleting the S equation and using (4.2) to replace S. In order to conserve notation, we retain the labels E_0, E_1, E_2 for the equilibria of (4.3). For convenience, and since we require some of the relations below, we restate the equilibrium conditions here. The steady state E_0 is given by

$$E_0 = (0, Q_1^0, 0, Q_2^0),$$

where the Q_i^0 are uniquely determined by $\rho_i(1, Q_i^0) = Q_i^0 \mu_i(Q_i^0)$. The steady state E_1 is given by

$$E_1 = (\hat{x}_1, \hat{Q}_1, 0, \hat{Q}_2),$$

provided that $\mu_1(Q_1) = 1$ has a solution $\hat{Q}_1 > 0$ and $\rho_1(1, \hat{Q}_1) > \hat{Q}_1$. In this case,

$$\mu_1(\hat{Q}_1) = 1,$$

$$\rho_1(1 - \hat{Q}_1 \hat{x}_1, \hat{Q}_1) = \hat{Q}_1,$$

$$\rho_2(1 - \hat{Q}_1 \hat{x}_1, \hat{Q}_2) = \hat{Q}_2 \mu_2(\hat{Q}_2).$$

Similarly, the steady state E_2 is given by

$$E_2 = (0, \tilde{Q}_1, \tilde{x}_2, \tilde{Q}_2),$$

provided that $\mu_2(Q_2) = 1$ has a solution $\tilde{Q}_2$ and $\rho_1(1, \tilde{Q}_2) > \tilde{Q}_2$. In this case,

$$\mu_2(\tilde{Q}_2) = 1,$$

$$\rho_2(1 - \tilde{x}_2 \tilde{Q}_2, \tilde{Q}_2) = \tilde{Q}_2,$$

$$\rho_1(1 - \tilde{x}_2 \tilde{Q}_2, \tilde{Q}_1) = \tilde{Q}_1 \mu_1(\tilde{Q}_1).$$

We continue to assume that if both E_1 and E_2 exist then (3.10) holds:

$$\tilde{S} = 1 - \tilde{x}_2 \tilde{Q}_2 < 1 - \hat{x}_1 \hat{Q}_1 = \hat{S}.$$

If (3.10) holds, then E_0, E_1, and E_2 are the only possible steady states of (4.3).

The local stability of E_0 is determined by $J_0 = [a_{ij}]$, the Jacobian matrix of (4.3) at E_0. The nonzero entries of J_0 are

$$a_{11} = \mu_1(Q_1^0) - 1, \quad a_{21} = -Q_1^0 \frac{\partial \rho_1}{\partial S},$$

$$a_{22} = -Q_1^0 \mu_1'(Q_1^0) - \mu_1(Q_1^0) + \frac{\partial \rho_1}{\partial Q_1}, \quad a_{23} = -Q_2^0 \frac{\partial \rho_1}{\partial S},$$

$$a_{33} = \mu_2(Q_2^0) - 1, \quad a_{41} = -Q_1^0 \frac{\partial \rho_2}{\partial S}, \quad a_{43} = -Q_2^0 \frac{\partial \rho_2}{\partial S},$$

and

$$a_{44} = -\mu_2(Q_2^0) - Q_2^0 \mu_2'(Q_2^0) + \frac{\partial \rho_2}{\partial Q_2}.$$

The arguments of the partial derivatives of ρ_i are $(1, Q_i^0)$. It follows that the eigenvalues of J_0 are its diagonal entries and that the two eigenvalues $\mu_i(Q_i^0) - 1$ ($i = 1, 2$) determine the stability of E_0, since the other two eigenvalues are negative.

PROPOSITION 4.1. *E_0 is locally asymptotically stable if both $\mu_i(Q_i^0) < 1$, $i = 1, 2$, and unstable if $\mu_i(Q_i^0) > 1$ for some i. Furthermore, $\mu_i(Q_i^0) > 1$ if and only if E_i exists.*

Proof. The first assertion has already been noted. If $\mu_1(Q_1^0) > 1$ then, by our assumptions about μ_1, $\hat{Q}_1$ exists such that $\mu_1(\hat{Q}_1) = 1$ and $\hat{Q}_1 < Q_1^0$. Therefore $\rho_1(1, Q_1^0) = Q_1^0 \mu_1(Q_1^0) > Q_1^0 > \hat{Q}_1$. This implies that E_1 exists. Conversely, if E_1 exists then $\rho_1(1, \hat{Q}_1) > \hat{Q}_1 = \hat{Q}_1 \mu_1(\hat{Q}_1)$, so

$$Q_1^0 \mu_1(Q_1^0) - \rho_1(1, Q_1^0) = 0 > \hat{Q}_1 \mu_1(\hat{Q}_1) - \rho_1(1, \hat{Q}_1).$$

Therefore, by monotonicity of $Q\mu_1(Q) - \rho_1(1, Q)$, $Q_1^0 > \hat{Q}_1$ and consequently

$$\mu_1(Q_1^0) > \mu_1(\hat{Q}_1) = 1. \qquad \square$$

The local stability of E_1 is determined by $J_1 = [c_{ij}]$, the Jacobian matrix of (4.3) at E_1. The nonzero entries of J_1 are

$$c_{12} = \hat{x}_1 \mu_1'(\hat{Q}_1), \quad c_{21} = -\hat{Q}_1 \frac{\partial \rho_1}{\partial S},$$

$$c_{22} = -1 - \hat{x}_1 \frac{\partial \rho_1}{\partial S} - \hat{Q}_1 \mu_1'(\hat{Q}_1) + \frac{\partial \rho_1}{\partial Q_1}, \quad c_{23} = -\hat{Q}_2 \frac{\partial \rho_1}{\partial S},$$

$$c_{33} = \mu_2(\hat{Q}_2) - 1, \quad c_{41} = -\hat{Q}_1 \frac{\partial \rho_2}{\partial S}, \quad c_{42} = -\hat{x}_1 \frac{\partial \rho_2}{\partial S},$$

$$c_{43} = -\hat{Q}_2 \frac{\partial \rho_2}{\partial S}, \quad c_{44} = \frac{\partial \rho_2}{\partial Q_2} - \mu_2(\hat{Q}_2) - \hat{Q}_2 \mu_2'(\hat{Q}_2),$$

where the argument of the partial derivatives of ρ_i is $(1 - \hat{x}_1 \hat{Q}_1, \hat{Q}_i)$. It is easy to see that J_1 has three eigenvalues with negative real part and $\lambda_1 = \mu_2(\hat{Q}_2) - 1$; the sign of λ_1 determines the stability of E_1. A parallel analysis shows that the stability of E_2, if it exists, is determined by the eigenvalue $\lambda_2 = \mu_1(\tilde{Q}_1) - 1$ of the Jacobian of (4.3) at E_2.

PROPOSITION 4.2. *If E_1 exists and E_2 does not exist, then $\lambda_1 < 0$ and E_1 is locally asymptotically stable. Similarly, if E_2 exists and E_1 does not exist, then $\lambda_2 < 0$ and E_2 is locally asymptotically stable. If E_1 and E_2 exist and (3.10) holds, then $\lambda_1 > 0$ and $\lambda_2 < 0$; hence E_1 is unstable and E_2 is locally asymptotically stable.*

Proof. Suppose E_1 exists and E_2 does not and that $\lambda_1 \geq 0$. Then $\mu_2(\hat{Q}_2) \geq 1$, so there exists a unique solution, $\tilde{Q}_2$, of $\mu_2(Q_2) = 1$. By monotonicity of μ_2 it follows that $\hat{Q}_2 \geq \tilde{Q}_2$. Since

$$\rho_2(1, \tilde{Q}_2) > \rho_2(1 - \hat{Q}_1 \hat{x}_1, \hat{Q}_2) = \mu_2(\hat{Q}_2)\hat{Q}_2 \geq \hat{Q}_2 \geq \tilde{Q}_2,$$

we conclude that E_2 exists, contradicting our hypothesis. Therefore, $\lambda_1 < 0$ if E_1 exists and E_2 does not.

Suppose that E_1 and E_2 exist and that (3.10) holds. Then

$$\tilde{Q}_2 \mu_2(\tilde{Q}_2) - \rho_2(\tilde{S}, \tilde{Q}_2) = \tilde{Q}_2 - \rho_2(\tilde{S}, \tilde{Q}_2)$$
$$= 0$$
$$= \hat{Q}_2 \mu_2(\hat{Q}_2) - \rho_2(\hat{S}, \hat{Q}_2)$$
$$< \hat{Q}_2 \mu_2(\hat{Q}_2) - \rho_2(\tilde{S}, \hat{Q}_2),$$

implying that $\tilde{Q}_2 < \hat{Q}_2$. Similar reasoning gives $\tilde{Q}_1 < \hat{Q}_1$. Therefore,

$$\lambda_2 = \mu_1(\tilde{Q}_1) - 1 < \mu_1(\hat{Q}_1) - 1 = 0$$

and

$$\lambda_1 = \mu_2(\hat{Q}_2) - 1 > \mu_2(\tilde{Q}_2) - 1 = 0. \qquad \square$$

In the next section, these local stability considerations will be shown to lead to corresponding global results. For this analysis, it will be important to approximate the one-dimensional unstable manifold of E_1 when both E_1 and E_2 exist and (3.10) holds. To this end, we provide information on an eigenvector corresponding to the eigenvalue λ_1 of J_1. Let $x = (x_1, Q_1, x_2, Q_2)$ denote such an eigenvector. We find that

$$x_1 = \lambda_1^{-1} \hat{x}_1 \mu_1'(\hat{Q}_1) Q_1,$$

$$\left[\lambda_1^{-1} \hat{Q}_1 \frac{\partial \rho_1}{\partial S} \hat{x}_1 \mu_1'(\hat{Q}_1) + \lambda_1 + \hat{x}_1 \frac{\partial \rho_1}{\partial S} + 1 + \hat{Q}_1 \mu_1'(\hat{Q}_1) - \frac{\partial \rho_1}{\partial Q_1} \right] Q_1 = -\hat{Q}_2 \frac{\partial \rho_1}{\partial S},$$

$$\left[-\frac{\partial \rho_2}{\partial Q_2} + \mu_2(\hat{Q}_2) + \hat{Q}_2 \mu_2'(\hat{Q}_2) \right] Q_2 = -\hat{Q}_1 \frac{\partial \rho_2}{\partial S} x_1 - \hat{x}_1 \frac{\partial \rho_2}{\partial S} Q_1 - \hat{Q}_2 \frac{\partial \rho_2}{\partial S},$$

$$x_2 = 1,$$

where the argument of the partial derivatives of ρ_i is $(1 - \hat{x}_1 \hat{Q}_1, \hat{Q}_i)$. If $\lambda_1 > 0$ then evidently

$$x_1 < 0, \quad Q_1 < 0, \quad x_2 = 1, \tag{4.4}$$

by our assumptions concerning μ_i and ρ_i.

5. Global Behavior of the Reduced System

The global asymptotic behavior of the reduced system, (4.3), is worked out in this section. The main result is stated immediately, for the convenience of the reader who may not wish to slog through the remainder of this mathematically technical section. Essentially, we claim that competitive exclusion holds as expected; the winner is the organism that can grow at the lowest nutrient concentration.

THEOREM 5.1.

   (i) *If E_0 is the only steady state, then all solutions tend to E_0 as $t \to \infty$.*
  (ii) *If E_0 and E_1 are the only steady states, then all solutions with $x_1(0) >$ 0 approach E_1 as $t \to \infty$.*
 (iii) *If E_0 and E_2 are the only steady states, then all solutions with $x_2(0) >$ 0 approach E_2 as $t \to \infty$.*
  (iv) *If E_0, E_1 and E_2 exist and* (3.10) *holds, then all solutions with $x_2(0) >$ 0 approach E_2 as $t \to \infty$.*

Case (iv) is the interesting one, since both organisms can survive in the absence of competition in the chemostat. Recall that (3.10) is simply our convention of labeling as x_2 the organism that can grow at the lowest nutrient concentration.

The proof will be divided into various cases and presented as separate propositions. The key to the proof is the use of new variables defined by

$$x_1 = x_1, \quad U_1 = x_1 Q_1,$$
$$x_2 = x_2, \quad U_2 = x_2 Q_2. \tag{5.1}$$

In the new variables (x_1, U_1, x_2, U_2), system (4.3) takes the form

$$x_1' = x_1(\mu_1(U_1/x_1) - 1),$$
$$U_1' = \rho_1(1 - U_1 - U_2, U_1/x_1)x_1 - U_1;$$
$$x_2' = x_2(\mu_2(U_2/x_2) - 1), \tag{5.2}$$
$$U_2' = \rho_2(1 - U_1 - U_2, U_2/x_2)x_2 - U_2.$$

As we will see, (5.2) is particularly useful as a tool in the analysis of those solutions of (4.3) for which $x_i > 0$. An appropriate domain for (5.2) is

$$\Delta = \{(x_1, U_1, x_2, U_2) \in \mathbb{R}_+^4 \mid x_i > 0, U_1 + U_2 \leq 1\},$$

which is positively invariant for (5.2). In fact, notice that $(U_1 + U_2)' = -1$ when $U_1 + U_2 = 1$, so this hyperplane repells in Δ.

Although (5.2) appears to be singular at $x_i = 0$, it is not hard to see that the functions $\mu_i(U_i/x_i)x_i$ and $\rho_i(1 - U_1 - U_2, U_i/x_i)x_i$ are locally Lipschitz, vanishing at $x_i = U_i = 0$, in a wedge shaped region $0 < c < U_i/x_i < C$ of the origin in the U_i-x_i plane. If $P_i > 0$, so that μ_i and ρ_i are defined for $Q_i \geq P_i$, then the lower bound c may be chosen as P_i. Therefore, we can view

$$E_0 = (0, 0, 0, 0),$$

$$E_1 = (\hat{x}_1, \hat{U}_1, 0, 0),$$

$$E_2 = (0, 0, \tilde{x}_2, \tilde{U}_2),$$

as steady states of (5.2), where $\hat{U}_1 = \hat{x}_1 \hat{Q}_1$ and $\tilde{U}_2 = \tilde{Q}_2 \tilde{x}_2$ - provided, of course, that E_0, E_1, E_2 exist for (4.3).

The principal reason that (5.2) is a useful way of viewing (4.3) is that (5.2) generates a strongly monotone dynamical system in Δ. Observe that for fixed U_1 (U_2), the (x_1, U_1) subsystem ((x_2, U_2) subsystem) is cooperative, while the two subsystems compete in the sense that an increase in U_2 (U_1) has a negative effect on U_1' (U_2'). By Theorem C.1 of Appendix C, (5.2) preserves the partial ordering defined by

$$(x_1, U_1, x_2, U_2) \leq_K (\bar{x}_1, \bar{U}_1, \bar{x}_2, \bar{U}_2)$$

if and only if $x_1 \leq \bar{x}_1$, $U_1 \leq \bar{U}_1$, $x_2 \geq \bar{x}_2$, and $U_2 \geq \bar{U}_2$. By this we mean that two solutions with initial data so related remain related in the future. Furthermore, as the variational matrix of (5.2) is irreducible in Δ, Theorem C.1 implies that if the initial data are distinct and ordered as shown then the strong order relation, denoted by

$$(x_1, U_1, x_2, U_2) <_K (\bar{x}_1, \bar{U}_1, \bar{x}_2, \bar{U}_2),$$

holds for all future times.

As a first use of (5.2), we apply Theorem B.1 to obtain bounds on solutions of (5.2) and, consequently, of (4.3). If $(x_1(t), U_1(t), x_2(t), U_2(t))$ is a solution of (5.2) in Δ then

$$x_i' = x_i(\mu_i(U_i/x_i) - 1),$$
$$U_i' \leq \rho_i(1 - U_i, U_i/x_i)x_i - U_i \tag{5.3}$$

for $i = 1, 2$. The solutions (x_i, U_i) can be compared to the solutions $(\bar{x}_i, \bar{U}_i)$ of

$$\bar{x}_i' = \bar{x}_i(\mu_i(\bar{U}_i/\bar{x}_i) - 1),$$
$$\bar{U}_i' = \rho_i(1 - \bar{U}_i, \bar{U}_i/\bar{x}_i)\bar{x}_i - \bar{U}_i \tag{5.4}$$

with $(x_i(0), U_i(0)) = (\bar{x}_i(0), \bar{U}_i(0))$, since (5.4) is a cooperative system. More precisely,

$$x_i(t) \le \bar{x}_i(t) \quad \text{and} \quad U_i(t) \le \bar{U}_i(t) \quad \text{for } t \ge 0, \ i = 1, 2. \tag{5.5}$$

Of course, (5.4) is just (3.4) transformed by (5.1). Therefore, either by a direct analysis of the cooperative planar system (5.4) or by appeal to Theorem 2.1, we know that

$$\lim_{t \to \infty} (\bar{x}_i(t), \bar{U}_i(t)) = \begin{cases} (0,0) & \text{if } E_i \text{ does not exist,} \\ (\hat{x}_1, \hat{U}_1) & \text{if } i = 1 \text{ and } E_1 \text{ exists,} \\ (\tilde{x}_2, \tilde{U}_2) & \text{if } i = 2 \text{ and } E_2 \text{ exists.} \end{cases} \tag{5.6}$$

In any case, (5.5) and (5.6) imply the boundedness of solutions of (5.2) and hence the boundedness of solutions of (4.3). Furthermore, (5.5) and (5.6) essentially imply the first assertion of Theorem 5.1.

The second and third assertions of Theorem 5.1 are symmetric, so it is sufficient to prove only the second one. It will often be necessary to use our knowledge of the behavior of solutions of (5.2) in order to draw conclusions about the corresponding solution of (4.3). In particular, it will be necessary to use knowledge of the behavior of $x_i(t)$ and $U_i(t)$ to determine the behavior of $Q_i(t) = U_i(t)/x_i(t)$. From (5.2), we find that $Q_i(t)$ satisfies

$$Q_i' = \rho_i(1 - U_1(t) - U_2(t), Q_i) - \mu_i(Q_i)Q_i, \tag{5.7}$$

where we have selectively introduced the argument t in order to make the point that we may view this equation as a non-autonomous equation for Q_i, particularly when we know the limiting behavior of the U_i. The next lemma addresses this issue.

LEMMA 5.1. *Let $(x_1(t), U_1(t), x_2(t), U_2(t))$ be a solution of (5.2) satisfying $x_i(0) > 0$ and $U_i(0) \ge 0$ for $i = 1, 2$. Then there exist constants c, C such that $0 < c < C$ and*

$$c < Q_i(t) = U_i(t)/x_i(t) < C$$

for all large t. If $U_i(t) \to U_i(\infty)$ as $t \to \infty$ for $i = 1, 2$, then $Q_i(t) \to Q_i(\infty)$ as $t \to \infty$ where $Q = Q_i(\infty)$ is the unique solution of

$$0 = \rho_i(1 - U_1(\infty) - U_2(\infty), Q) - \mu_i(Q)Q.$$

Proof. The upper bound for Q_i is obvious from (5.7), which implies that $Q_i' < 0$ whenever Q_i is large, so it suffices to show that $\liminf_{t \to \infty} Q_i(t) > 0$. If not, then there exists $t_n \to \infty$ such that $Q_i(t_n) \to 0$ and $Q_i'(t_n) \le 0$. Now $0 < U_1(t) + U_2(t) < 1$ for $t > 0$, and we may assume that $U_1(t_n) + U_2(t_n) \to c \in [0, 1]$. Furthermore, $c < 1$ since (as noted previously) the line $U_1 + U_2 = 1$ is repelling for (5.2). Therefore, from (5.7) we have

$$\lim_{n \to \infty} Q_i'(t_n) = \rho_1(1 - c, 0) \le 0,$$

contradicting that $\rho_1(1 - c, 0) > 0$ since $1 - c > 0$.

The second assertion of the lemma follows easily because the right side of (5.7) is strictly decreasing in the variable Q_i. If the limit superior of $Q_i(t)$ as $t \to \infty$ differed from the limit inferior, then we could find two sequences of times t_n and s_n tending to infinity along which $Q_i(t)$ approaches distinct limits and along which $(Q_i)'(t)$ vanishes. Taking the limit as $n \to \infty$ of (5.7) along each sequence produces a contradiction to the monotonicity mentioned previously. $\qquad \square$

The importance of the first assertion of Lemma 5.1 lies in the local Lipschitz continuity of (5.2) in the domain $\{(x_1, U_1, x_2, U_2) \in \mathbb{R}_+^4 : U_1 + U_2 \le 1,$ $x_1 \le \hat{x}_1, x_2 \le \bar{x}_2,$ and either $(x_i, U_i) = 0$ or $c < x_i/U_i < C\}$. This domain contains the positive limit set of any solution satisfying the hypotheses of Lemma 5.1. Therefore, this limit set is invariant for (5.2).

The next result establishes part (ii) of the theorem.

PROPOSITION 5.2. *If E_0 and E_1 are the only steady states, then all solutions of (4.3) with $x_1(0) > 0$ approach E_1 as $t \to \infty$.*

Proof. If $x_2(0) = 0$ then the result follows from Theorem 2.1. Therefore, assume that $x_2(0) > 0$. By Proposition 4.2, E_1 is locally asymptotically stable. Conditions (5.5) and (5.6) together imply that $(x_2(t), U_2(t)) \to 0$ as $t \to \infty$ and

$$\limsup_{t \to \infty} x_1(t) \le \hat{x}_1, \qquad \limsup_{t \to \infty} U_1(t) \le \hat{U}_1.$$

Suppose that $x_1(t)$ does not converge to 0 as $t \to \infty$. Then the omega limit set of the solution $x(t) \equiv (x_1(t), Q_1(t), x_2(t), Q_2(t))$ contains a point $(\bar{x}_1, \bar{Q}_1, 0, \bar{Q}_2)$ with $\bar{x}_1 > 0$. As this point belongs (by Theorem 2.1) to the domain of attraction of E_1, and since E_1 is locally asymptotically stable (by Proposition 4.2), it follows that $x(t) \to E_1$ as $t \to \infty$ and we are done.

If $x_1(t) \to 0$ as $t \to \infty$ then, by Lemma 5.1, $U_1(t) \to 0$ and $Q_1(t) \to Q_1^0$ as $t \to \infty$. Now $Q_1^0 > \hat{Q}_1$ (see proof of Proposition 4.1) and so, taking $\bar{Q}$ with $\hat{Q}_1 < \bar{Q} < Q_1^0$, we conclude that $Q_1(t) \ge \bar{Q}$ for all $t \ge t_0$, for some $t_0 > 0$. Therefore

$$x_1' \ge x_1(\mu_1(\bar{Q}_1) - 1)$$

holds for $t \ge t_0$. By comparison, therefore,

$$x_1(t) \ge x_1(t_0) \exp[(\mu_1(\bar{Q}) - 1)(t - t_0)].$$

Because $\mu_1(\bar{Q})-1 > \mu_1(\hat{Q}_1)-1 = 0$, $x_1(t) \to \infty$ as $t \to \infty$, a contradiction. Consequently, our assumption that $x_1(t) \to 0$ as $t \to \infty$ cannot hold. This completes our proof. □

We begin the treatment of Theorem 5.1(iv) by showing that the part of the one-dimensional unstable manifold of E_1 contained in Ω is a "hetero-clinic" orbit connecting E_1 to E_2.

PROPOSITION 5.3. *Let E_1 and E_2 exist and let* (3.10) *hold. Then there exists a solution* $(x_1^*(t), Q_1^*(t), x_2^*(t), Q_2^*(t)) = x^*(t)$ *of* (4.3) *satisfying*

(a) $x^*(t) \to E_1$ *as* $t \to -\infty$;
(b) $x^*(t) \to E_2$ *as* $t \to +\infty$;
(c) $x_1^*(t)$ *and* $U_1^*(t) = x_1^*(t)Q_1^*(t)$ *are monotone decreasing;*
(d) $x_2^*(t)$ *and* $U_2^*(t) = x_2^*(t)Q_2^*(t)$ *are monotone increasing.*

Proof. By (4.4), an eigenvector $x = (x_1, Q_1, x_2, Q_2)$ of the Jacobian matrix J_1, at the steady state E_1 of (4.3) corresponding to the positive eigenvalue λ_1, can be shown to satisfy $x_1, Q_1 < 0$ and $x_2 = 1 > 0$; Q_2 is unimportant, as we shall see. It follows that the part of the unstable manifold of E_1 belonging to Ω is a one-dimensional curve that can be expressed, near E_1, in terms of a parameter r as

$$x(r) = E_1 + rx + o(r)$$

as $r \to 0^+$, where $o(r)$ represents a term satisfying $o(r)/r \to 0$ as $r \to 0$.
In the new coordinates (5.1), $x(r)$ takes the form

$$y(r) = (\hat{x}_1, \hat{x}_1\hat{Q}_1, 0, 0) + r(x_1, \hat{Q}_1x_1 + \hat{x}_1Q_1, 1, \hat{Q}_2) + o(r).$$

We will follow the orbit of (4.3) through $x(r)$, for small $r > 0$, by considering the orbit of (5.2) through $y(r)$. Let $F = (F_1, F_2, F_3, F_4)$ denote the right side of (5.2). Then straightforward calculation shows that

$$F_1(y(r)) = r\hat{x}_1 \frac{d\mu_1}{dQ_1}(\hat{Q}_1)Q_1 + o(r),$$

$$F_2(y(r)) = r\hat{x}_1Q_1[\lambda_1 + \hat{Q}_1\mu_1'(\hat{Q}_1)] + o(r),$$

$$F_3(y(r)) = r\lambda_1 + o(r),$$

$$F_4(y(r)) = r\hat{Q}_2\lambda_1 + o(r).$$

Therefore,

$$F(y(r)) <_K 0$$

for all sufficiently small $r > 0$. It follows from Theorem C.2 that the solution $(x_1^*(t), U_1^*(t), x_2^*(t), U_2^*(t))$, starting at $t = 0$ and at such a point $y(r)$, satisfies assertions (c) and (d) of the proposition. Therefore, the $x_i^*(t)$ and $U_i^*(t)$ have limits as $t \to \infty$. By (5.5) and (5.6), these limits are finite. In fact, since $(x_2^*(0), U_2^*(0)) \le (\tilde{x}_2, \tilde{U}_2)$, it follows that $(x_2^*(t), U_2^*(t)) \le (\tilde{x}_2, \tilde{U}_2)$ for all $t \ge 0$.

Since $x_2^*(t)$ is monotone increasing to a positive limit, it follows from (4.3) that $(x_2^*)'(t) \to 0$ as $t \to \infty$ and therefore that $Q_2^*(t) \to \tilde{Q}_2$. If $x_1^*(t)$ has a positive limit as $t \to \infty$ then, since $(x_1^*)'(t) \to 0$ as $t \to \infty$, it follows that $Q_1^*(t) \to \hat{Q}_1$. But then $(x_1^*(t), Q_1^*(t), x_2^*(t), Q_2^*(t))$ has the limit $(x_1^*(\infty), \hat{Q}_1, x_2^*(\infty), \tilde{Q}_2)$ as $t \to \infty$, which is not a rest point of (4.3). This is impossible, so we conclude that $x_1^*(t) \to 0$ as $t \to \infty$. Now, using that $Q_2^*(t) \to \tilde{Q}_2$ (and therefore $(Q_2^*)'(t) \to 0$) as $t \to \infty$, we conclude that

$$\rho_2(1 - \tilde{Q}_2 x_2^*(\infty), \tilde{Q}_2) - \mu_2(\tilde{Q}_2)\tilde{Q}_2 = 0.$$

Obviously $x_2^*(\infty) = \tilde{x}_2$, since $\mu_2(\tilde{Q}_2) = 1$. Thus we have shown that

$$(x_1^*(t), Q_1^*(t), x_2^*(t), U_2^*(t)) \to E_2 \quad \text{as } t \to \infty.$$

Since $(x_1^*(0), Q_1^*(0), x_2^*(0), U_2^*(0)) = x(r)$, a point of the unstable manifold of E_1, $(x_1^*(t), Q_1^*(t), x_2^*(t), Q_2^*(t)) \to E_1$ as $t \to -\infty$. This concludes our proof. $\qquad\square$

We can determine the asymptotic behavior of solutions of (4.3) for which $x_i(0) > 0$, $i = 1, 2$, by considering solutions of (5.2) in Δ. We begin with the following preliminary result.

LEMMA 5.4. *Every solution of* (5.2) *with initial condition belonging to*

$$R = \{(x_1, U_1, x_2, U_2) \in \mathbb{R}_+^4 : x_i, U_i > 0 \ (i = 1, 2) \ and$$

$$(0, 0, \tilde{x}_2, \tilde{U}_2) \le_K (x_1, U_1, x_2, U_2) \le_K (\hat{x}_1, \hat{U}_1, 0, 0)\}$$

remains in R for all $t \ge 0$ and satisfies

$$(x_1(t), U_1(t), x_2(t), U_2(t)) \to (0, 0, \tilde{x}_2, \tilde{U}_2) \tag{5.8}$$

and

$$Q_1(t) = U_1(t)/x_1(t) \to \tilde{Q}_1, \qquad Q_2(t) = U_2(t)/x_2(t) \to \tilde{Q}_2 \tag{5.9}$$

as $t \to \infty$.

Proof. By strong monotonicity, any solution starting at $t = 0$ in R satisfies

$$(0, 0, \tilde{x}_2, \tilde{U}_2) <_K (x_1(t), U_1(t), x_2(t), U_2(t)) <_K (\hat{x}_1, \hat{U}_1, 0, 0)$$

for all $t > 0$. This assertion also follows from the comparison arguments (5.5) and (5.6). Noting that $y(r)$ satisfies $y(r) <_K (\hat{x}_1, \hat{U}_1, 0, 0)$ and that $y(r) \to (\hat{x}_1, \hat{U}_1, 0, 0)$ as $r \to 0$, it follows that we can find $r > 0$ such that

$$(x_1(1), U_1(1), x_2(1), U_2(1)) <_K y(r).$$

Monotonicity of (5.2) implies that

$$(x_1(t+1), U_1(t+1), x_2(t+1), U_2(t+1)) \leq_K (x_1^*(t), U_1^*(t), x_2^*(t), U_2^*(t))$$

for all $t \geq 0$, where $(x_1^*(t), U_1^*(t), x_2^*(t), U_2^*(t))$ is the solution of (5.2) starting at $y(r)$ and described in Proposition 5.3. This implies

$$x_1(t+1) \leq x_1^*(t) \to 0 \quad \text{and} \quad U_1(t+1) \leq U_1^*(t) \to 0 \quad \text{as } t \to \infty,$$

so it follows that $(x_1(t), U_1(t)) \to 0$ as $t \to \infty$. Since

$$x_2(t+1) \geq x_2^*(t) \to \bar{x}_2 \quad \text{and} \quad U_2(t+1) \geq U_2^*(t) \to \bar{U}_2,$$

we conclude that

$$\liminf_{t \to \infty} x_2(t) \geq \bar{x}_2 \quad \text{and} \quad \liminf_{t \to \infty} U_2(t) \geq \bar{U}_2.$$

On the other hand, the comparison arguments (5.5) and (5.6) imply that

$$\limsup_{t \to \infty} x_2(t) \leq \bar{x}_2 \quad \text{and} \quad \limsup_{t \to \infty} U_2(t) \leq \bar{U}_2.$$

Therefore, (5.8) follows; (5.9) follows from the second assertion of Lemma 5.1. $\square$

We complete the proof of Theorem 5.1(iv) in the next result.

PROPOSITION 5.5. *Every solution of* (5.2) *satisfying* $x_i(0) > 0$ *and* $U_i(0) > 0$ *for* $i = 1, 2$ *must satisfy* (5.8) *and* (5.9).

Proof. This is obvious if the solution ever meets the positively invariant set R, which by Lemma 5.4 lies in the basin of attraction of E_2. On the other hand, the omega limit set $\L$ consists of points x satisfying $(0, 0, \bar{x}_2, \bar{U}_2) \leq_K x \leq_K ((\hat{x}_1, \hat{U}_1, 0, 0)$, by the comparison arguments (5.5) and (5.6). We may suppose that the solution remains outside R for all t, and that therefore an omega limit point $x = (x_1, U_1, x_2, U_2)$ must satisfy one or more of $x_1 = \hat{x}_1$, $x_2 = \bar{x}_2$, $U_2 = \bar{U}_2$, or $U_1 = \hat{U}_1$.

If $x_1 = \hat{x}_1$ then $x_1' = 0$ at (x_1, U_1, x_2, U_2), since $\L$ is invariant and $\bar{x}_1 \leq \hat{x}_1$ for all points $(\bar{x}_1, \bar{U}_1, \bar{x}_2, \bar{U}_2) \in \L$. Therefore, $U_1 = \hat{U}_1$.

If $U_1 = \hat{U}_1$, then $U_1' = 0$ at (x_1, U_1, x_2, U_2) and so

$$\hat{U}_1 = \rho_1(1 - \hat{U}_1 - U_2, \hat{U}_1/x_1)x_1 \leq \rho_1(1 - \hat{U}_1, \hat{U}_1/\hat{x}_1)\hat{x}_1 = \hat{U}_1.$$

Therefore, $x_1 = \hat{x}_1$ and $U_2 = 0$. By Lemma 5.1, $U_2 = 0$ implies $x_2 = 0$. It follows that if either $x_1 = \hat{x}_1$ or $U_1 = \hat{U}_1$ then $(x_1, U_1, x_2, U_2) = (\hat{x}_1, \hat{U}_1, 0, 0)$. Similarly, if $x_2 = \tilde{x}_2$ or $U_2 = \tilde{U}_2$ then $(x_1, U_1, x_2, U_2) = (0, 0, \tilde{x}_2, \tilde{U}_2)$. Consequently, either $\text{Ł} = (\hat{x}_1, \hat{U}_1, 0, 0)$ or $\text{Ł} = (0, 0, \tilde{x}_2, \tilde{U}_2)$. As the proof is complete in the second case, assume that $\text{Ł} = (\hat{x}_1, \hat{U}_1, 0, 0)$. Equation (5.7) is satisfied by $Q_2(t) = U_2(t)/x_2(t)$. Since $U_1(t) \to \hat{U}_1$ and $U_2(t) \to 0$ as $t \to \infty$, we conclude from Lemma 5.1 that $Q_2 \to \hat{Q}_2$ as $t \to \infty$. But then $\mu_2(U_2(t)/x_2(t)) - 1 \to \mu_2(\hat{Q}_2) - 1 > 0$, implying that x_2 grows exponentially. This contradiction rules out the case $\text{Ł} = (\hat{x}_1, \hat{U}_1, 0, 0)$, and the proof is complete. $\qquad\square$

6. Competitive Exclusion

In Section 5, the global behavior for the reduced system (4.3) was determined. It remains to show that the results obtained for this system carry over to the original model system (3.4). This will be done by making a change of variables in (3.4) and using the results of Appendix F.

Under the conventions and assumptions described previously, our main result is the following.

THEOREM 6.1. *Assume that the steady states of (3.4) are nondegenerate. Then the following assertions hold.*

(i) *If (3.6) and (3.8) do not hold, then E_0 is the only steady state and every solution of (3.4) satisfies*

$$(x_1(t), Q_1(t), x_2(t), Q_2(t), S(t)) \to E_0 \quad as \ t \to \infty.$$

(ii) *If (3.6) holds and (3.8) does not hold then E_0 and E_1 are the only steady states and every solution for which $x_1(0) > 0$ satisfies*

$$(x_1(t), Q_1(t), x_2(t), Q_2(t), S(t)) \to E_1 \quad as \ t \to \infty.$$

(iii) *If (3.8) holds and (3.6) does not hold then E_0 and E_2 are the only steady states and every solution for which $x_2(0) > 0$ satisfies*

$$(x_1(t), Q_1(t), x_2(t), Q_2(t), S(t)) \to E_2 \quad as \ t \to \infty.$$

(iv) *If (3.6) and (3.8) hold then E_0, E_1, E_2 exist; if also (3.10) holds then every solution for which $x_2(0) > 0$ satisfies*

$$(x_1(t), Q_1(t), x_2(t), Q_2(t), S(t)) \to E_2 \quad as \ t \to \infty.$$

The first three assertions of the theorem describe outcomes in which one or both populations are eliminated from the chemostat – owing not to competition but to the chemostat's environment, which is hostile for one or both populations. For example, in case (ii) the x_2 population is unable to survive in the chemostat, even in the absence of its rival x_1, and therefore has no chance for survival in competition with another population. These cases, then, have nothing to do with competition.

The most interesting case is, of course, the last one. If both (3.6) and (3.8) hold, then the chemostat environment is sufficiently benign that each competitor can survive in it in the absence of the other competitor. The theorem states that the competitor capable of reducing the nutrient concentration to the lowest level is the winner. From a different viewpoint, the competitor that can grow at the lower nutrient concentration is the winner. Competitive exclusion holds, with the weaker competitor being washed out of the chemostat.

The assumption that the steady states are nondegenerate is a mild one. In fact, for the last assertion of the theorem it is an empty assumption (it holds automatically). For the first assertion of the theorem, nondegeneracy holds for E_0 if and only if $\mu_i(Q_i^0) \neq D$, $i = 1, 2$. For the second (third) assertion, only the single condition $\mu_2(Q_2^0) \neq D$ $(\mu_1(Q_1^0) \neq D)$ is needed to ensure that the nondegeneracy assumption holds for both steady states. As all eigenvalues are real, nondegeneracy is equivalent to hyperbolicity.

The argument for Theorem 6.1 follows the familiar pattern, using Appendix F. The main ideas are briefly sketched here. Set

$$Z = 1 - S - Q_1 x_1 - Q_2 x_2$$

in (3.4) and note that $Z' = -Z$. Replace S by $1 - Z - Q_1 x_1 - Q_2 x_2$ in (3.4) to obtain the new system

$$x_1' = x_1(\mu_1(Q_1) - 1),$$

$$Q_1' = \rho_1(1 - Z - Q_1 x_1 - Q_2 x_2, Q_1) - \mu_1(Q_1)Q_1,$$

$$x_2' = x_2(\mu_2(Q_2) - 1), \qquad\qquad (6.1)$$

$$Q_2' = \rho_2(1 - Z - Q_1 x_1 - Q_2 x_2, Q_2) - \mu_2(Q_2)Q_2,$$

$$Z' = -Z.$$

It suffices to determine the global behavior of solutions of (6.1). Clearly $Z(t) \to 0$ as $t \to \infty$, so (6.1) converges to (4.3) as $t \to \infty$. In order to apply the theorem of Appendix F, we note that the equilibria of (4.3) are isolated and that, by Theorem 5.1, every solution converges to an equilibrium of (4.3). Furthermore, there are no cycles of equilibria for (4.3).

Table 6.1

Equilibria present	Conditions for hyperbolicity
E_0	$\mu_i(Q_i^0) \neq 1,$ $i = 1, 2$
E_0, E_1	$\mu_2(Q_2^0) \neq 1$
E_0, E_2	$\mu_1(Q_1^0) \neq 1$
E_0, E_1, E_2	$\tilde{S} < \hat{S}$

Therefore, we conclude from Theorem F.1 that every solution of (6.1), and therefore every solution of (3.4), converges to an equilibrium as $t \to \infty$. However, in order to determine the global behavior of (3.4), we must determine which initial data are attracted to which equilibria. That is, we must determine the stable manifolds of each equilibrium of (3.4). Fortunately, owing to the invariance of the $x_1 = 0$ and $x_2 = 0$ hyperplanes, it is possible to locate these (global) stable manifolds provided we assume that each equilibrium is hyperbolic. From Propositions 4.1 and 4.2, we easily verify the hyperbolicity conditions in Table 6.1. From the form of Jacobian matrices J_i, $i = 0, 1, 2$ (see Section 4), the dimensions of the stable manifolds of each equilibria for (3.4), (4.3), and (6.1) are easily deduced. The form of (6.1) implies that the dimension of the stable manifold for each equilibrium of (3.4) is one more than the dimension of the stable manifold for the corresponding equilibrium of (4.3). Consequently, assuming the hyperbolicity assumptions of Table 6.1, the dimension of the stable manifold of each equilibrium is as follows:

dim $M^+(E_0) = 5$ when E_0 is the only equilibrium;
dim $M^+(E_0) = 4$ when exactly one of the single-population equilibria E_1, E_2 exists;
dim $M^+(E_0) = 3$ when both E_1 and E_2 exist;
dim $M^+(E_1) = 5$ if only E_0 and E_1 exist;
dim $M^+(E_1) = 4$ if E_0, E_1, and E_2 exist;
dim $M^+(E_2) = 4$ if only E_0 and E_2 exist;
dim $M^+(E_2) = 5$ if E_0, E_1, E_2 exist.

The arguments in the four cases of Theorem 6.1 are very similar, so we present only one case, the last and most interesting one. When E_0, E_1, and E_2 exist and $\tilde{S} < \hat{S}$, then E_2 is a local attractor for (3.4) and E_1 is unstable, with a four-dimensional stable manifold consisting of that portion of the $x_2 = 0$ invariant hyperplane for (3.4) with $x_1 > 0$. The one-dimensional unstable manifold of E_1 connects to E_2, by Proposition 5.3. The equilibrium E_0 has a three-dimensional stable manifold contained in the region where $x_1 = 0$ and $x_2 = 0$. Since every solution of (3.4) converges, we conclude that E_2 attracts all solutions corresponding to initial data satisfying $x_2(0) > 0$. This establishes the last case of the main theorem. The other cases follow similar arguments.

7. Discussion

The conclusions of Theorem 6.1 correspond precisely to those of Theorem 5.1 of Chapter 1 and Theorem 3.2 of Chapter 2. In fact, following Grover [G2], a constant-yield model can be associated with (3.1) in such a way that both models give the same predictions (this is not proved in [G2]). Consider the case where both E_1 and E_2 exist. Omit from (3.1) the equations for Q_i and substitute

$$\mu_1(Q_1) = \hat{Q}_1^{-1}\rho_1(S, \hat{Q}_1),$$

$$\mu_2(Q_2) = \tilde{Q}_2^{-1}\rho_2(S, \tilde{Q}_2)$$

for $\mu_i(Q_i)$ in the equations for x_i, $i = 1, 2$. Replace Q_i by the equilibrium values $\hat{Q}_1$ and $\tilde{Q}_2$ in the equation for S. This results in the system

$$x_1' = x_1(\hat{Q}_1^{-1}\rho_1(S, \hat{Q}_1) - D),$$

$$x_2' = x_2(\tilde{Q}_2^{-1}\rho_2(S, \tilde{Q}_2) - D), \tag{7.1}$$

$$S' = D(S^0 - S) - x_1\rho_1(S, \hat{Q}_1) - x_2\rho_2(S, \tilde{Q}_2),$$

which can be viewed as the constant-yield model corresponding to (3.1). Its global behavior is determined by the break-even nutrient concentrations for each population, that is, the value of S at which $x_i' = 0$. By (3.5) and (3.7) these are $S = \hat{S}$ for $i = 1$ and $S = \tilde{S}$ for $i = 2$. The main results of Chapters 1 and 2 show that the winner is the population with the smaller break-even concentration – provided, of course, that it is smaller than S^0. This is precisely the conclusion of Theorem 6.1. Furthermore, the equilibria of (7.1) are obtained from those of (3.1) by deleting the Q_i components.

The predictions of the variable-yield model (3.1) and the corresponding constant-yield model (7.1) are identical. Typical solutions of each model approach the corresponding equilibrium in a monotone fashion (see Proposition 5.3).

In one respect, the variable-yield model has been a disappointment in the sense that it was hoped that the transient behavior of its solutions would better fit the transient behavior seen in experiments with certain algae [CN1]. The experiments, described in [CM], involved the growth of a *Chlamydomonas reinhardii* population on a nitrogen substrate. Following a step increase in the dilution rate, damped oscillations were observed in cell numbers. Cunningham and Nisbet [CN1] note that the single-population variable-yield model could not reproduce these oscillations without the introduction of time delays into the equations. See also the monograph [NG].

The variable-yield model is not the only model proposed for uncoupling the per-capita nutrient uptake rate from the per-capita growth rate. Tang and Wolkowicz [TaW] formulate a model in which the nutrient substrate is converted by the organism to an intermediate product, on the external surface of the cell, which is subsequently released. The per-capita growth rate of the organism is assumed to depend on the concentration of this intermediate product, which in turn is assumed to be uniformly distributed throughout the chemostat. This model gives different predictions than both the variable-yield and constant-yield models. The asymptotic behavior of both variable- and constant-yield models is practically independent of the initial conditions. This is not true for the models developed in [TaW].

9

A Size-Structured Competition Model

1. Introduction

The models considered in previous chapters of this monograph have ignored the fact that populations of microorganisms contain individuals with differing body size and that individuals of different size have different characteristics. Body size is clearly an important factor in determining an organism's energy requirements and its ability to uptake resources. Furthermore, if an organism can grow as well as reproduce, then it becomes important to determine how the organism allocates its energy resources between growth and reproduction.

Data from [Wi] – particularly figures 3, 6, 8, 10, 18, 19, and 21 – leave no doubt that, at least for certain populations of algae, individual cell volume varies significantly during the course of experiments in the chemostat. These data also suggest that steady-state size distributions are reached which have remarkably stable shapes with respect to changes in the control parameters for the chemostat (flow rates, temperature, CO_2).

The purpose of this chapter is to present a model of competition in the chemostat for a single resource that takes these factors into account, and to determine the extent to which these factors influence the outcome of competition. The model presented here is a special case of a more general class of models formulated in [DMKH] and treated in [MD, sec. I.3]. This simplified model is one of several considered in an elegant paper by Cushing [Cu2]. Most of the results of this chapter are taken from [Cu2].

A model that accounts for individual variation in one or more characteristics – such as age, size, or class – is often called a *structured* population model, and the particular characteristics allowed to vary are called the *structure variables*. In this chapter, a size-structured population model is presented. There is a large and rapidly developing literature on structured population modeling, much of which is reviewed in the important book of Metz and Diekmann [MD].

The central theme of structured population modeling is to begin with a careful description of the life cycle of an individual (how it changes in size, reproduces, dies, etc.) and then to infer the aggregate behavior of the population by simply summing over all individuals. This is how we begin. In Section 2, the model of a single size-structured population in a chemostat is formulated. The model takes the form of an ordinary differential equation for the resource coupled with a hyperbolic partial differential equation for the size density function of the population, together with appropriate initial and boundary conditions. In Section 3, these equations are reduced to a system of ordinary differential equations for the resource, the scaled total surface area, total length, and total number of the population. A further reduction is made to a system of two equations which can be directly compared to the equation for single population growth derived in Chapter 1.

In Section 4, competition between two populations is analyzed. Again, the equations can be reduced to a system that can be directly compared to the systems derived in Chapters 1 and 2. Section 5 explores the evolution in time of the population average length, surface area, and volume; in Section 6 we formulate the conservation principle, which played such a crucial role in earlier chapters. The steady-state size distribution of a population is determined in Section 7. Our findings are summarized in a discussion section, where a comparison is made between the conclusions derived from the size-structured model and the unstructured models considered in Chapters 1 and 2.

2. The Single-Population Model

An individual in the population is characterized by its length l. It is assumed that all individuals have the same length (l_b) at birth and that they do not shrink, so that $l_b \le l$. If nutrient is available at concentration S in the chemostat, then the rate of nutrient uptake by an individual of length l is given by

$$\text{nutrient uptake rate} = l^2 f(S),$$

where $f(S)$, the uptake rate per unit surface area, satisfies:

(a) f is continuously differentiable;
(b) $f(0) = 0$; and
(c) $f'(S) > 0$.

The Monod function

$$f(S) = \frac{mS}{a+S}$$

is the primary example. The factor l^2 in the uptake rate reflects the assumption that, for fixed nutrient concentration, uptake is proportional to surface area. We suppose that a fraction κ of the energy derived from the ingested nutrient is used for growth of the organism and the remaining fraction, $1-\kappa$, is channeled into reproduction. It is assumed that the amount of energy required for maintenance of the organism can be neglected. If η is the conversion factor relating nutrient to biomass, then the rate of growth of the organism is given by

$$\frac{d(l^3)}{dt} = \eta^{-1}\kappa l^2 f(S)$$

or, in terms of length, by

$$\frac{dl}{dt} = \frac{\kappa}{3\eta} f(S).$$

The remaining fraction of uptake, $(1-\kappa)l^2 f(S)$, is used for producing offspring, each requiring ωl_b^3 units of resource, where ω is a conversion factor relating nutrient units to the weight of offspring. Therefore, the birth rate for individuals of length l is taken to be

$$\text{birth rate} = \frac{1-\kappa}{\omega l_b^3} f(S)l^2.$$

Finally, the washout rate D in the chemostat and the population death rate d are assumed to be constant, independent of l. Therefore, the removal rate of the organism is given by $D_1 = D+d$.

Having described the behavior of individuals, we now focus attention at the population level. Let $\rho(t, l)$ be the density of individuals of size l at time t, so that

$$\int_a^b \rho(t, l)\, dl$$

is the number of individuals with lengths l satisfying $a \le l \le b$. Fix a time $t_0 > 0$ and consider the fate of the cohort of individuals with length between a and b at time t_0. At time $t > t_0$, this cohort occupies the size range $a(t) \le l \le b(t)$, where

$$a(t) = a + \int_{t_0}^{t} \frac{\kappa}{3\eta} f(S(r)) \, dr,$$

$$b(t) = b + \int_{t_0}^{t} \frac{\kappa}{3\eta} f(S(r)) \, dr.$$

Here, it is assumed that the nutrient $S(t)$ is given for $t > t_0$. The number of individuals in this cohort can change only owing to mortality and wash-out, so

$$\frac{d}{dt} \int_{a(t)}^{b(t)} \rho(t, l) \, dl = -D_1 \int_{a(t)}^{b(t)} \rho(t, l) \, dl.$$

Using the Leibniz rule for the derivative on the left, this expression becomes

$$\int_{a(t)}^{b(t)} \frac{\partial \rho}{\partial t} (t, l) \, dl + b'(t) \rho(t, b(t)) - a'(t) \rho(t, a(t)) = -D_1 \int_{a(t)}^{b(t)} \rho(t, l) \, dl.$$

Since $a'(t) = b'(t) = (\kappa/3\eta) f(S(t))$, the fundamental theorem of calculus can be applied to the previous equality, resulting in

$$\int_{a(t)}^{b(t)} \left[\frac{\partial \rho}{\partial t} (t, l) + \frac{\kappa}{3\eta} f(S(t)) \frac{\partial \rho}{\partial l} (t, l) + D_1 \rho(t, l) \right] dl = 0.$$

This holds at $t = t_0$, so the limits of integration can be taken to be arbitrary numbers a, b satisfying $l_b < a < b$. If the integrand (the term in brackets) is not identically zero as a function of l, for arbitrary fixed t_0, then there is a point l_0 where it is (say) positive. It would then be positive in some interval containing l_0, since the integrand is continuous in l. Taking a and b to be the endpoints of such an interval in the integral would result in a contradiction. Therefore, the integrand must be identically zero:

$$\frac{\partial \rho}{\partial t} + \frac{\kappa}{3\eta} f(S(t)) \frac{\partial \rho}{\partial l} = -D_1 \rho$$

holds for $l_b < l$ and $t > 0$. This equation describes how ρ changes with time. It must, of course, be supplemented with appropriate boundary and initial conditions. For example, it is necessary to specify the initial density ρ_0 of the population at time $t = 0$:

$$\rho(0, l) = \rho_0(l), \quad l \geq l_b.$$

The rate at which offspring of size l_b are added to the population must also be determined. The number of offspring born in the time interval from t to $t+\Delta t$, where $\Delta t > 0$ is small, is given by adding the individual contribution of each size class:

$$\frac{1-\kappa}{\omega l_b^3} f(S(t)) \int_{l_b}^{\infty} l^2 \rho(t,l)\, dl\, \Delta t.$$

Here, Δt must be small so that offspring born of offspring in this period can be neglected. At time $t+\Delta t$, the offspring born at time t have length given approximately by

$$l_{\Delta t} = l_b + \frac{\kappa}{3\eta} f(S(t))\Delta t;$$

hence, at time $t+\Delta t$, the newborns accounted for in the preceding expression occupy the size range $l_b \le l \le l_{\Delta t}$. It therefore follows that

$$\int_{l_b}^{l_{\Delta t}} \rho(t+\Delta t, l)\, dl = \frac{1-\kappa}{\omega l_b^3} f(S(t)) \int_{l_b}^{\infty} l^2 \rho(t,l)\, dl\, \Delta t$$

or

$$\frac{3\eta}{\kappa f(S(t))\Delta t} \int_{l_b}^{l_{\Delta t}} \rho(t+\Delta t, l)\, dl = \frac{3\eta(1-\kappa)}{\kappa \omega l_b^3} \int_{l_b}^{\infty} l^2 \rho(t,l)\, dl.$$

Letting $\Delta t \to 0$ in this last expression leads to:

$$\rho(t, l_b) = \frac{3\eta(1-\kappa)}{\kappa \omega l_b^3} \int_{l_b}^{\infty} l^2 \rho(t,l)\, dl.$$

This equation gives the desired boundary condition at $l = l_b$.

Finally, an expression for the nutrient S is needed. We assume as usual that the nutrient is supplied to the chemostat at constant concentration $S^{(0)}$, and that the flow rate into and out of the vessel is at the constant rate D. The rate of uptake of nutrient by the individuals at time t is given by summing over individuals of all sizes, yielding

$$f(S(t)) \int_{l_b}^{\infty} l^2 \rho(t,l)\, dl.$$

This leads immediately to an equation for S of the form

$$S' = D(S^{(0)} - S) - f(S(t)) \int_{l_b}^{\infty} l^2 \rho(t,l)\, dl.$$

In summary, S and ρ satisfy

$$S' = D(S^{(0)} - S) - f(S(t)) \int_{l_b}^{\infty} l^2 \rho(t,l)\, dl, \qquad (2.1a)$$

$$\frac{\partial \rho}{\partial t} = -\frac{\kappa}{3\eta} f(S(t)) \frac{\partial \rho}{\partial l} - D_1 \rho, \qquad (2.1b)$$

$$\rho(t, l_b) = \frac{3\eta(1-\kappa)}{\kappa \omega l_b^3} \int_{l_b}^{\infty} l^2 \rho(t, l)\, dl, \qquad (2.1c)$$

$$\rho(0, l) = \rho_0(l), \qquad (2.1d)$$

$$S(0) = S_0. \qquad (2.1e)$$

It might appear unnatural to allow unbounded individual length in the model (2.1), inasmuch as the algae or bacteria being modeled are quite small. However, it will be shown in Section 5 that the average individual length at time t approaches a constant value as $t \to \infty$ which is of order l_b in magnitude. Furthermore, the standard deviation in length at time t approaches a constant value of order l_b as $t \to \infty$. Therefore, while the model may allow some individuals to attain unnaturally large sizes, these individuals represent a negligible fraction of the total population. Another fact bears mentioning in this regard. It is expected that the initial length distribution of the population, $\rho_0(l)$, vanishes for large lengths – say, for $l \ge l_m$ where $l_m > l_b$. It then follows that

$$\rho(t, l) \equiv 0, \quad l \ge l_m(t), \qquad (2.2)$$

where

$$l_m(t) = l_m + \int_0^t \frac{\kappa}{3\eta} f(S(r))\, dr.$$

Before proceeding further, it seems appropriate to take a critical look at the assumptions of the model that lead to the equations (2.1). From a general perspective, the model appears to handle nutrient uptake and growth quite reasonably. The scaling of nutrient uptake by cell surface area and growth in terms of cell volume is reasonable. Neglecting the energy required for cell maintenance is a drawback of the model. However, the most serious deficiency of the model is its description of reproduction. The model does not adequately reflect the cell division process, at least for bacteria, where a mother cell divides into two daughter cells of comparable size. The size of the daughter cell is very much of interest. One approach is to consider the ratio of the size of the daughter to the mother as a random variable and treat the problem in a stochastic way. See for example the discussion in Harvey [Ha, sec. 4.2]. Metz and Diekmann [MD, p. 237] analyze just such a model, although they scale uptake in a different way (see Section 8). Despite the obvious deficiencies in the model, it has the important advantage of being mathematically tractable, as we shall see. It is highly likely that a model that more accurately accounts for reproduction will be much more complicated or impenetrable to mathematical analysis. In what follows, we explore the predictions of the present model, cautiously optimistic that some of these predictions may hold up in a more accurate model.

3. Reduction to Ordinary Differential Equations

Systems containing coupled partial differential equations and integro-differential equations, such as (2.1), present significant challenges to mathematical analysis. Much progress on these difficult equations is presented in [MD]. Following Cushing [Cu2] with only minor differences, we assume that (2.1) defines a unique solution $S(t)$ and $\rho(t, l)$ for $t > 0$ and introduce the moment functions:

$$A(t) = l_b^{-2} \int_{l_b}^{\infty} \rho(t, l) l^2 \, dl,$$

$$L(t) = l_b^{-1} \int_{l_b}^{\infty} \rho(t, l) l \, dl, \qquad (3.1)$$

$$P(t) = \int_{l_b}^{\infty} \rho(t, l) \, dl.$$

In [Cu2], the factors l_b^{-2} and l_b^{-1} of $A(t)$ and $L(t)$ are omitted. We introduce them to make for cleaner expressions. Ignoring a scaling factor, $A(t)$ is the total surface area of the population, $L(t)$ is the total length of the population, and $P(t)$ is the total number of individuals in the population at time t.

The immediate goal is to obtain differential equations for these new variables. Multiplying (2.1b) by $(l/l_b)^2$ and integrating from l_b to infinity results in

$$A'(t) + \frac{\kappa}{3\eta l_b^2} f(S(t)) \int_{l_b}^{\infty} l^2 \frac{\partial \rho}{\partial l}(t, l) \, dl = -D_1 A(t).$$

Integrating by parts in the integral, and requiring that $l^2 \rho(t, l) \to 0$ (see (2.2) for justification) as $l \to \infty$, leads to

$$A'(t) + \frac{\kappa}{3\eta l_b^2} f(S(t))[-l_b^2 \rho(t, l_b) - 2l_b L(t)] = -D_1 A(t).$$

The boundary conditions (2.1c) can be used to obtain

$$A'(t) = \tfrac{2}{3} \beta l_b^{-1} f(S) L + \alpha l_b^{-1} f(S) A - D_1 A,$$

where

$$\alpha = (1 - \kappa)/\omega \quad \text{and} \quad \beta = \kappa/\eta.$$

In a similar way, multiplying (2.1b) by (l/l_b) (or by 1), and using the boundary condition (2.1c) and requiring $l^i \rho(t, l) \to 0$ as $l \to \infty$, leads to

$$L' = -D_1 L + \alpha l_b^{-1} f(S) A + \tfrac{1}{3} \beta l_b^{-1} f(S) P,$$

$$P' = -D_1 P + \alpha l_b^{-1} f(S) A.$$

The net result is that we can trade (2.1) for the following system of ordinary differential equations:

$$S' = D(S^{(0)} - S) - f(S) l_b^2 A,$$

$$A' = -D_1 A + \alpha l_b^{-1} f(S) A + \tfrac{2}{3} \beta l_b^{-1} f(S) L,$$

$$L' = -D_1 L + \alpha l_b^{-1} f(S) A + \tfrac{1}{3} \beta l_b^{-1} f(S) P, \tag{3.2}$$

$$P' = -D_1 P + \alpha l_b^{-1} f(S) A.$$

Initial conditions for (3.2) are obtained from (2.1e) and by putting $t = 0$ into (3.1) and using (2.1d).

In [Cu2] α is called the *reproductive efficiency* of the organism, since it is a ratio of the fraction of energy derived from uptake that is allocated to reproduction to the conversion factor relating food units to weight for reproduction (ωl_b^3 is the amount of nutrient needed to produce one offspring). For similar reasons, β is called the *growth efficiency* of the organism.

It will be useful to write the last three equations of (3.2) in vector form. Let $p = \text{col}(A, L, P)$ and let

$$M = \begin{pmatrix} \alpha & \alpha & \alpha \\ \tfrac{2}{3}\beta & 0 & 0 \\ 0 & \tfrac{1}{3}\beta & 0 \end{pmatrix}.$$

Denoting the transpose of M by M^t, the last three equations of (3.2) take the form

$$p' = -D_1 p + l_b^{-1} f(S) M^t p. \tag{3.3}$$

Equation (3.3) can be simplified further by the change of variables $q = T^t p$, where T^t is the transpose of a nonsingular matrix T to be determined shortly. Introducing this change in (3.3) results in

$$q' = -D_1 q + l_b^{-1} f(S) (T^{-1} M T)^t q.$$

The problem is to determine the matrix T in such a way that the new variable $q = \text{col}(x, y, z)$ retains some biological interpretation while at the same time the new system becomes more tractable. The next result describes how to do this; it is based on the Perron–Frobenius theorem (see Appendix A, Theorem A.4).

LEMMA 3.1. *The matrix M has a positive eigenvalue μ and a corresponding eigenvector*

$$v = \mathrm{col}(\tfrac{9}{2}(\mu/\beta)^2, 3(\mu/\beta), 1).$$

The eigenvalue $\mu = \mu(\alpha, \beta)$ is strictly increasing in α and β. Furthermore, it can be expressed as either

$$\mu = \alpha\delta(\beta/\alpha) \quad or \quad \mu = \beta\epsilon(\alpha/\beta),$$

where $\delta(r)$ is strictly increasing in $r \geq 0$ and satisfies $\delta(0) = 1$ and where $\epsilon(s)$ is strictly increasing in $s \geq 0$ and satisfies $\epsilon(0) = 0$. The functions $\mu(\alpha, \beta), \delta(r), \epsilon(s)$ are smooth. Corresponding to the eigenvalue μ, M^t also has a positive eigenvector $w > 0$ satisfying $w \cdot v = 1$, the first component of which is given by

$$w_1 = \frac{2\mu\beta^2}{9\mu^3 + 6\alpha\beta\mu + 4\alpha\beta^2}.$$

The remaining eigenvalues of M are $\gamma \pm i\nu$, where $\gamma < 0$ and $\nu > 0$. There is a nonsingular matrix T such that

$$T^{-1}MT = \begin{pmatrix} \mu & 0 & 0 \\ 0 & \gamma & \nu \\ 0 & -\nu & \gamma \end{pmatrix}.$$

The first column of T is the eigenvector v, and the first row of T^{-1} is the transpose of the eigenvector w.

Proof. Both M and M^t are irreducible, nonnegative matrices, so Theorem A.4 implies that the spectral radius $\mu = \mu(M)$ is a positive eigenvalue and that there is a corresponding positive eigenvector v. As $\mu(M) = \mu(M^t)$, μ is an eigenvector of M^t with a corresponding positive eigenvector w. The eigenvectors v and w are easily calculated in terms of μ.

The characteristic polynomial of M is

$$-\lambda^3 + \alpha\lambda^2 + \tfrac{2}{3}\alpha\beta\lambda + \tfrac{2}{9}\alpha\beta^2 = 0. \tag{3.4}$$

It is easy to show that (3.4) has only one real root, which must be μ. If we denote the eigenvalues of M by $\lambda_1 = \mu$, $\lambda_2 = \gamma + i\nu$, and $\lambda_3 = \bar{\lambda}_2$, then

$$\lambda_1\lambda_2 + \lambda_1\lambda_3 + \lambda_2\lambda_3 = -\tfrac{2}{3}\alpha\beta \quad or \quad \gamma^2 + \nu^2 + 2\gamma\mu = -\tfrac{2}{3}\alpha\beta.$$

It follows that $\gamma < 0$.

By factoring α (resp. β) out of the matrix M resulting in $M = \alpha M'$ ($M = \beta M''$), we see that $\mu(M) = \alpha\mu(M') = \alpha\delta(\beta/\alpha)$ ($\mu = \beta\mu(M'') =$

$\beta\epsilon(\alpha/\beta)$). Theorem A.4(v) implies that $\delta(r)$ is strictly increasing in r, $\mu(\alpha,\beta)$ is strictly increasing in α and β, and $\epsilon(s)$ is strictly increasing in s. That each of these functions is smooth in its arguments follows from the implicit function theorem and the fact that they are simple roots of (3.4). One can easily obtain expressions for their derivatives.

Finally, the canonical form $T^{-1}MT$ expressed in the lemma is just the real Jordan canonical form of M. □

Despite its slightly different scaling, our μ agrees with the μ in [Cu2]. Cushing calls μ the "physiological efficiency coefficient" of the population, since it reflects both the reproductive efficiency and the growth efficiency of the organism.

In the new variables $q = \text{col}(x,y,z)$,

$$x = v \cdot p = \tfrac{9}{2}(\mu/\beta)^2 A + 3(\mu/\beta)L + P > 0$$

is a weighted average of A, L, and P that will serve as a measure of population size. Furthermore,

$$A = c \cdot q = w_1 x + c_2 y + c_3 z,$$

where $c = \text{col}(w_1, c_2, c_3)$ is the first column of T^{-1} containing the first component $w_1 > 0$ of the positive eigenvector w. Consequently, the equations for S, x, y, z take the form

$$S' = D(S^{(0)} - S) - f(S)l_b^2(w_1 x + c_2 y + c_3 z),$$
$$x' = -D_1 x + \mu l_b^{-1} f(S) x,$$
$$y' = -D_1 y + l_b^{-1} f(S)(\gamma y - \nu z),$$
$$z' = -D_1 z + l_b^{-1} f(S)(\nu y + \gamma z).$$

(3.5)

Introducing the complex variable

$$\eta = y + iz,$$

the last two equations can be expressed as

$$\eta' = [-D_1 + \gamma l_b^{-1} f(S(t)) + i\nu l_b^{-1} f(S(t))]\eta.$$

Consequently,

$$\frac{d}{dt}|\eta|^2 = 2[-D_1 + \gamma l_b^{-1} f(S(t))]|\eta|^2.$$

Since $\gamma < 0$, it follows that $\eta(t) \to 0$ as $t \to \infty$ at an exponential rate. Therefore, it suffices to consider the system (3.5) with $y = z = 0$:

$$S' = D(S^{(0)} - S) - w_1 l_b^2 f(S)x,$$

$$x' = -D_1 x + \mu l_b^{-1} f(S)x.$$

To compare this equation with the standard model of Chapter 1, let

$$\bar{f}(S) = \mu l_b^{-1} f(S) \quad \text{and} \quad \gamma^{-1} = w_1 l_b^3 / \mu.$$

Then

$$S' = D(S^{(0)} - S) - \gamma^{-1} \bar{f}(S)x,$$

$$x' = (\bar{f}(S) - D_1)x. \tag{3.6}$$

The key parameter associated with (3.6), as observed in Chapters 1 and 2, is the break-even nutrient concentration λ, defined to be the solution of

$$\bar{f}(\lambda) = \mu l_b^{-1} f(\lambda) = D_1$$

or (equivalently)

$$f(\lambda) = l_b D_1 / \mu, \tag{3.7}$$

where $\lambda = \infty$ if no such solution exists. Since f is strictly increasing, λ is uniquely defined. If insufficient nutrient is supplied to the chemostat (i.e. if $S^{(0)} \le \lambda$) then

$$\lim_{t \to \infty} x(t) = 0 \quad \text{and} \quad \lim_{t \to \infty} S(t) = S^{(0)}.$$

However, if adequate nutrient is supplied ($\lambda < S^{(0)}$) then

$$\lim_{t \to \infty} x(t) = \gamma(S^{(0)} - \lambda)\frac{D}{D_1} \quad \text{and} \quad \lim_{t \to \infty} S(t) = \lambda$$

hold for all solutions of (3.6) for which $x(0) > 0$.

In Section 5, we show that if $\lambda < S^{(0)}$ and $P(0) > 0$ (equivalently, if $x(0) > 0$) then

$$\lim_{t \to \infty} P(t) = \frac{1 - \kappa}{\omega l_b^3}(S^{(0)} - \lambda)\frac{D}{D_1}. \tag{3.8}$$

Obviously, $P(t) \to 0$ when $S^{(0)} \le \lambda$.

Whether or not the population described by the model (2.1) can survive in the chemostat is determined solely by the break-even concentration λ. It is evident from (3.7) that λ increases with increasing l_b and decreases with increasing μ. Recall that smaller is "better" when it comes to λ. A population that can grow at low nutrient levels is more likely to survive and be a strong competitor. Decreasing l_b has the effect of making off-spring cheaper to produce – since each costs ωl_b^3 in nutrient units – so decreasing l_b should have a positive effect on a population's ability to survive in the chemostat. Increasing μ also has the effect of increasing the ability of the population to survive in the chemostat, and since $\mu = \mu(\alpha, \beta)$ is increasing in α and in β, an increase in either the reproductive efficiency $\alpha = (1 - \kappa)/\omega$ or the growth efficiency $\beta = \kappa/\eta$ has the effect of increasing the population's ability to survive in the chemostat.

A natural question to ask is whether a population is better off devoting all, or nearly all, of its resources to reproduction at the expense of growth. Why should an organism grow at all? More explicitly, how does the ability of an organism to survive in the chemostat change as we vary the parameter κ, the fraction of energy derived from nutrient uptake that is channeled to growth? Mathematically, we ask: How does μ depend on κ? Expressing α and β in terms of κ and differentiating (3.4) with respect to κ leads to

$$\frac{\partial \mu}{\partial \kappa} = \frac{9\eta^2\mu^2 + 6\eta(2\kappa - 1)\mu + 2\kappa(3\kappa - 2)}{-27\omega\eta^2\mu^2 + 18\eta^2\mu(1 - \kappa) + 6\eta\kappa(1 - \kappa)}.$$

Now, as $\kappa \to 0$, $\mu \to \omega^{-1}$ and therefore

$$\frac{\partial \mu}{\partial \kappa}(0^+) = \eta^{-1}\left(\frac{2}{3} - \frac{\eta}{\omega}\right).$$

Consequently, if $\eta < \frac{2}{3}\omega$ then an organism that devotes nearly all its energy resources to reproduction will do better if it allocates more to growth, but if the reverse inequality holds then it will do better by devoting even more resources to reproduction.

It is quite remarkable – and not at all obvious from the equations (2.1) – that the question of survival of the population in the chemostat should depend on a single number.

4. Competition

The equations describing the competition between two size-structured populations for a single nutrient can easily be inferred from (2.1). A subscript i is used for variables and parameters associated with the ith population, $i = 1, 2$. In particular, let l_i denote the length at birth of an individual of the ith population. We consider only two populations for simplicity; a similar analysis can be carried out for any number of competitors [Cu2]. The equations are

$$S' = D(S^{(0)} - S) - f_1(S(t))\int_{l_1}^{\infty} l^2\rho_1(t, l)\,dl$$

$$-f_2(S(t))\int_{l_2}^{\infty} l^2\rho_2(t, l)\,dl, \tag{4.1a}$$

$$\frac{\partial \rho_i}{\partial t} = -\frac{\kappa_i}{3\eta_i} f_i(S(t)) \frac{\partial \rho_i}{\partial l} - D_i \rho_i, \tag{4.1b}$$

$$\rho_i(t, l_i) = \frac{3\eta_i(1-\kappa_i)}{\kappa_i \omega_i l_i^3} \int_{l_i}^{\infty} l^2 \rho_i(t, l) \, dl, \tag{4.1c}$$

$$\rho_i(0, l) = \rho_{i0}(l), \tag{4.1d}$$

$$S(0) = S_0. \tag{4.1e}$$

Introducing the moment functions A_i, L_i, P_i as in (3.1), we may reduce (4.1) to

$$S' = D(S^{(0)} - S) - f_1(S)l_1^2 A_1 - f_2(S)l_2^2 A_2,$$

$$A_i' = -D_i A_i + \alpha_i l_i^{-1} f_i(S) A_i + \tfrac{2}{3}\beta_i l_i^{-1} f_i(S) L_i,$$

$$L_i' = -D_i L_i + \alpha_i l_i^{-1} f_i(S) A_i + \tfrac{1}{3}\beta_i l_i^{-1} f_i(S) P_i, \tag{4.2}$$

$$P_i' = -D_i P_i + \alpha_i l_i^{-1} f_i(S) A_i,$$

$$i = 1, 2,$$

where

$$\alpha_i = (1-\kappa_i)/\omega_i \quad \text{and} \quad \beta_i = \kappa_i/\eta_i.$$

The quantity α_i is the reproductive efficiency and β_i is the growth efficiency of the ith organism.

In the same way that (3.2) was reduced to (3.6) by a change of variables, (4.2) can be reduced to

$$S' = D(S^{(0)} - S) - w_1 l_1^2 f_1(S) x_1 - w_2 l_2^2 f_2(S) x_2,$$

$$x_i' = -D_i x_i + \mu_i l_i^{-1} f_i(S) x_i, \quad i = 1, 2, \tag{4.3}$$

where

$$x_i = \tfrac{9}{2}(\mu_i/\beta_i)^2 A_i + 3(\mu_i/\beta_i) L_i + P_i \tag{4.4}$$

and $\mu_i > 0$ is the spectral radius of

$$M_i = \begin{pmatrix} \alpha_i & \alpha_i & \alpha_i \\ \tfrac{2}{3}\beta_i & 0 & 0 \\ 0 & \tfrac{1}{3}\beta_i & 0 \end{pmatrix}.$$

The term μ_i will also be referred to as the "physiological efficiency coefficient" of the ith organism; $w_i > 0$ is the first component of the positive eigenvector W_i of M_i^t and is given by

$$w_i = \frac{2\mu_i \beta_i^2}{9\mu_i^3 + 6\alpha_i \beta_i \mu_i + 4\alpha_i \beta_i^2}. \tag{4.5}$$

The following scaling will allow a direct comparison of (4.3) to the equations immediately preceding (3.1) or (4.1) in Chapter 2:

$$\bar{f}_i(S) = \mu_i l_i^{-1} f_i(S), \qquad \gamma_i^{-1} = w_i l_i^3/\mu_i.$$

Then (4.3) becomes

$$S' = D(S^{(0)} - S) - \gamma_1^{-1}\bar{f}_1(S)x_1 - \gamma_2^{-1}\bar{f}_2(S)x_2,$$

$$x_1' = (\bar{f}_1(S) - D_1)x_1, \qquad\qquad\qquad (4.6)$$

$$x_2' = (\bar{f}_2(S) - D_2)x_2.$$

Define the break-even nutrient concentration λ_i to be the solution of

$$\bar{f}_i(\lambda_i) = \mu_i f_i(\lambda_i)/l_i = D_i,$$

or $\lambda_i = \infty$ if no such solution exists. Then we have the following result.

THEOREM 4.1. *Suppose that either*

(i) $f_i(S) = m_i S/(a_i + S)$ *or*
(ii) *the f_i are restricted as in Section 2 and $D_1 = D_2 = D$.*

Assume also that

$$\lambda_1 < \lambda_2 \quad and \quad \lambda_1 < S^{(0)}.$$

If $x_1(0) > 0$ in (4.6), then

$$\lim_{t \to \infty} x_1(t) = \gamma_1(S^{(0)} - \lambda_1)\frac{D}{D_1},$$

$$\lim_{t \to \infty} x_2(t) = 0, \qquad\qquad\qquad (4.7)$$

$$\lim_{t \to \infty} S(t) = \lambda_1.$$

Proof. If (i) holds then this is Theorem 4.1 of Chapter 2. If (ii) holds then this is Theorem 3.2 of Chapter 2. $\qquad\square$

Competitive exclusion holds, and the winner is the population able to grow at the lowest nutrient concentration. It is interesting to examine the condition $\lambda_1 < \lambda_2$ with respect to the new parameters l_i and μ_i. From

$$f_i(\lambda_i) = l_i D_i/\mu_i$$

it is evident that the length at birth or the physiological efficiency coefficient can decide the winner under certain conditions. All else (i.e. $f_i, D_i,$ μ_i) being equal, the winner is the population with the smaller length at birth; if f_i, D_i, l_i are equal then the winner is the population with the larger physiological efficiency coefficient. As μ_i depends on the growth efficiency α and the reproductive efficiency β, it follows that either can be of decisive importance under suitable conditions. In fact, it is possible that the winner could be the population with a smaller uptake function for all values of S in the range $0 < S < S^{(0)}$, provided its length at birth is suitably smaller – or its physiological efficiency coefficient suitably larger – than that of its rival. This is not possible for the models of Chapters 1 and 2. Typically, however, all things are not equal and the winner will be decided by a complicated weighting of the form of the uptake functions (Michaelis–Menten parameters), the death rates, length at birth, and the growth and reproductive efficiencies of the organisms.

In Section 5 we show that, under the hypotheses of Theorem 4.1,

$$\lim_{t \to \infty} P_1(t) = \frac{1 - \kappa_1}{\omega_1 l_1^3} (S^{(0)} - \lambda_1) \frac{D}{D_1}. \tag{4.8}$$

Obviously, $P_2(t) \to 0$ as $t \to \infty$. It is also possible to obtain expressions for the limiting values of $A_i(t)$ and $L_i(t)$.

5. Average Cell Size

A remarkable feature of the model (4.1) of size-structured competitors is that the average individual length and the average individual surface area of the ith population at time t approaches a constant value as $t \to \infty$, and this value is independent of initial conditions and independent of whether or not a competing population is present. In order to see this, let

$$\bar{A}(t) = A_i(t)/P_i(t),$$

$$\bar{L}(t) = L_i(t)/P_i(t)$$

denote the (scaled) averages corresponding to the ith population at time t. Direct calculation gives

$$\bar{A}' = l_b^{-1} f(S(t))[\alpha \bar{A} + \tfrac{2}{3}\beta \bar{L} - \alpha \bar{A}^2],$$

$$\bar{L}' = l_b^{-1} f(S(t))[\alpha \bar{A} + \tfrac{1}{3}\beta - \alpha \bar{A}\bar{L}],$$

where we have omitted the subscript i (from f_i, α_i, and β_i) and replaced l_i with l_b. Passing to the new time variable τ given by

$$\tau = \frac{l_b}{\alpha} \int_0^t \frac{dr}{f(S(r))},$$

the system now becomes

$$\frac{d\bar{A}}{d\tau} = \bar{A} + \frac{2\beta}{3\alpha}\bar{L} - \bar{A}^2,$$

$$\frac{d\bar{L}}{d\tau} = \bar{A} + \frac{\beta}{3\alpha} - \bar{A}\bar{L}.$$

(5.1)

Analysis of (5.1) leads to the following result.

THEOREM 5.1. *For any solution of* (4.2),

$$\lim_{t\to\infty} \frac{A_i(t)}{P_i(t)} = \frac{\mu_i}{\alpha_i} = \delta_i\left(\frac{\beta_i}{\alpha_i}\right),$$

(5.2a)

$$\lim_{t\to\infty} \frac{L_i(t)}{P_i(t)} = 1 + \frac{\beta_i}{3\mu_i} = 1 + \left(3\epsilon_i\left(\frac{\alpha_i}{\beta_i}\right)\right)^{-1}.$$

(5.2b)

Proof. It suffices to show that $(\bar{A}, \bar{L}) = (\mu/\alpha, 1 + \beta/3\mu)$ is a globally attracting equilibrium for (5.1). Setting the derivatives to zero and solving for $\bar{L}$ in terms of $\bar{A}$ in the second equation leads to the equation for $\bar{A}$:

$$-\bar{A}^3 + \bar{A}^2 + \frac{2\beta}{3\alpha}\bar{A} + \frac{2\beta^2}{9\alpha^2} = 0.$$

Comparing this with (3.4), it is clear that $\alpha\bar{A}$ must satisfy (3.4). Therefore $\alpha\bar{A} = \mu$, since μ is the only real root of (3.4). It follows that there is only one equilibrium of (5.1) and it is given in the first line of the proof. A direct calculation shows that the trace of the variational matrix at this equilibrium is negative and the determinant is positive. Therefore, the equilibrium is locally asymptotically stable.

The Dulac criterion of Chapter 1 can be used to rule out periodic orbits for (5.1). In fact, setting $\beta(\bar{A}, \bar{L}) = (\bar{A}\bar{L})^{-1}$ and computing the divergence gives

$$\frac{\partial}{\partial\bar{A}}\left[\left(\frac{1}{\bar{A}\bar{L}}\right)\left(\bar{A} + \frac{2\beta}{3\alpha}\bar{L} - \bar{A}^2\right)\right] + \frac{\partial}{\partial\bar{L}}\left[\left(\frac{1}{\bar{A}\bar{L}}\right)\left(\bar{A} + \frac{\beta}{3\alpha} - \bar{A}\bar{L}\right)\right]$$

$$= -\frac{2\beta}{3\alpha}\bar{A}^{-2} - \bar{L}^{-1} - \bar{L}^{-2} - \frac{\beta}{3\alpha}\bar{A}^{-1}\bar{L}^{-2} < 0$$

in the open first quadrant.

Finally, it is easy to see from Proposition B.7 that solutions of (5.1) remain nonnegative if their initial conditions are nonnegative. A tedious analysis shows that solutions are bounded. An application of the Poincaré–Bendixson theorem completes the proof. □

Keeping in mind the scale factors l_i^2 in A_i and l_i in L_i, Theorem 5.1 implies that the asymptotic value of the average individual surface area is $l_i^2\mu_i/\alpha_i$ and of the average individual length $l_i(1+\beta_i/(3\mu_i))$. Since δ_i is monotone increasing, the asymptotic average length increases with β_i and decreases with increasing α_i. Similarly, since ϵ_i is monotone increasing, the asymptotic average surface area increases with β_i and decreases with α_i.

The reader might wonder why the average individual volume has been ignored. As in (3.1), it can be defined by

$$V = l_b^{-3} \int_{l_b}^{\infty} l^3 \rho(t,l)\, dl.$$

Direct computation gives

$$V' = (\alpha+\beta)l_b^{-1} f(S)A - D_1 V. \tag{5.3}$$

The average individual volume $\bar{V} = V/P$ satisfies

$$\bar{V}' = \alpha l_b^{-1} f(S)\bar{A}\left[\frac{\alpha+\beta}{\alpha} - \bar{V}\right]. \tag{5.4}$$

Consequently, the limiting value is

$$\lim_{t\to\infty} \frac{V_i(t)}{P_i(t)} = 1 + \frac{\beta_i}{\alpha_i}. \tag{5.5}$$

As before, this value should be multiplied by l_i^3 to account for our scaling.

From (5.5) and (5.1) we may conclude that any reasonable measure of average individual size for the ith population, whether it be the asymptotic value of $L_i(t)/P_i(t)$, $A_i(t)/P_i(t)$, $V_i(t)/P_i(t)$ or some weighted average of these, is increasing in β_i and decreasing in α_i.

Finally, we aim to verify that (4.8) holds under the hypotheses of Theorem 4.1. From (4.4) it follows that

$$P_1(t) = \frac{x_1(t)}{\frac{9}{2}(\mu_1/\beta_1)^2 \bar{A}_1(t) + (3\mu_1/\beta_1)\bar{L}_1(t) + 1}.$$

The limit (4.8) results from taking the limit as $t\to\infty$ and using (4.5), (4.7), and (5.1).

The standard deviation $\sigma(t)$ in the length at time t also approaches an asymptotic value. This can be seen from the formula for the variance,

$$\sigma^2(t) = l_1^2(\bar{A}_1(t) - \bar{L}_1^2(t)).$$

Using (3.4) and (5.2), we obtain

$$\lim_{t\to\infty} \sigma(t) = \frac{l_1\beta_1}{3\mu_1}.$$

6. The Conservation Principle

In the previous chapters, the conservation principle has often played a decisive role in allowing for the reduction of the system of equations to one of lower dimension. Thus it is natural to ask what form the principle takes in the present context and whether there is additional information to be gained from it. This question, not addressed in [Cu2], is the subject of this brief section.

Suppose that the individual death rates d_i are so small that they can be ignored. That is, suppose

$$D_1 = D_2 = D.$$

What is required is an expression for the amount of nutrient that is stored by each population in the form of offspring and in the form of biomass derived from growth. Such an expression for the ith population is given by

$$U_i(t) = \frac{l_i^3}{\alpha_i + \beta_i} V_i(t) = \frac{\omega_i \eta_i}{(1 - \kappa_i)\eta_i + \kappa_i \omega_i} \int_{l_i}^{\infty} l^3 \rho_i(t, l) \, dl.$$

The total amount of nutrient (in all its forms) in the chemostat at time t is

$$T(t) = S(t) + U_1(t) + U_2(t).$$

Direct calculation using (4.2), (5.3), and $D_i = D$ gives

$$T' = D(S^{(0)} - T). \tag{6.1}$$

Here is the desired conservation principle: T approaches $S^{(0)}$ at an exponential rate,

$$S(t) + U_1(t) + U_2(t) \to S^{(0)}, \tag{6.2}$$

as $t \to \infty$.

The conservation principle leads to a plausible explanation of (4.8) in the case where $D = D_1 = D_2$. Asymptotically, as $t \to \infty$, the total amount of nutrient (in all its forms) is given by

$$S(\infty) + U_1(\infty) + U_2(\infty) = S^{(0)}.$$

If $\lambda_1 < \lambda_2$ and $\lambda_1 < S^{(0)}$ then (by Theorem 4.1) $S(\infty) = \lambda_1$ and $U_2(\infty) = 0$, since the second population is eliminated from the chemostat. Therefore, the amount of nutrient stored in the form of offspring and in biomass derived from growth for the first population is

$$U_1(\infty) = S^{(0)} - \lambda_1.$$

The fraction $1 - \kappa_1$ of this is in the form of offspring, each worth $\omega_1 l_1^3$ nutrient units, so

$$P_1(\infty) = \frac{(1 - \kappa_1)(S^{(0)} - \lambda_1)}{\omega_1 l_1^3}.$$

7. The Steady-State Size Distribution

The steady-state size distribution for a single population obeying equations (2.1), or of the surviving population for two competing populations obeying (4.1), can be readily computed. The case of a single population will be treated, with an appropriate subscript on the result yielding the distribution of the surviving population in the case of competition. Assume that $0 < \lambda < S^{(0)}$. Then set S' and $\partial \rho / \partial t$ equal to zero in (2.1a,b); using the fact that the steady-state value of S is λ, we have $f(S) = l_b D_1 / \mu$ in (2.1b). This leads to

$$\rho(l) = \rho(l_b) \exp\left[-\frac{3\eta\mu}{\kappa l_b}(l - l_b)\right], \tag{7.1}$$

where $\rho(l_b)$ must be determined either from (2.1a) or (more efficiently) from (3.8). The latter yields

$$\rho(l_b) = 3\frac{\mu\eta}{\kappa l_b}(1 - \kappa)\frac{S^{(0)} - \lambda}{\omega l_b^3}\frac{D}{D_1}. \tag{7.2}$$

The data in [Wi] are in terms of cell volume, so for comparison purposes it is necessary to convert the distribution by length, (7.1), to one by volume. Let $R(v)$ be the steady-state cell-volume distribution corresponding to the distribution (7.1). Then the number of individuals with cell volume in the range v_1 to v_2, where $v_b = l_b^3 \le v_1 < v_2$, is given by

$$\int_{v_1}^{v_2} R(v)\, dv.$$

If $v = l^3$, then differentiating the identity

$$\int_{v_b}^{v} R(v)\, dv = \int_{l_b}^{l} \rho(\eta)\, d\eta$$

with respect to l leads to

$$R(v) = \tfrac{1}{3} v^{-2/3} \rho(v^{1/3})$$

or

$$R(v) = \frac{\mu\eta}{\kappa l_b}(1-\kappa)\frac{S^{(0)}-\lambda}{\omega v_b}\frac{D}{D_1}v^{-2/3}\exp\left[-\frac{3\eta\mu}{\kappa}\left(\left(\frac{v}{v_b}\right)^{1/3}-1\right)\right]. \qquad (7.3)$$

As this distribution monotonically decreases with v for $v \geq v_b$, it fails to account for the peak in observed cell volume seen in Figure 7.1 (taken from [Wi, fig. 19]).

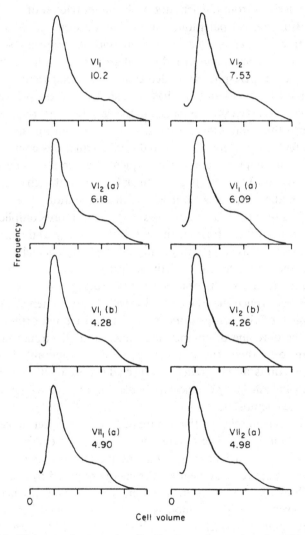

Figure 7.1. Eight steady-state size distributions observed under different experimental conditions (flow rate, temperature, CO_2), scaled for equal means and areas. The mean cell size for each graph is indicated next to the graph. (From [Wi, fig. 19], Copyright 1971, Academic Press. Reproduced by permission.)

8. Discussion

The system of equations (4.1), like those in Chapters 1 and 2, models competition between two populations competing for a single nutrient in the chemostat. The difference is that for the model introduced in this chapter, each population contains individuals having different lengths. The energy derived from nutrient uptake by an individual of the ith population is assumed to be partitioned into a fraction κ_i channeled to growth and a fraction $1 - \kappa_i$ channeled to reproduction. Appropriate scaling of uptake (by surface area) and of individual growth (by volume) has been taken into account. Despite this added complexity, relative to the models considered in Chapters 1 and 2, it has been shown that the principal prediction of competitive exclusion made in the earlier model remains valid. The winner is the population able to grow at the lowest nutrient concentration. In fact, the system (4.1) was reduced to equations comparable to those treated in Chapters 1 and 2. Consequently, we could define a break-even concentration λ for each population and conclude that the population having the smaller value of λ is the winner in competition. The novelty in the case of size-structured competitors is that λ is more complicated; it not only depends on the form of the uptake function f and the removal rate but also (inversely) on the physiological efficiency coefficient μ of the population and on the length at birth. In turn, μ depends on the growth efficiency and the reproductive efficiency. In the discussion following Theorem 4.1, it was argued that the physiological efficiency coefficients or the length at birth of the competitors can be of decisive importance in determining the outcome of competition under suitable circumstances. It would seem to be theoretically possible and of considerable interest to test the predictions of Theorem 4.1, especially those relating to the role of physiological efficiency coefficients in determining the winner of competition in the chemostat.

Like all models in science, the one treated in this chapter makes many unrealistic assumptions. These were pointed out at the end of Section 2. The most notable deficiencies are that the model inadequately reflects the cell division process and neglects the energy required for cell maintenance. It should be pointed out, however, that the main predictions of the simple (even less realistic) models of Chapters 1 and 2 survive intact in the more complex model treated in this chapter. Therefore, it is not unreasonable to expect that many of the predictions made on the basis of Theorem 4.1 will continue to hold for more realistic models.

As noted in Section 2, Metz and Diekmann [MD, p. 237] describe a different size-structured model, one that reflects the cell-division process quite well. They assume that cell size x varies among the individual cells of the population, from a minimum value x_{min} to a maximum value that is normalized to 1. A function $b(x)$ gives the per-unit time probability of a cell of size x dividing. Small cells are not allowed to divide ($b(x) = 0$, $x \leq a$). A mother cell of size x is assumed to divide into two daughter cells, one of size px and one of size $(1-p)x$, with probability $d(p)$, $0 < p < 1$. Of course, $d(p) = d(1-p)$ and $\int_0^1 d(p)\,dp = 1$. The unit of size x – whether length, area, or volume – is not specified in [MD]. This makes their assumption that the growth rate of a cell of size x is proportional to x (and to $f(S)$) subject to different interpretations. The reader is referred to [MD, p. 238] for the equations and hypotheses. Their model also can be reduced to the equations considered in Chapter 1.

It would be of considerable interest to construct and analyze a model that treats growth and consumption as in this chapter (following [Cu2], i.e., proportional to surface area) and that treats cell division as in [MD]. It seems unlikely that this marriage of the two approaches would yield a model that can easily be reduced to the ordinary differential equation models of Chapter 1.

Cushing's study [Cu2] was motivated by the so-called size-efficiency hypothesis formulated in [BD]. Based on studies of zooplankton communities, Brooks and Dodson proposed that (1) larger individuals are more efficient at exploiting resources, which provides the potential for competitive exclusion of the smaller individuals in the population, and (2) size-selective predation by predators, which falls more heavily upon the larger individuals, can allow for the survival of smaller individuals or in some cases can even result in the elimination of larger individuals. To test hypothesis (1), it is natural to ask whether, in the model treated in Section 4, the superior competitor is necessarily the larger competitor. To answer this question, some measure of population size must be chosen. In Section 5, it was argued that suitable choices are the asymptotic values of average individual length, surface area, volume, or some weighted average of these. Each of these measures of average individual size shares the common feature that it is an increasing function of growth efficiency κ_i/η_i and the length l_i at birth, and a decreasing function of reproductive efficiency $(1 - \kappa_i)/\omega_i$.

On the other hand, competitive success is determined solely by having the smaller break-even concentration λ_i. Since λ_i decreases (and so the ith population becomes a stronger competitor) as κ_i/η_i or $(1 - \kappa_i)/\omega_i$ increases and as l_i decreases, it is not evident that a clear-cut relationship exists between competitive success and average individual size. For example, in the case where the two populations have identical per-unit surface area uptake rates $f_1 = f_2$ and identical death rates $D_1 = D_2$, the superior competitor is the one with the larger value μ_i since that populatioin has the smaller λ_i (see (3.7)). However, from (5.2), the population with larger μ_i need not have the larger average size, by any of the measures of size. Consequently, the model considered here does not appear to support the size-efficiency hypothesis. For more on this interesting subject the reader is referred to [Cu2] and the references therein.

10

New Directions

1. Introduction

In this chapter, several recent models that make use of the chemostat are described. The situations that occur are not as fully understood as those in the previous chapters, and no attempt will be made to present a detailed analysis. The mathematical results are only partial and, in fact, in some cases the modeling is clearly inadequate at this time. The title of the chapter is intended to suggest that further work is needed. The problems are important and interesting, and it is hoped that the reader might find something of interest here and contribute to the development of the theory.

Three types of new directions are discussed. In two of these, ordinary differential equations are not an adequate model to describe the phenomenon of interest; functional differential equations and partial differential equations provide the appropriate setting. In the remaining case ordinary differential equations are appropriate but the modeling is not complete. Improving the model would result in a larger system for which the techniques of monotone dynamical systems are inappropriate. The problems will be described and results indicated, but no proofs are given. In all cases, much more work needs to be done before the problem is appropriately modeled and analyzed.

2. The Unstirred Chemostat

The gradostat was an attempt to create a nutrient gradient in a piece of laboratory apparatus; it was discussed in detail in Chapters 5 and 6. An alternative to the gradostat is to remove the "well mixed" hypothesis – that is, to think of the chemostat but without mixing the vessel. If the nutrient is added to the vessel at one point, allowed to diffuse, and then removed at a different point, a gradient will occur. The nutrient concentration will become spatially dependent, and the organisms will compete at different nutrient levels at different locations. (There is a tacit assumption that the turnover of the chemostat – typically 12–24 hours – is so slow that there is no relevant transport.)

Reformulating the chemostat with diffusion introduces a new level of difficulty into the modeling. First of all, the equations will become nonlinear partial differential equations with all of their attendant complexity. The input and the output now occur at the boundary, so the boundary conditions for the system of partial differential equations must be formulated with care. If nutrient and organisms are diffusing, a new constant occurs: the diffusion coefficient. To simplify matters, we will assume that all the quantities diffuse with the same constant, an assumption that is mathematically convenient but not biologically rigorous. The analysis is limited to only one space dimension, so one must perform a thought experiment to visualize a one-dimensional chemostat when the real one is three-dimensional. A tubular reactor is an approximation. This will have the mathematical consequence that the rest states will be solutions of a boundary value problem for ordinary differential equations. With these limiting assumptions, the problem was considered in [HSW], [HW2], and [SoW] with an attempt to recover the standard gradostat results in this setting. The resulting boundary value problems were also considered in [BT].

We do not give the derivation here, but the equations take the form

$$\frac{\partial S}{\partial t} = d\frac{\partial^2 S}{\partial x^2} - \frac{m_1 S u}{a_1 + S} - \frac{m_2 S v}{a_2 + S},$$

$$\frac{\partial u}{\partial t} = d\frac{\partial^2 u}{\partial x^2} + \frac{m_1 S u}{a_1 + S}, \qquad (2.1)$$

$$\frac{\partial v}{\partial t} = d\frac{\partial^2 v}{\partial x^2} + \frac{m_2 S v}{a_2 + S},$$

$$0 < x < 1,$$

with boundary conditions

$$\frac{\partial S}{\partial x}(t, 0) = -S^{(0)},$$

$$\frac{\partial u}{\partial x}(t, 0) = \frac{\partial v}{\partial x}(t, 0) = 0,$$

$$\frac{\partial S}{\partial x}(t,1) + rS(t,1) = 0, \qquad (2.2)$$

$$\frac{\partial u}{\partial x}(t,1) + ru(t,1) = 0,$$

$$\frac{\partial v}{\partial x}(t,1) + rv(t,1) = 0$$

and initial conditions

$$S(0,x) = S_0(x) \geq 0;$$

$$u(0,x) = u_0(x) \geq 0, \quad u_0(x) \not\equiv 0; \qquad (2.3)$$

$$v(0,x) = v_0(x) \geq 0, \quad v_0(x) \not\equiv 0.$$

The boundary conditions in (2.2) are fairly intuitive and appropriate for this type of equation. However, the boundary conditions are not defined in terms of the operating parameters of the simple chemostat. The problem will be considered in a heuristic way to see how the units compare between the simple chemostat and the chemostat without the assumption of well mixing. To keep matters simple, we work with the nutrient equation without consumption (equivalently, zero initial conditions for the microorganisms); the other cases will be clear by analogy. Under these circumstances, the simple chemostat takes the form

$$S'(t) = (S^{(0)} - S(t))D.$$

The units of S are concentration, mass/volume $= m/l^3$. The total mass of substrate is VS, where V is the volume of the vessel; if F is the flow rate (the rate of the pump operating the chemostat) then the parameter D is defined as F/V. Rewriting the equation just displayed for the mass of the substrate in the vessel yields

$$VS'(t) = FS^{(0)} - FS(t). \qquad (2.4)$$

Equation (2.4) states that the rate of change in mass is proportional to the difference between the incoming flux and the outgoing flux.

When considering the partial differential equation, the basic quantity $S(t,x)$ becomes a density, measured in units of mass per unit length. The nutrient equation, using a subscript to denote differentiation, is

$$S_t = dS_{xx}. \qquad (2.5)$$

If one integrates over the interval $[0,1]$, an equation for the total mass of the nutrient is obtained:

$$\frac{d}{dt} \int_0^1 S(t,x)\,dx = dS_x(t,1) - dS_x(t,0). \tag{2.6}$$

The two terms on the right-hand side of the equation represent the flux at the right and left endpoints, so equation (2.6) is the counterpart of (2.4). These quantities must be determined from the boundary conditions. The flux at the left end is given by $\bar{S}^{(0)}F$, where $\bar{S}^{(0)}$ corresponds to the $S^{(0)}$ of the basic chemostat (as a density; i.e., the units are m/l). The condition at the left endpoint can be written

$$dS_x(t,0) = -\bar{S}^{(0)}F;$$

if one defines

$$S^{(0)} = \bar{S}^{(0)}F/d$$

then the first boundary condition in (2.2) is obtained. Similarly, the flux at the right-hand end is given by

$$dS_x(t,1) = -FS(t,1).$$

Thus, if r is defined by $r = F/d$ then the second boundary condition holds.

Equation (2.6) states that the rate of change of the mass of the nutrient in the vessel is proportional to the difference between the input nutrient flux and the output nutrient flux, as in the basic chemostat. The diffusion coefficient d has units of length squared over time, l^2/t; thus the units are appropriate.

The following basic lemma will allow the problem to be simplified. (Note that this simplification depends upon the fact that the diffusion coefficients are the same.)

LEMMA 2.1. *The solutions $S(t,x), u(t,x), v(t,x)$ of (2.1)–(2.3) exist for all $t > 0$ with $0 < x < 1$. The solutions are nonnegative and bounded, and*

$$\sup_{[0,1]} |S(t,\cdot) + u(t,\cdot) + v(t,\cdot) - \phi| = O(e^{-\alpha t}) \quad \text{as } t \to \infty \tag{2.7}$$

for some $\alpha > 0$, where

$$\phi = \phi(x) = S^{(0)}\left(\frac{1+r}{r} - x\right), \quad 0 < x < 1.$$

The function $\phi(x)$ represents the distribution of nutrient for the case of no consumption ($u_0(x) \equiv 0$, $v_0(x) \equiv 0$). The lemma reflects the fact that the total nutrient and equivalent organism biomass equilibrate to this function as well. As noted frequently in this work, this is essentially a definition of the chemostat if all variables are taken into account. The parameters $S^{(0)}$ and r are reflected in the function $\phi(x)$; these are the operating parameters of the apparatus.

Solutions of (2.1)–(2.3) generate a semidynamical system on

$$C_+ \times C_+ \times C_+,$$

where C_+ is the set of nonnegative continuous functions on $[0, 1]$ with the usual supremum norm. The setting is far more abstract than any considered in the body of this book, but the basic definitions of dynamical systems theory and persistence carry over. This semidynamical system is denoted by $\pi(x, t)$, where $t \geq 0$ and x represents the triple of initial conditions given by (2.3). For $t > 0$, the semiflow is compact [H3]. The lemma shows that the system is dissipative, and that all omega limit sets are in the subset given by $S + u + v - \phi = 0$. On this set, (2.1)–(2.3) become

$$\frac{\partial u}{\partial t} = d\frac{\partial^2 u}{\partial x^2} + f_1(\phi - u - v)u,$$

$$\frac{\partial v}{\partial t} = d\frac{\partial^2 v}{\partial x^2} + f_2(\phi - u - v)v; \tag{2.8}$$

$$\frac{\partial u}{\partial x}(t, 0) = 0, \qquad \frac{\partial u}{\partial x}(t, 1) + ru(t, 1) = 0,$$

$$\frac{\partial v}{\partial x}(t, 0) = 0, \qquad \frac{\partial v}{\partial x}(t, 1) + rv(t, 1) = 0; \tag{2.9}$$

$$u(0, x) = u_0(x) \geq 0, \qquad v(0, x) = v_0(x) \geq 0,$$

$$u_0(x) \not\equiv 0, \quad v_0(x) \not\equiv 0, \qquad \phi(x) - u_0(x) - v_0(x) \geq 0, \tag{2.10}$$

where

$$f_i(S) = \begin{cases} m_i S/(a_i + S) & \text{if } S \geq 0, \\ 0 & \text{if } S \leq 0 \end{cases}$$

for $i = 1, 2$. ([HSW] allows for a more general functional response than this.)

This system can be investigated as a dynamical system on the space $C_+ \times C_+$. Note that the theorem used in Appendix F is not adequate to show that the dynamics are the same as those of the original equations; however, the orbits in the omega limit set are solutions of these equations. The steps in the analysis generally follow those of the gradostat, although the investigation is much more technical. For example, one establishes the existence of a steady state (temporally independent but spatially dependent) for each population growing in the absence of the other. The parameter conditions involve the eigenvalues of linear problems. We state two results to give the flavor of one-population growth problems.

For example, if only the u population is growing in the chemostat, the equation is

$$\frac{\partial u}{\partial t} = d\frac{\partial^2 u}{\partial x^2} + \frac{m_1(\phi(x)-u)}{a_1+\phi(x)-u}u;$$

(2.11)

$$\frac{\partial u}{\partial x}(t,0) = 0, \qquad \frac{\partial u}{\partial x}(t,1)+ru(t,1) = 0.$$

The following lemma provides conditions under which an organism cannot survive in the given environment – that is, with the given r and the given input concentration $S^{(0)}$.

LEMMA 2.2. *If $m_1 < \lambda_0 d$ then $u(x,t)$ decays to zero exponentially as $t \to \infty$, where $\lambda_0 > 0$ is the smallest eigenvalue of*

$$\psi'' + \lambda\left(\frac{\phi(x)}{a_1+\phi(x)}\right)\psi = 0;$$

$$\psi'(0) = 0, \qquad \psi'(1) + r\psi(1) = 0.$$

This lemma states that if the maximum growth rate is small then the organism will tend to extinction as time becomes large.

LEMMA 2.3. *If $m_1 > \lambda_0 d$ and $u(t,x)$ is the solution of (2.11), then*

$$\lim_{t \to \infty} u(t,x) = \hat{u}(x)$$

uniformly in x, where $\hat{u}(x)$ is the unique positive steady-state solution of

$$u'' + \frac{m_1(\phi(x)-u)}{a_1+\phi(x)-u}u = 0;$$

$$u'(0) = 0, \qquad u'(1)+ru(1) = 0.$$

These lemmas and their counterpart for v (labeling the eigenvalue as μ_0) establish the existence of the rest points on the boundary of $C_+ \times C_+$. As before, we label these rest points E_0, E_1, E_2. As with the gradostat, the condition for coexistence is tied to the question of invasiveness. Now, however, the conditions take the form of comparison with the eigenvalues of certain Sturm–Liouville problems rather than with the stability modulus of matrices, as was the case in Chapter 6. We describe just enough of this to show the parameters on which the result depends.

First consider the following boundary value problem:

$$\lambda p(x) = dp'' + \left[\frac{-m_1 a_1 \hat{u}}{(a_1 + \phi - \hat{u})^2} + \frac{m_1(\phi - \hat{u})}{a_1 + \phi - \hat{u}} \right] p + \frac{-m_1 a_1 \hat{u}}{(a_1 + \phi - \hat{u})^2} q,$$

$$\lambda q(x) = dq'' + \frac{m_2(\phi - \hat{u})}{a_2 + \phi - \hat{u}} q;$$

$$p'(0) = 0, \qquad q'(0) = 0,$$

$$p'(1) + rp(1) = 0, \qquad q'(1) + rq(1) = 0.$$

Since the second equation is independent of the first, the set of eigen-values for the full system is a subset of the set of eigenvalues of the second equation and therefore they are real.

Think of m_2 as a parameter and let $\hat{\lambda}(m_2)$ be the largest eigenvalue of the Sturm–Liouville problem just displayed. The eigenvalue $\hat{\lambda}(m_2)$ is a strictly increasing function of m_2 satisfying

$$\hat{\lambda}(m_2) < 0 \quad \text{if } m_2 \text{ is small}$$

and

$$\lim_{m_2 \to \infty} \hat{\lambda}(m_2) = \infty.$$

Since $\hat{\lambda}(m_2)$ is monotone, there is a unique value m_2^* such that $\hat{\lambda}(m_2^*) = 0$. If $m_2 > m_2^*$ then the largest eigenvalue is positive and E_1 is unstable; there is a similar result for E_2. Of course, given λ and q, one must solve the remaining boundary value problem for p. This is dealt with in varying degrees of generality in [BT], [HSW], and [SoW]. Table 2.1 describes the situation.

The theorem on coexistence can be stated in terms of the instability of the boundary rest points.

THEOREM 2.4. *Fix m_1 and m_2 such that $m_1 > \lambda_0 d$ and $m_2 > \mu_0 d$. If, in addition, $m_1 > m_1^*$ and $m_2 > m_2^*$, then the semidynamical system gener-ated by (2.8)–(2.10) is uniformly persistent.*

Although this is a coexistence result, it contains much less information than the previous results on models that involve ordinary differential equa-tions. To apply the theorem, one must fix the parameters and then com-pute the m_i^*s to see if both inequalities hold.

One can proceed further and show the existence of an interior order interval to which all solutions converge. In fact, from monotonicity, al-most all solutions must converge to a rest point in this order interval [HSW]. Finally, we note that the system of equations in this section is

Table 2.1

	Existence	Instability
$E_0 = (0, 0)$	Always	E_1 or E_2 exists
$E_1 = (\hat{u}, 0)$	$m_1 > \lambda_0 d$	$m_2 > m_2^*$
$E_2 = (0, \bar{v})$	$m_2 > \mu_0 d$	$m_1 > m_1^*$

similar in form to those studied in chemotaxis; see Lauffenberger and Calcagno [LC].

Many questions remain open. Key among these are the questions of uniqueness of the interior steady state in the case of Michaelis–Menten dynamics and of sufficient conditions for uniqueness in the general case. The model does not contain any transport terms, although their inclusion would be important to model a moving stream. Periodic coefficients, as discussed in Chapter 7, are certainly relevant to this model and have not been considered. In addition, the case of non-equal diffusion is not considered at all by these methods. Hence further modeling and mathematics are still needed.

3. Delays in the Chemostat

Delays occur naturally in biological systems. In the chemostat, the use of ordinary differential equations carries the implication that changes occur instantaneously, so this is a deficiency of the model. There are two obvious sources of delays: delays due to the possibility that the organism stores the nutrient (so that the "free" nutrient concentration does not reflect the nutrient available for growth); and delays due to the cell cycle. An internal-stores model was considered in Chapter 8 without delays; ones that include delays appear in [Cap], [McD], and [NG]. Bush and Cook [BC] investigated the growth of one organism in a chemostat with a delay term that reflected the delay between consumption and growth. (As a consequence there is a delay in the growth equation but no delay in the consumption term in the nutrient equation.) This approach was extended to the competitive situation in [FSW1]. This model is discussed first; the form of the equations here is very much like that of those discussed in Chapters 1 and 2.

The model is written for Michaelis–Menten kinetics but holds in greater generality at the expense of a more complicated statement for the theorems. It is supposed that each competitor has a delay τ_i which affects its growth rate; that is, the quantity τ_i is the time delay in nutrient conversion. The specific growth rate is assumed to be a function of the nutrient level at time $t - \tau_i$. The model takes the form of a system of differential difference equations:

$$S'(t) = 1 - S(t) - x_1(t)f_1(S(t)) - x_2(t)f_2(S(t)),$$

$$x_1'(t) = x_1(t)[f_1(S(t-\tau_1)) - 1], \tag{3.1}$$

$$x_2'(t) = x_2(t)[f_2(S(t-\tau_2)) - 1]$$

with

$$f_i(S) = \frac{m_i S}{a_i + S},$$

where $\tau_1, \tau_2 \geq 0$, $S(t) = \phi(t) \geq 0$ on $[-\tau, 0]$, $\tau = \max(\tau_1, \tau_2)$, and $x_i(0) = x_{i0} \geq 0$ $(i = 1, 2)$.

The last two equations can be written in integral form as

$$x_i(t) = x_i(0) \exp\left[\int_0^t f_i(S(\theta - \tau_i) - 1)\, d\theta\right].$$

This illustrates that the proper initial value problem is the one indicated by the initial conditions just listed. The theory for such delay differential equations is much more complicated than that for ordinary differential equations, and is not so widely known among nonspecialists. The basic reference is Hale [H1]; see also Kuang [K2].

Let C denote the space of continuous functions on $[-\tau_2, 0]$ equipped with the sup norm. We will tacitly assume the labeling is such that $\tau_2 > \tau_1$. Using our integral representation and a simple inequality argument for S', it is not difficult to show that solutions of the system (3.1) are nonnegative for all positive time; (3.1) defines a semidynamical system on $C_+ \times C_+ \times C_+$ (C_+ was defined in Section 2). The "conservation" argument used previously to obtain boundedness (and to reduce the complexity of the problem) is no longer valid, since the uptake and the consumption terms do not cancel. This fact alone casts suspicion on the model as a description of the chemostat. The model does, however, produce oscillations, which makes it very interesting. The boundedness and the continuability of solutions of the system (3.1) can be established, but it is not quite as easy as with the previous chemostat problems.

The investigation of solutions takes the following form. First, one population growing on the nutrient is analyzed (after some scaling) and a bifurcation (with the delay as parameter) is shown to exist, establishing the existence of a periodic solution $(\hat{S}(t), \hat{x}_1(t), 0)$. For one population of microorganisms, the two-dimensional system governing growth is

$$S'(t) = 1 - S(t) - x_1(t)f_1(S(t)),$$

$$x_1'(t) = x_1(t)[f_1(S(t-\tau)) - 1]. \tag{3.2}$$

After scaling, one has

$$S'(t) = \tau[1 - S(t) - x_1(t)f_1(S(t))],$$

$$x_1'(t) = \tau x_1(t)[f_1(S(t-1)) - 1]. \tag{3.3}$$

System (3.3) has a unique equilibrium point $E_* = (S^*, x^*)$, since f_1 is strictly increasing. A periodic orbit bifurcates from E_* for τ large.

THEOREM 3.1. *There exists $\tau_0 > 0$ such that a family of periodic solutions of (3.2) bifurcates from the equilibrium point E^* for τ near τ_0.*

Although stability may in principle be computed, the calculation is extremely complicated. Numerical calculations suggest the asymptotic stability of the limit cycle, but the stability has not been rigorously established. Assuming that the solution is asymptotically stable, a secondary bifurcation can be shown to occur. The argument is quite technical and requires a form of a Poincaré map in the appropriate function space; it is analogous to the bifurcation theorem used in Chapter 3 for bifurcation from a simple eigenvalue. The principal theorem takes the form of a bifurcation statement.

THEOREM 3.2. *Suppose that the parameters a_1, m_1, and τ_1 are chosen so that (3.2) has a (linearly) asymptotically orbitally stable periodic solution $(\hat{S}(t), \hat{x}(t))$ with period $T > 0$. Fix a_2 and $\tau_2 > \tau_1$. Then there exist a critical value m_2^* and a branch of periodic orbits of (3.1), with positive x_2 component, bifurcating from the hypothesized orbit for m_2 near m_2^*.*

All of the comments raised in Chapter 3 in connection with a similar bifurcation apply. The computations needed to determine the direction of bifurcation (which side of m_2^*) and the stability are formidable. However, for particular parameter values one can solve the differential equations numerically and exhibit the periodic orbit. Figure 3.1 shows the time course of a sample problem, and Figure 3.2 shows the projection of the periodic or the coexisting periodic orbits onto each of the possible pairs of variables.

Ellermeyer [E] takes a different approach to modeling the internal storage of nutrient. With our preceding notation let y_i, $i = 1, 2$, be the nutrient

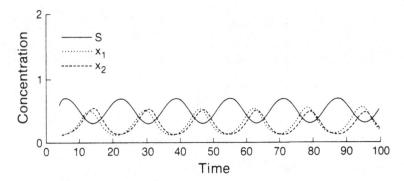

Figure 3.1. Plot of 100 time steps in the case of oscillatory coexistence. Parameters are $a_1 = 1.0$, $a_2 = 1.0$, $m_1 = 3.1$, $m_2 = 3.09$, $\tau_1 = 3.0$, $\tau_2 = 4.0$. (From [FSW1], reprinted with permission from the *SIAM Journal on Applied Mathematics,* volume 49, number 3, pp. 859–70. Copyright 1989 by the Society for Industrial and Applied Mathematics, Philadelphia, Pennsylvania. All rights reserved.)

stored internally by population i. Taking the input and washout into consideration, this quantity is given by

$$y_i(t) = \int_{t-\tau_i}^{t} e^{-D(t-u)} f_i(S(u)) x_i(u)\, du, \quad i = 1, 2.$$

The exponential accounts for the stored nutrient, which washes out of the vessel – along with the cells containing it – during the storage period. Balancing input and output with consumption and washout yields integral equations of the form

$$S(t) = S(0) + \int_0^t DS^{(0)}\, du - \int_0^t DS(u)\, du - \int_0^t \sum_{i=1}^{2} f_i(S(u)) x_i(u)\, du,$$

$$x_i(t) + y_i(t) = x_i(0) + y_i(0) + \int_0^t f_i(S(u)) x_i(u)\, du - D\int_0^t [x_i(u) + y_i(u)]\, du.$$

Differentiation produces the following system of differential difference equations:

$$S'(t) = D(S(0) - S(t)) - \sum_{i=1}^{2} f_i(S(t)) x_i(t),$$

$$x_1'(t) = -Dx_1(t) + e^{-D\tau_1} f_1(S(t - \tau_1)) x_1(t - \tau_1),$$

$$x_2'(t) = -Dx_2(t) + e^{-D\tau_2} f_2(S(t - \tau_2)) x_2(t - \tau_2).$$

(3.4)

The initial conditions now take the form

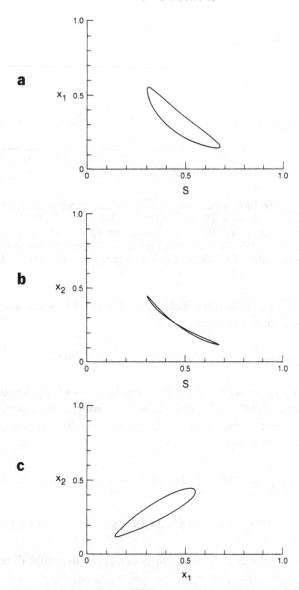

Figure 3.2. Plot of projections onto two dimensions of the solution given in Figure 3.1: **a** S-x_1; **b** S-x_2; **c** x_1-x_2. (From [FSW1], reprinted with permission from the *SIAM Journal on Applied Mathematics,* volume 49, number 3, pp. 859–70. Copyright 1989 by the Society for Industrial and Applied Mathematics, Philadelphia, Pennsylvania. All rights reserved.)

$$S(t) = \phi_0(t),$$

$$x_1(t) = \phi_1(t),$$

$$x_2(t) = \phi_2(t),$$

$$-\tau_2 \le t \le 0,$$

where again one assumes that $\tau_2 \ge \tau_1$ and that the initial conditions are continuous functions. These same equations occurred also in [FSW2].

Although the equations look quite similar to (3.1), the behavior of solutions is quite different. There is a conservation principle but, as we would expect owing to the delay, it takes a rather different form.

LEMMA 3.3. *For any fixed initial condition* $\phi = (\phi_0, \phi_1, \phi_2)$, *the corresponding trajectory satisfies*

$$\lim_{t \to \infty}[S(t) + e^{D\tau_1}x_1(t+\tau_1) + e^{D\tau_2}x_2(t+\tau_2)] = S^{(0)}.$$

As in the model of Section 2, the problem can be studied on its omega limit set with three rest points E_0, E_1, E_2. A local stability analysis and, for some special cases, the asymptotic behavior of solutions were given in [E]. However, the populations cannot invade each other simultaneously (E_1 and E_2 cannot be simultaneously unstable), so the persistence theory does not hold [E]. Moreover, for Michaelis–Menten dynamics, when one of the boundary rest points is locally stable and the other unstable, the locally stable one is globally stable [HWE]. In particular, the oscillation observed in the case of system (3.2) does not occur with (3.4). Indeed, the delayed system seems to behave much like the simple chemostat.

The totally different behavior of the two models illustrates the importance of the modeling process. Both models appear reasonable. Neither uses any cell physiology in deriving the delay term. The cell cycle does not appear in either. Clearly, more work in this direction is needed, with particular emphasis on more careful modeling of the delay.

4. A Model of Plasmid-Bearing, Plasmid-Free Competition

Genetically altered organisms are used in industry to manufacture a desired product, for example, a pharmaceutical. The alteration is accomplished by introducing DNA into the cell, frequently in the form of a plasmid. Plasmids contain bits of DNA which exist separately from the chromosome and replicate independently; the plasmid codes for the added

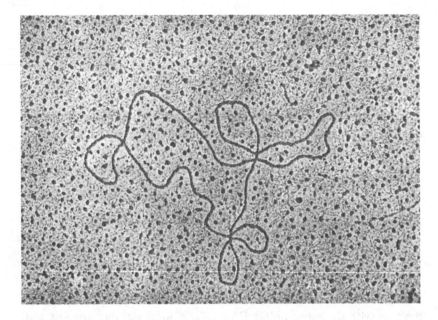

Figure 4.1. A plasmid. Photograph courtesy of Mervyn Bibb, John Innes Institute, Norwich, England. Reproduced by permission.

production. A micrograph of a plasmid is shown in Figure 4.1. A good description of plasmids can be found in Postgate [Pos]. The burden imposed on the cell by the task of production can result in the genetically altered (the plasmid-bearing) organism being a less able competitor than the plasmid-free organism. Unfortunately, the plasmid can be lost in the reproductive process; that is, it may not be passed to the daughter cells, producing a plasmid-free organism (the "wild" type). Since commercial production can take place on a scale of many generations, it is important to understand the asymptotic behavior of this system.

A model of competition between plasmid-bearing and plasmid-free organisms in a chemostat was proposed by Stephanopoulos and Lapidus [SLa], who give a local analysis of various models. A global analysis of the behavior of system trajectories was presented in [HWW], and a portion of that analysis is sketched here. The major remaining problem is discussed after the description of the known results. There are other models of plasmid loss (and conjugation), for example, Stewart and Levin [SL2] and Macken, Levin, and Waldstätter [MLW]. The survey article of Simonsen [Si] contains a discussion of the experiments and the theory.

The model of Stephanopoulos and Lapidus takes the form

$$S' = (S^{(0)} - S)D - f_1(S)\frac{x_1}{\gamma} - f_2(S)\frac{x_2}{\gamma},$$

$$x_1' = x_1(f_1(S)(1-q) - D),$$

$$x_2' = x_2(f_2(S) - D) + qx_1 f_1(S),$$

$$S(0) = S_0 \geq 0, \qquad x_i(0) > 0, \quad i = 1, 2,$$

(4.1)

where for descriptive purposes we will take

$$f_i(S) = \frac{m_i S}{a_i + S} \quad \text{and} \quad 0 < q < 1.$$

The variables and the units are those which have been used since Chapter 1: $S(t)$ is the nutrient concentration at time t, $x_1(t)$ is the concentration of plasmid-bearing organisms at time t, and $x_2(t)$ is the concentration of plasmid-free organisms at time t; $S^{(0)}$ is the input concentration of the nutrient, and D is the washout rate of the chemostat. These are the operating parameters. The m_i term is the maximal growth rate of x_i, and a_i is the Michaelis–Menten (or half-saturation) constant of x_i. These are assumed to be known (measurable) properties of the organism that characterize its growth and reproduction. A plasmid is lost in reproduction with probability q, and γ is the yield constant.

We proceed as before to obtain dimensionless variables by measuring concentrations in units of $S^{(0)}$, time in units of $1/D$, and x_i in units of $1/\gamma_i S^{(0)}$. The number of parameters can be reduced, and the equations then take the form

$$S' = 1 - S - x_1 f_1(S) - x_2 f_2(S),$$

$$x_1' = x_1(f_1(S)(1-q) - 1),$$

$$x_2' = x_2(f_2(S) - 1) + qx_1 f_1(S).$$

(4.2)

Although the system (4.2) looks similar to the equations for the chemostat in Chapter 1, the analysis is more difficult because the system is no longer competitive. Stephanopoulos and Lapidus used a very clever index argument to generate phase portraits. However, such arguments are only local; [HWW] determined the global asymptotic behavior.

Let $\Sigma(t) = 1 - S(t) - x_1(t) - x_2(t)$. The system (4.2) may be written

$$\Sigma' = -\Sigma,$$

$$x_1' = x_1[f_1(1 - \Sigma - x_1 - x_2)(1 - q) - 1],$$

$$x_2' = x_2[f_2(1 - \Sigma - x_1 - x_2) - 1] + qx_1 f_1(1 - \Sigma - x_1 - x_2).$$

Table 4.1

I	$f_1(1)(1-q) < 1$	$f_2(1) < 1$	
II	$f_1(1)(1-q) > 1$	$f_2(1) < 1$	
III	$f_1(1)(1-q) < 1$	$f_2(1) > 1$	
IV	$f_1(1)(1-q) > 1$	$f_2(1) > 1$	(a) $f_1(\lambda_2)(1-q) < 1$
			(b) $f_1(\lambda_2)(1-q) > 1$

Table 4.2

I	$\{E_1\}$
II	$\{E_1, E_*\}$
III	$\{E_1, E_2\}$
IV	(a) $\{E_1, E_2\}$
	(b) $\{E_1, E_2, E_*\}$

Clearly $\lim_{t \to \infty} \Sigma(t) = 0$ and trajectories on the omega limit set satisfy $\Sigma = 0$. If all of the rest points of the limiting system are hyperbolic (which will be implied by the conditions stated) and if there are no periodic orbits (which needs to be proved), then the results of Appendix F apply. The limiting system is

$$x_1' = x_1[f_1(1-x_1-x_2)(1-q)-1],$$
$$x_2' = x_2[f_2(1-x_1-x_2)-1]+qx_1f_1(1-x_1-x_2). \tag{4.3}$$

The equations, of course, are restricted to the region

$$\Omega = \{(x_1, x_2) \mid x_i \geq 0, x_1+x_2 \leq 1\};$$

Ω is a positively invariant region. Note that the system is not a competitive one.

The analysis breaks conveniently into four cases, which are given in Table 4.1. The rest point sets are shown in Table 4.2. The rest points are defined as $E_1 = (0,0)$, $E_2 = (0, 1-\lambda_2)$, and $E_* = (x_1^*, x_2^*)$ where the quantities λ_2, x_1^*, x_2^* are defined next.

The λ_2 term is defined to be the unique value such that $f_2(\lambda_2) = 1$ and λ^* is defined by $f_1(\lambda^*) = 1/(1-q)$, if such a λ^* exists. A necessary and sufficient condition for there to be such a λ^* is that $f_1(1) > 1/(1-q)$. If one assumes $\lambda^* \neq \lambda_2$ (a hyperbolicity assumption), then the interior rest point E_* will exist and have coordinates

$$x_1^* = \frac{(1-\lambda^*)(1-f_2(\lambda^*))}{f_1(\lambda^*)-f_2(\lambda^*)},$$

$$x_2^* = 1-\lambda^*-x_1^*$$

$$= \frac{(1-\lambda^*)(f_1(\lambda^*)-1)}{f_1(\lambda^*)-f_2(\lambda^*)},$$

provided the right-hand side is positive and less than 1.

The principal global result takes the following form.

THEOREM 4.1.

(i) *In case I, E_1 is a global attractor of the interior of Ω.*
(ii) *In cases II and IV(b), E_* is a global attractor of the interior of Ω.*
(iii) *In cases III and IV(a), E_2 is a global attractor of the interior of Ω.*

The theorem shows clearly that plasmid loss is detrimental (or fatal) to the production of the chemostat. To compensate for this possibility, in commercial production a plasmid that codes for resistance to an antibiotic is added to the DNA that codes for the item to be produced. Thus, if the plasmid is lost then the "wild" type is susceptible to (inhibited by) the antibiotic. The antibiotic is introduced into the feed bottle along with the nutrient. The dynamics produced by adding an inhibitor to the chemostat was modeled in Chapter 4. A new direction for research on chemostat models would be to include the inhibitor, as in Chapter 4, and the plasmid model of this section (or one of the more general models) into the same model. This is a mathematically more difficult problem to analyze, since the reduced system will not be planar. Moreover, because the methods of monotone dynamical systems do not apply, other techniques would need to be found in order to obtain global results. The model also assumes extremely simple behavior for the plasmid; more could be included in a model.

11

Open Questions

In this brief final chapter we collect in one place the main questions that remain unanswered concerning the models explored in this work. We proceed more or less in the order of the chapters. In many cases, the open problems mentioned here have already been identified and discussed in the discussion section of the corresponding chapter, so the reader may wish to check there as well.

As noted in the discussion section of Chapter 2, there remains a gap in our knowledge of the basic chemostat model in the case of differing removal rates for the competitors. The principal open problem is to extend the result of [Hsu1], described in Section 4 of Chapter 2, to general monotone uptake functions. It would also be desirable to include the class of not necessarily monotone uptake functions identified in [BWo1]. The recent work [WLu] represents a major step in this direction.

In Chapter 3, two limit cycles played a prominent role. The first occurred in the planar system, representing oscillations in the simple food chain. Neither the stability nor the uniqueness of the limit cycle was established. Kuang [K1] shows that the limit cycle is unique and asymptotically stable – for parameter values near where the rest point loses stability – by examining the Hopf bifurcation from the rest point. For other values of parameters, the uniqueness and stability questions remain open.

The second limit cycle was obtained as a simple bifurcation from the first limit cycle. It represents the oscillations of two competing populations preying on the population consuming the nutrient. The direction of bifurcation ($a_3 < a_3^*$ or $a_3 > a_3^*$) and the stability of this limit cycle were left undetermined, although numerical simulations suggest that it is stable and exists for $a_3 < a_3^*$. Numerical simulations suggest the possibility of establishing a global bifurcation such that, as parameters are adjusted, the cycle "lifts off" from the cycle in the x–y plane of the octant (one

248

food chain) and moves toward and eventually coalesces with the cycle in the x-z plane (second food chain). This was established in [Ke2] under an additional condition. The limit cycle continues, so long as hyperbolicity is maintained, since there is strong dissipation (solutions are bounded by a uniform constant). The problem seems to be bounding the period of the limit cycle from above. Establishing the global bifurcation rigorously would be of interest and, along with uniqueness, would help to complete the theory. The proof of Lemma 5.1 in Chapter 3 is very long and inelegant. A simpler proof would be of interest.

Limit cycles also appear in Chapter 4, but no bifurcation theorems were used (although the Hopf bifurcation theorem could have been used). Uniqueness of these cycles is a question of major interest and importance. To more accurately model the chemostat as it is used in commercial production, the plasmid model discussed in Chapter 10 should be combined with the inhibitor model of Chapter 4. More specifically, consider two organisms – differing only by the presence or absence of a plasmid that confers immunity to the inhibitor – competing in a chemostat (equation (4.2) of Chapter 9 modified for the presence of the inhibitor). The techniques of analysis used in Chapter 4 do not apply, since the system is not competitive in the mathematical sense. Yet an understanding of this system would be very important.

The model in Chapter 4 had only one competitor taking up the inhibitor, but this is only a first approximation. One could justify this assumption only by showing that nothing changes if the system is modified to allow for uptake by both competitors. However, such modification complicates the analysis. Finally, could it be that a threshold amount needs to be consumed before the inhibitor is effective? Adding thresholds to the model would be an interesting modeling task.

In Theorem 7.2 of Chapter 4, we assumed there were no limit cycles and showed that the positive rest point was globally attracting. It might be conjectured that this rest point is globally attracting when it is locally asymptotically stable. Establishing this conjecture would require criteria for eliminating limit cycles for competitive systems. The development of such criteria would be very important for applications to many ecological systems. Remark 7.4 of Chapter 4 provided an interesting case where the stability computations for the positive rest point could be carried out and the existence of a limit cycle established. Are there more general cases where the computations can be carried out?

In Chapter 5, a complete analysis of competition between two microorganisms with Michaelis–Menten uptake functions in a two-vessel gradostat was given. This analysis relied on two calculations which established that if a positive coexistence rest point exists then it is both unique and asymptotically stable. As noted in Section 4 of Chapter 6, more general conditions are known [HSo] for two monotone uptake functions in order that these two results hold. Furthermore, counterexamples are given in [HSo] where these conditions fail and there exists an unstable positive rest point for some two-vessel gradostat (not necessarily the same one considered in Chapter 5). As a result of our ignorance of general sufficient conditions for these two results to hold for monotone uptake functions and for more general n-vessel gradostats, the analysis of Chapter 6 is less complete than that of Chapter 5. The main difficulty seems to be the possible non-uniqueness of the positive rest point. Sufficient conditions for the uniqueness of this rest point would greatly simplify the main result (Theorem 4.4) of Chapter 6, since in that case $E_* = E_{**}$ and so E_* would be globally attracting. Such sufficient conditions on the monotone uptake functions, even if limited to the case of the standard n-vessel gradostat, would be highly desirable.

One might ask if the results of Chapter 5 hold for the case of nonnegligible death rates. There seems to be no work in this direction.

A more interesting question from the ecological point of view is that of how many different populations can coexist in an n-vessel gradostat. [JST] shows that this number cannot exceed n. Some numerical simulations and conjectures appear in [BWu; CB], but very little is known about this question. New techniques must be developed to handle this problem, since the resulting equations do not generate a monotone dynamical system when the number of competitors exceeds two (see [JST]).

The principal open questions that arise in the treatment of the chemostat with periodic washout rate, discussed in Chapter 7, are analogous to those mentioned in connection with Chapter 6. Namely, can sufficient conditions be given for the uniqueness of the positive periodic solution (fixed point of the Poincaré map) that represents the coexistence of the two populations? What can be said of the case of more than two competitors?

Similarly, our analysis of the variable-yield model in Chapter 8 is limited to two competing populations because we rely on the techniques of monotone dynamical systems theory. One would expect the main result of Chapter 8 to remain valid regardless of the number of competitors, just as it did for the simpler constant-yield model treated in Chapters 1 and 2. Perhaps the LaSalle corollary of Chapter 2 can be used to carry out such an extension, using arguments similar to those used in [AM] (described in Chapter 2). As noted in [NG], a structured model in which individual organisms vary in the amount of nutrient stored would be more appropriate, but such models are mathematically less tractable and require knowledge of various parametric functions which is generally unavailable.

A size-structured model of the growth in a chemostat of a single population, and the competition between two populations, was analyzed in Chapter 9. The model erred principally in its description of the cell-division process. Metz and Diekmann describe a size-structured model in [MD, p. 237] which more accurately accounts for the cell-division process. A brief description of their assumptions follows. The cell size x is assumed to vary among individuals of the population, from a minimum value x_{min} to a maximum value which is normalized to 1. A function $b(x)$ gives the per-unit time probability of a cell of size x dividing. Small cells are not allowed to divide ($b(x) = 0$, $x \le a$). A mother cell of size x divides into two daughter cells, one of size px and one of size $(1-p)x$, with probability $d(p)$, $0 < p < 1$. One must assume that $d(p) = d(1-p)$ and $\int_0^1 d(p)\,dp = 1$. The unit of size x – whether length, area, or volume – is not specified in [MD], and this makes their assumption that the growth rate of a cell of size x is proportional to x (and to $f(S)$) subject to different interpretations. Their model also can be reduced to the equations considered in Chapter 1.

An interesting open problem is to construct and analyze a model which, following [Cu2], treats growth and consumption as in Chapter 9 (i.e., as proportional to surface area) and which treats cell division as in [MD]. Questions of existence, uniqueness, and continuous dependence of solutions on initial data were not considered for the model discussed in Chapter 9. This problem seems to be an open one. The analysis should follow along the lines of arguments given in [MD, p. 238].

All of the open problems for the standard gradostat system of Chapter 6 are open problems for the unstirred chemostat model discussed in Chapter 10. It can be shown [HSW] that the dynamics of the unstirred chemostat system mirror those of the gradostat in the sense that there is an order interval, bounded by two (possibly identical) positive rest points, that attracts all solutions. Furthermore, an open and dense set of initial data generates solutions that converge to a stable rest point. The question of the uniqueness of the interior rest point is a major open problem. Another is how to handle the case where the diffusion coefficients of the competitors and nutrient are distinct. Although there must still be conservation of total nutrient, it is no longer a pointwise conservation relation and the reduction to two equations is not clear. Even if accomplished, it may be difficult to exploit. If one is forced to analyze the full system (three equations), then the corresponding dynamical system is no longer monotone and different techniques need to be applied. From the biological point of view, this is important because it is expected that the nutrient, being a molecule, will diffuse differently than the organisms, which are cells.

One of the basic assumptions in the unstirred chemostat was that the turnover rate was so slow that any transport (perhaps induced by pumps operating the chemostat) was negligible. If one thinks of a model for a flowing stream, or a lake with circulation, this assumption is unwarranted. Hence a mathematical analysis of the case where transport has been added to the model equations would be an important contribution. The steady-state case (with equal diffusion) was considered in [JW], but the dynamical model is the important one.

Delay models were discussed in Chapter 10. We repeat here that the most interesting problem is a modeling one. Since the problem is sensitive to how the delay is introduced, care must be taken in the modeling. A physical delay is caused by the physiology of the cell, so model equations must be modified to consider or approximate the cell physiology. Once a model is known, analysis of the corresponding system of equations (either functional differential equations or hyperbolic partial differential equations of a structured model) would be an important contribution. It is likely, however, that the delay will be state-dependent, and the theory for such equations is not well developed. A model with delays due to both cell physiology and diffusion in an unstirred chemostat would also be of interest.

Finally, as has been noted repeatedly, most of the material after Chapter 2 is limited to two competing populations. Obviously, in nature one is concerned with greater numbers of competitors. The models are straightforward – one just adds equations – but the techniques for analysis, namely monotone dynamics, fail here. This remains a major consideration in the development of the theory. One final point deserves mention here, since it is not addressed adequately in the text. How robust or how extensive is the parameter region where coexistence occurs in the models discussed in Chapters 3–7 and 9? The theory developed in these chapters points to the calculations required to address this question, but the parameter space is multidimensional and exploring all of it will not be possible. Operating diagrams for specific competitors – where the range of dilution rates, reservoir nutrient concentration, and other control parameters that result in coexistence of the populations are displayed – will probably remain the most useful answers.

Appendices

A Matrices and Their Eigenvalues

B Differential Inequalities

C Monotone Systems

D Persistence

E Some Techniques in Nonlinear Analysis

F A Convergence Theorem

A

Matrices and Their Eigenvalues

Matrices are encountered on a number of occasions in the text, particularly in arguments for the linearization about rest points or periodic orbits. On some occasions we encounter matrices of dimension large enough that direct computation of the eigenvalues may not be feasible. Fortunately, the matrices are often of a special form and there are theorems to cover such cases. In this appendix we list some of the useful theorems in the analysis of these special systems, along with appropriate references.

For stability at a rest point one wishes to show that the eigenvalues of the linearization lie in the left half of the complex plane. There is a totally general result, the Routh–Hurwitz criterion, that can determine this. It is an algorithm for determining the signs of the real parts of the zeros of a polynomial. Since the eigenvalues of a matrix A are the roots of a polynomial

$$f(z) = a_0 z^n + a_1 z^{n-1} + \cdots + a_n$$

obtained from

$$f(z) = \det(A - zI),$$

this theory applies. Unfortunately, the explanation of the algorithm depends on describing a certain index for quotients of polynomials, and the explanation of the necessary computations is also very complex. Thus, although the question of the signs of the real parts of the roots can be answered theoretically, practical application of the theory is very difficult. A complete explanation of the theory can be found in the appendix to [Co]. When the degree of the polynomial is small, however, the computations can be carried out. If

$$a_0 z^3 + a_1 z^2 + a_2 z + a_3 = 0$$

with $a_0 > 0$, then the relevant condition is

$$a_3 > 0, \quad a_1 > 0, \quad a_1 a_2 > a_0 a_3.$$

The next result is also general and, while the conditions are not often met, it is an important tool when it can be applied. The theorem is called the Gerschgorin circle theorem. An excellent general reference on matrices is Lancaster and Tismenetsky [LT], and most of the results here are quoted from that source. Another important source is Berman and Plemmons [BP], particularly for special results on nonnegative matrices.

THEOREM A.1 [LT, p. 371]. *Let A be an $n \times n$ matrix with elements denoted by a_{ij}, and let ρ be defined by $\rho_i = \Sigma'_k |a_{ik}|$ where Σ'_k denotes the sum on k from $k = 1$ to n with the term $k = i$ omitted. (The diagonal element of the matrix is omitted from the sum.) Then every eigenvalue of A lies in at least one of the disks*

$$\{z \,|\, |z - a_{ii}| \leq \rho_i\}, \quad i = 1, 2, \ldots, n,$$

in the complex plane. Furthermore, a set of m of these disks having no point in common with the remaining $n - m$ disks contains m and only m eigenvalues of A.

As generally used in stability theory, the diagonal element is negative, and Theorem A.1 is used in attempts to show that the radius of the disk (called a *Gerschgorin disk*) is smaller in absolute value. There are many generalizations that yield finer results at the expense of a more complicated criterion. One of these, useful for our work, involves the concept of an irreducible matrix.

A matrix is said to be *irreducible* if it cannot be put into the form

$$\begin{bmatrix} A & B \\ 0 & C \end{bmatrix}$$

(where A and C are square matrices) by reordering the standard basis vectors. In applications, not being irreducible often leads to the interpretation that the object being studied can be separated into two compartments, with at least one of them not influencing the other.

There is a simple test to see if an $n \times n$ matrix A is irreducible. Write n points $P_1, P_2, \ldots, P_n$ in the plane. If $a_{ij} \neq 0$, draw a directed line segment $P_i P_j$ connecting P_i to P_j. The resulting graph is said to be *strongly connected* if, for each pair (P_i, P_j), there is a directed path $P_i P_{k_1}, P_{k_1} P_{k_2}, \ldots,$ $P_{r-1} P_j$. A square matrix is irreducible if and only if its directed graph is strongly connected [LT, p. 529].

This test reflects the preceding interpretation of "separated compartments." When there is no path between two vertices of the graph, no material may pass from the first to the second vertex. Hence there is no influence of the first on the second.

THEOREM A.2 [LT, p. 376]. *Let A be an irreducible matrix, and let λ be an eigenvalue of A lying on the boundary of the union of the Gerschgorin disks. Then λ is on the boundary of each of the Gerschgorin disks.*

The set of all eigenvalues of the matrix A is called the *spectrum* of A and is denoted by $\sigma(A)$. The *stability modulus* of a matrix A, denoted by $s(A)$, is defined by

$$s(A) = \max\{\operatorname{Re}\lambda \mid \lambda \in \sigma(A)\}.$$

The traditional statement in stability theory that all of the eigenvalues have negative real part becomes $s(A) < 0$. The *spectral radius* of a matrix, denoted $\mu(A)$, is defined by

$$\mu(A) = \max\{|\lambda| \mid \lambda \in \sigma(A)\}.$$

For special matrices there are theorems that give information about the stability modulus. A matrix is said to be *positive* if all of the entries are positive; this is written $A > 0$. (Similarly, a matrix is nonnegative if all of the entries are nonnegative.) The very elegant Perron–Frobenius theory applies to such matrices.

THEOREM A.3. *If the $n \times n$ matrix A is nonnegative, then*

 (i) *$\mu(A)$ is an eigenvalue; and*
(ii) *there is a nonnegative eigenvector v associated with $\mu(A)$.*

Since the hypotheses allow for A to be a null matrix, the theorem is of limited value. If irreducibility is added, a stronger theorem applies.

THEOREM A.4 [LT, p. 536]. *If the $n \times n$ matrix A is nonnegative and irreducible, then:*

  (i) *the matrix A has a positive eigenvalue $r = \mu(A)$;*
 (ii) *there is a positive eigenvector v associated with r;*
(iii) *the eigenvalue r has algebraic multiplicity 1;*
(iv) *any nonnegative eigenvector of A is a positive multiple of v; and*
 (v) *if $B \geq A$ but $B \neq A$ then $\mu(B) > \mu(A)$.*

The following theorem is a consequence of Theorem A.4 and the observation that if A is a matrix such that $A + cI \geq 0$, then $\mu(A + cI) = s(A) + c$.

THEOREM A.5. *If A is irreducible and has nonnegative off-diagonal elements, then $s(A)$ is an eigenvalue with algebraic multiplicity 1, and $\operatorname{Re}\lambda < s(A)$ for $\lambda \in \sigma(A)$ and $\lambda \neq s(A)$. Moreover, there exists an eigenvector $v > 0$ such that $Av = s(A)v$. Any nonnegative eigenvector of A is a positive multiple of v. If B is an $n \times n$ matrix satisfying $B \geq A$ and $B \neq A$, then $s(B) > s(A)$.*

The following consequence is also useful.

THEOREM A.6. *Let*

$$A = \begin{bmatrix} B & C \\ D & E \end{bmatrix},$$

where B and E are square matrices of dimension k and l (respectively) with nonnegative off-diagonal elements, $D \leq 0$ and $C \leq 0$, and where A is irreducible. Then $s(A)$ is an eigenvalue of multiplicity 1 and $\operatorname{Re}\lambda < s(A)$ for $\lambda \neq s(A)$. Moreover, there is an eigenvector $v = (v_1, v_2)$, $v_1 \in R^k$ and $v_2 \in R^l$ satisfying $v_1 > 0$ and $v_2 < 0$, such that $Av = s(A)v$.

If A is a positive matrix, some of the theorems may be strengthened.

THEOREM A.7 [LT, p. 542]. *If A is a positive matrix then, in addition to the conclusions in Theorem A.4, r exceeds the modulus of all other eigenvalues of A.*

Another concept that is important is that of "positive (or negative) definite." For this, it is required that A be a symmetric matrix, that is $a_{ij} = a_{ji}$. An important theorem is the following.

THEOREM A.8. *The spectrum of a symmetric matrix is real.*

Thus, one may order the eigenvalues of a symmetric matrix

$$\lambda_1 \leq \lambda_2 \leq \cdots \leq \lambda_n.$$

A symmetric matrix A is said to be *positive definite* if all of the eigenvalues are positive; it is said to be *negative definite* if all of the eigenvalues are negative. Semidefinite is similarly defined. There is a simple test to determine if a symmetric matrix is positive or negative definite.

The principal minors are those determinants whose upper left element is a_{11} and have contiguous rows and columns. For example, the first principal minor, d_1, is just a_{11}. The second is given by

$$d_2 = \det \begin{bmatrix} a_{11} & a_{21} \\ a_{21} & a_{22} \end{bmatrix},$$

and the last by

$$d_n = \det \begin{bmatrix} a_{11} & a_{12} & \cdots & a_{1n} \\ a_{21} & a_{22} & \cdots & a_{2n} \\ \vdots & \vdots & \ddots & \vdots \\ a_{n1} & a_{n2} & \cdots & a_{nn} \end{bmatrix}.$$

THEOREM A.9. *A symmetric matrix is positive definite if and only if*

$$d_1 > 0, d_2 > 0, \ldots, d_n > 0.$$

THEOREM A.10. *A symmetric matrix is negative definite if and only if*

$$d_1 < 0, d_2 > 0, \ldots, (-1)^n d_n > 0.$$

Theorem A.10 allows us to conclude stability if the matrix is the variational matrix evaluated at a rest point. An important result is that the test in Theorem A.10 will work for some matrices that are not symmetric, not in the sense of being negative definite but in the sense of yielding stability based on the sign of the real parts of the eigenvalues. The type of matrix is closely associated with the orderings and monotone flow discussed in Appendices B and C.

THEOREM A.11 [S6, thm. 2.7; BP, chap. 6]. *Let A be as in Theorem A.6. Let $\tilde{A}$ be defined by*

$$\tilde{A} = \begin{bmatrix} B & -C \\ -D & E \end{bmatrix}.$$

Then $s(A) < 0$ if and only if $(-1)^k d_k > 0$, $k = 1, 2, \ldots, n$, where d_k is the kth principal minor of $\tilde{A}$.

Finally, we will have need of the following.

THEOREM A.12. *Let A have nonnegative off-diagonal elements and be irreducible. Then*

(i) *if $s(A) < 0$ then $-A^{-1} > 0$; and*

(ii) *if $s(A) > 0$ and $r > 0$, then $Ax + r = 0$ has no solution x satisfying $x > 0$.*

Proof. (i) follows from the identity $-A^{-1} = \int_0^\infty e^{At}\, dt$ and (B.7) of Theorem B.3, which implies that $e^{At} > 0$ for $t > 0$.

If (ii) is false then there exists an $x > 0$ such that $-Ax > 0$. According to [BP, p. 134, I_{27}], $-A = sI - (sI + A)$, where $s > 0$ is chosen so that $B = sI + A \geq 0$. The spectral radius of B, $\mu(B)$, satisfies $\mu(B) = \mu(sI + A) = s + s(A) > s$, so $-A$ cannot be an M matrix. This contradiction establishes (ii). $\qquad\square$

B

Differential Inequalities

In this appendix, basic theorems on differential inequalities are stated and interpreted. The main theorem is usually attributed to Kamke [Ka] but the work of Müller [Mü] is prior. A more general version due to Burton and Whyburn [BWh] is also needed. We follow the presentation in Coppel [Co, p. 27] and Smith [S2; S6]. The nonnegative cone in $\mathbb{R}^n$, denoted by $\mathbb{R}^n_+$, is the set of all n-tuples with nonnegative coordinates. One can define a partial order on $\mathbb{R}^n$ by $y \le x$ if $x - y \in \mathbb{R}^n_+$. Less formally, this is true if and only if $y_i \le x_i$ for all i. We write $x < y$ if $x_i < y_i$ for all i. The same notation will be used for matrices with a similar meaning.

Let $f : \mathbb{R} \times D \to \mathbb{R}^n$, where D is an open subset of $\mathbb{R}^n$, be a vector-valued function, $f = (f_1, f_2, ..., f_n)$. We first give the general form of the needed condition. The function f is said to be of *type K* in D if, for each i and all t, $f_i(t, a) \le f_i(t, b)$ for any two points a and b in D satisfying $a \le b$ and $a_i = b_i$.

The object is to compare solutions of the system of differential equations

$$x' = f(t, x), \tag{B.1}$$

with solutions of the system of differential inequalities

$$z' \le f(t, z) \tag{B.2}$$

or

$$y' \ge f(t, y) \tag{B.3}$$

on an interval. We always assume that solutions of initial value problems for (B.1) are unique.

THEOREM B.1. *Let f be continuous on $\mathbb{R} \times D$ and of type K. Let $x(t)$ be a solution of (B.1) defined on $[a, b]$. If $z(t)$ is a continuous function on $[a, b]$ satisfying (B.2) on (a, b) with $z(a) \le x(a)$, then $z(t) \le x(t)$ for all*

261

t in [*a*, *b*]. If *y*(*t*) is continuous on [*a*, *b*] satisfying (B.3) on (*a*, *b*) with *y*(*a*) ≥ *x*(*a*), then *y*(*t*) ≥ *x*(*t*) for all *t* in [*a*, *b*].

Proof. For $m = 1, 2, \ldots$, let $x_m(t)$ be a solution of the initial value problem

$$x' = f(t, x) + (1/m)e$$

satisfying $x_m(a) = x(a)$, where $e = (1, 1, \ldots, 1)$. Then [H2, chap. 1, lemma 3.1], $x_m(t)$ is defined on [*a*, *b*] for all sufficiently large *m* and $x_m(t) \to x(t)$ as $m \to \infty$, uniformly on [*a*, *b*]. We show that $z(t) < x_m(t)$, $a < t < b$, for all large *m*, from which the first assertion of the theorem follows by taking limits as $m \to \infty$. The second assertion is proved in a similar manner.

Let $m \geq 1$ be fixed such that $x_m(t)$ is defined on [*a*, *b*]. As $z_i(a) = x_{mi}(a)$ (the latter is the *i*th component of $x_m(a)$) and $z_i'(a) < x_{mi}'(a)$ for $1 \leq i \leq n$, it follows that $z_i(t) < x_{mi}(t)$ for $t > a$ and $t - a$ small. Consequently, if $z(t) < x_m(t)$ does not hold for some $t \in (a, b)$ then there exist *j* and $t_0 \in (a, b)$ such that $z_i(t) < x_{mi}(t)$ for $a < t < t_0$ and $1 \leq i \leq n$ and such that $z_j(t_0) = x_{mj}(t_0)$. Therefore,

$$f_j(t_0, z(t_0)) \geq z_j'(t_0) \geq x_{mj}'(t_0) = f_j(t_0, x_m(t_0)) + (1/m) > f_j(t_0, x_m(t_0)).$$

But $z(t_0) \leq x_m(t_0)$ and $z_j(t_0) = x_{mj}(t_0)$ implies, by the type-*K* condition, that $f_j(t_0, z(t_0)) \leq f_j(t_0, x_m(t_0))$. This contradiction proves the theorem. □

See [Co] for a more general result. The theorem is traditionally used when the solution of (B.1) is known, or a bound on it is known, and $z(t)$ or $y(t)$ arises from some more complicated differential equation whose right-hand side can be bounded by *f*.

The type-*K* condition can be expressed in terms of the partial derivatives of *f* on suitable domains. We say that *D* is *p-convex* if $tx + (1-t)y \in D$ for all $t \in [0, 1]$ whenever $x, y \in D$ and $x \leq y$. In all applications in this book, *D* is a convex set and so *D* is also *p*-convex. In the next result, we compare two solutions of the same equation which have related initial values.

COROLLARY B.2. *Let* $f(t, x)$ *and* $(\partial f/\partial x)(t, x)$ *be continuous on* $\mathbb{R} \times D$, *where D is a p-convex subset of* $\mathbb{R}^n$. *Let*

$$\frac{\partial f_i}{\partial x_j}(t, x) \geq 0, \quad i \neq j, \ (t, x) \in D, \tag{B.4}$$

hold. If $y(t)$ *and* $z(t)$ *are two solutions of* (B.1) *defined for* $t \geq t_0$ *satisfying* $y(t_0) \leq z(t_0)$, *then* $y(t) \leq z(t)$ *for all* $t \geq t_0$.

Proof. Conditions (B.4), together with the fundamental theorem of calculus, imply that f is of type K in $\mathbb{R} \times D$. In fact, if $a \le b$ and $a_i = b_i$ then

$$f_i(t, b) - f_i(t, a) = \int_0^1 \sum_{j \ne i} \frac{\partial f_i}{\partial x_j}(t, a + r(b-a))(b_j - a_j)\, dr \ge 0,$$

by (B.4). The result then follows by applying Theorem B.1. $\qquad\square$

The conclusion of Corollary B.2 holds if f is merely continuous and of type K in D. We have stated the special case of this more general result, since we will always have the additional smoothness and since (B.4) is easy to check in applications.

The system (B.1) is said to be a *cooperative* system if the hypotheses of Corollary B.2 hold; (B.1) is said to be *cooperative and irreducible* if it is a cooperative system and if $(\partial f/\partial x)(t, x)$ is an irreducible matrix (see Appendix A) for each $(t, x) \in \mathbb{R} \times D$.

We write $x(t, s, \xi)$ for the solution of (B.1) satisfying $x(s) = \xi$. Recall that if f has a continuous derivative with respect to x then $x(t, s, \xi)$ is continuously differentiable,

$$X(t) = \frac{\partial x}{\partial \xi}(t, s, \xi)$$

satisfies $X(s) = I$ (the identity matrix), and $X(t)$ is the fundamental matrix solution of

$$z'(t) = \frac{\partial f}{\partial x}(t, x(t))z(t), \tag{B.5}$$

where $x(t) = x(t, s, \xi)$. If (B.1) is a cooperative system, then Corollary B.2 can be applied to (B.5) since it is also a cooperative system. Indeed, if $x_i(t)$ denotes the ith column of $X(t)$ then $x_i(s)$ is the ith element of the standard basis for $\mathbb{R}^n$, which is nonnegative. Therefore, Corollary B.2 applied to $x_i(t)$ and 0 implies that $x_i(t) \ge 0$ for $t \ge s$. Therefore, if (B.1) is cooperative then

$$\frac{\partial x}{\partial \xi}(t, s, \xi) \ge 0, \quad t \ge s. \tag{B.6}$$

Recall that the inequality (B.6) means that each entry in the matrix is nonnegative. A stronger conclusion holds if (B.1) is cooperative and irreducible.

THEOREM B.3. *Let* (B.1) *be a cooperative and irreducible system in* $\mathbb{R} \times D$. *Then*

$$\frac{\partial x}{\partial \xi}(t, s, \xi) > 0 \tag{B.7}$$

for $t > s$. Furthermore, if $\xi_1, \xi_2 \in D$ are distinct points satisfying $\xi_1 \le \xi_2$, then

$$x(t, s, \xi_1) < x(t, s, \xi_2) \tag{B.8}$$

for $t > s$.

Proof. Inequality (B.8) follows from (B.7) and the formula

$$x(t, s, \xi_2) - x(t, s, \xi_1) = \int_0^1 \frac{\partial x}{\partial \xi}(t, s, \xi_1 + r(\xi_2 - \xi_1))(\xi_2 - \xi_1) \, dr.$$

Let

$$X(t) = \frac{\partial x}{\partial \xi}(t, s, \xi) \quad \text{and} \quad A(t) = \frac{\partial f}{\partial x}(t, x(t)).$$

Equations (B.4) and (B.6) imply that if $X(t) = (x_{ij}(t))$ then

$$x'_{ij}(t) \ge a_{ii}(t) x_{ij}(t), \quad t \ge s.$$

If $t_1 \ge s$ and $x_{ij}(t_1) > 0$, it follows that $x_{ij}(t) > 0$ for $t \ge t_1$. Consequently, $Z_{ij} \equiv \{t : t > s \text{ and } x_{ij}(t) = 0\}$ is either empty or is an interval $(s, e_{ij}]$, $s < e_{ij} \le \infty$. Obviously, Z_{ii} is empty since $x_{ii}(s) = 1$, $1 \le i \le n$. Suppose that Z_{ij} is not empty for some pair of indices i, j. Let $S = \{k : x_{kj}(t) = 0 \text{ for all } t \in (s, e_{ij}]\}$. Since $i \in S$, S is not empty, and as $X(t)$ is nonsingular, $S \ne \{1, 2, \ldots, n\}$. Let S^c denote the complement of S in $\{1, 2, \ldots, n\}$. For $t \in (s, e_{ij}]$ and $k \in S$,

$$0 = x'_{kj}(t) = \sum_l a_{kl}(t) x_{lj}(t) = \sum_{l \in S^c} a_{kl}(t) x_{lj}(t).$$

But, by the definition of S, it follows that $x_{lj}(t) > 0$ for t near e_{ij}, $l \in S^c$. Consequently there exists $t_1 < e_{ij}$ such that $x_{lj}(t) > 0$ for $t_1 < t$ and all $l \in S^c$. It follows from the equality just displayed that $a_{kl}(t) = 0$ for $t_1 < t \le e_{ij}$ and for all $l \in S^c$. As $k \in S$ was arbitrary, we have that $a_{kl}(t) = 0$ for $t_1 < t \le e_{ij}$ for all $k \in S$ and all $l \in S^c$. This contradicts the irreducibility of $A(t)$ and so completes the proof. $\square$

Another kind of system encountered frequently in our applications has the form

$$x' = F(t, x, y), \quad y' = G(t, x, y), \tag{B.9}$$

where $x, y \in \mathbb{R}^n$, $H = (F, G) : \mathbb{R} \times D \to \mathbb{R}^{2n}$, and $D \subset \mathbb{R}^{2n}$ is open. The function H, or (B.9), is said to be of *generalized type K* in D if (a) for each i

$(1 \le i \le n)$ $F_i(t, a, c) \ge F_i(t, b, d)$ whenever $a \ge b$, $c \le d$, $t \in \mathbb{R}$, and $a_i = b_i$; and (b) for each j $(n+1 \le j \le 2n)$, $G_j(t, a, c) \le G_j(t, b, d)$ whenever $a \ge b$, $c \le d$, $t \in \mathbb{R}$, and $c_j = d_j$. It is assumed that (a, c) and (b, d) belong to D. We write $(b, d) \le_K (a, c)$ if $b \le a$ and $d \ge c$; if $b < a$ and $d > c$ then we write $(b, d) <_K (a, c)$. The reader should be cautioned that the notations $<$ and $<_K$ are used in different ways in the literature.

There is no particular reason why the dimension of the vectors x and y appearing in (B.9) need to be the same. See Burton and Whyburn [BWh] and Smith [S2; S6] for the more general case. In all the systems considered in this work, the two components have the same dimension, so we will restrict our attention to this case even though all our results hold for the general case.

Let $P: \mathbb{R}^{2n} \to \mathbb{R}^{2n}$ be defined by $P(u, v) = (u, -v)$. If (B.9) is of generalized type K in D, let $D' = PD$. The change of variables $(u, v) = P(x, y)$ in (B.9) yields the system

$$u' = F(t, u, -v), \qquad v' = -G(t, u, -v). \tag{B.10}$$

It is easy to see that if (B.9) is of generalized type K in D then (B.10) is of type K in D', and conversely. Also observe that $(b, d) \le_K (a, c)$ if and only if $P(b, d) \le P(a, c)$. These simple observations allow us to state analogs of each of the preceding results.

Consider the differential inequality

$$u' \le F(t, u, v), \qquad v' \ge G(t, u, v). \tag{B.11}$$

Observe that if $z = (u, v)$ then (B.11) is just $z' \le_K H(t, z)$.

The following result is the analog of Theorem B.1. There is, of course an analogous result for the reverse set of inequalities in (B.11).

THEOREM B.4. *Let H be continuous on $\mathbb{R} \times D$ and suppose that H is of generalized type K in D. Let $(x(t), y(t))$ be a solution of (B.9) on an interval $[a, b]$. If $(u(t), v(t))$ is continuous on $[a, b]$ and satisfies (B.11) on (a, b) and if $(u(a), v(a)) \le_K (x(a), y(a))$, then $(u(t), v(t)) \le_K (x(t), y(t))$ for all $t \in [a, b]$.*

Theorem B.4 follows directly by an application of Theorem B.1 to (B.10), and interpreting this result by way of the transformation P. The next result is the analog of Corollary B.2 for (B.9).

COROLLARY B.5. *Let $H = (F, G)$ be continuous and have a continuous derivative with respect to (x, y) in $\mathbb{R} \times D$, where D is a convex subset of $\mathbb{R}^{2n}$. Assume that*

$$\frac{\partial F_i}{\partial x_j} \geq 0, \quad \frac{\partial G_i}{\partial y_j} \geq 0, \quad i \neq j;$$

$$\frac{\partial F_i}{\partial y_j} \leq 0, \quad \frac{\partial G_i}{\partial x_j} \leq 0, \quad all \ i, j.$$

(B.12)

Let $(x(t), y(t))$ *and* $(u(t), v(t))$ *be solutions of* (B.9) *defined for* $t \geq t_0$ *satisfying* $(x(t_0), y(t_0)) \leq_K (u(t_0), v(t_0))$. *Then* $(x(t), y(t)) \leq_K (u(t), v(t))$ *for all* $t \geq t_0$.

The convexity of D is a stronger requirement than necessary, but will be satisfied for all cases of interest here.

The same argument used to conclude (B.6) from (B.4) shows, when applied to (B.10), that if $z(t, s, \alpha) = (x(t, s, \alpha), y(t, s, \alpha))$ is the solution of (B.9) satisfying $z(s) = \alpha \equiv (\xi, \eta) \in D$ and if (B.12) holds, then

$$\frac{\partial z}{\partial \alpha}(t, s, \alpha) = \begin{pmatrix} \dfrac{\partial x}{\partial \xi} & \dfrac{\partial x}{\partial \eta} \\[2ex] \dfrac{\partial y}{\partial \xi} & \dfrac{\partial y}{\partial \eta} \end{pmatrix}$$

satisfies

$$\frac{\partial x}{\partial \xi} \geq 0, \quad \frac{\partial y}{\partial \eta} \geq 0,$$

$$\frac{\partial x}{\partial \eta} \leq 0, \quad \frac{\partial y}{\partial \xi} \leq 0.$$

(B.13)

The irreducibility assumption implies that the strong inequalities hold in (B.13).

THEOREM B.6. *Let* (B.9) *satisfy* (B.12) *and suppose that* $(\partial H/\partial z)(t, z)$ *is irreducible for each* $(t, z) \in \mathbb{R} \times D$, *where* D *is a convex subset of* $\mathbb{R}^{2n}$ *and* $H = (F, G)$, $z = (x, y)$. *Then strict inequality* ($>$ *or* $<$) *holds in each of the four inequalities of* (B.13). *Furthermore, if* (ξ_i, η_i) *for* $i = 1, 2$ *are distinct points of* D, $s \in \mathbb{R}$, *and* $(\xi_1, \eta_1) \leq_K (\xi_2, \eta_2)$, *then*

$$z(t, s, \xi_1, \eta_1) <_K z(t, s, \xi_2, \eta_2)$$

(B.14)

for $t > s$.

In the systems treated in this work, each component represents the concentration of a nutrient or microbial population and hence must be nonnegative. Therefore, in order to be biologically meaningful, the systems treated here must have the property that $\mathbb{R}_+^n$ is positively invariant. The next result supplies sufficient conditions for this basic property to hold.

PROPOSITION B.7. *Suppose that f in* (B.1) *has the property that solutions of initial value problems $x(t_0) = x_0 \geq 0$ are unique and, for all i, $f_i(t, x) \geq 0$ whenever $x \geq 0$ satisfies $x_i = 0$. Then $x(t) \geq 0$ for all $t \geq t_0$ for which it is defined, provided $x(t_0) \geq 0$.*

Proof. The assertion is obvious when f satisfies the stronger condition that $f_i(t, x) > 0$ whenever $x \geq 0$ satisfies $x_i = 0$. The general case can be treated by a limiting argument. For $s > 0$, define $f_s(t, x) = f(t, x) + sv$, where v is the vector with all entries equal to 1. Then f_s satisfies the stronger condition, and so solutions of $x' = f_s(t, x)$ which start nonnegative remain nonnegative in the future. Since the solution of (B.1) and $x(t_0) = x_0 \geq 0$ can be approximated at any fixed $t \geq t_0$ by solutions of the corresponding initial value problem for f_s for sufficiently small $s > 0$, by continuity of solutions with respect to parameters [H2] it follows that $x(t) \geq 0$. $\square$

The condition on f in Proposition B.7 is clearly necessary for the result to hold. An analogous condition (with inequalities reversed) suffices to ensure that $x(t) \leq 0$ for all $t \geq t_0$ provided $x(t_0) \leq 0$. An immediate corollary of the proposition and the previous remark is the following.

COROLLARY B.8. *Let f satisfy the hypotheses of Corollary B.2 and suppose that there exists $x_0 \in D$ such that $f(t, x_0) \geq 0$ for all $t \geq t_0$. Then the solution of* (B.1) *and $x(t_0) = x_0$ satisfies $x(t) \geq x_0$ for all $t \geq t_0$. A similar conclusion holds if the inequalities between vectors are reversed.*

Let $H = (F, G)$ satisfy the hypotheses of Corollary B.5, and suppose there exists $z_0 \in D$ such that $0 \leq_K H(t, z_0)$ for all $t \geq t_0$. Then the solution of (B.9) and $z(t_0) = z_0$ satisfies $z_0 \leq_K z(t)$ for all $t \geq t_0$. A similar conclusion holds if the inequalities between vectors are reversed.

Proof. Only the first assertion regarding (B.1) will be treated in detail. Following the change of variables $y = x - x_0$, this assertion is equivalent to showing that solutions of $y' = f(t, x_0 + y)$ which begin nonnegative remain nonnegative. Because f is type K, we have that if $y \geq 0$ and $y_i = 0$ for some i then

$$f_i(t, x_0 + y) \geq f_i(t, x_0) \geq 0.$$

Therefore, Proposition B.7 implies the desired conclusion. $\square$

C

Monotone Systems

Let $\pi(x, t)$ denote the dynamical system generated by the autonomous system of differential equations

$$x' = f(x), \tag{C.1}$$

where f is continuously differentiable on a subset $D \subset \mathbb{R}^n$. Recall that $\pi(x, t)$ denotes the solution of (C.1), which starts at position x at $t = 0$. It will be assumed that solutions of (C.1) extend to all $t \geq 0$. The basic concepts and definitions of a dynamical system were given in Chapter 1. The notation for the various partial order relations was given in Appendix B. The main results on monotone dynamical systems are stated here in terms of the dynamical system to suggest their generality, although most of the dynamical systems encountered in this work will be generated by differential equations of the form (C.1). A dynamical system will be called a *monotone dynamical system* with respect to $\leq_K$ if it has the property that $\pi(x, t) \leq_K \pi(y, t)$ for $t \geq 0$ whenever $x \leq_K y$. The dynamical system will be called *strongly monotone* with respect to $\leq_K$ if $x \leq_K y$ and $x \neq y$ implies that $\pi(x, t) <_K \pi(y, t)$ for all $t > 0$. A dynamical system is a monotone (resp. strongly monotone) dynamical system with respect to $\leq$ ($<$) when these conditions hold with $\leq$ replacing $\leq_K$ ($<$ replacing $<_K$). In order to simplify the statement of results, we assume that D is convex throughout this appendix.

Sufficient conditions for (C.1) to generate a dynamical system which is monotone (strongly monotone) are given in Appendix B. Corollary B.2, Theorem B.3, Corollary B.5, and Theorem B.6 imply the following result.

THEOREM C.1. *If (C.1) is cooperative in D, then π is a monotone dynamical system with respect to $\leq$ in D. If (C.1) is cooperative and irreducible in D, then π is a strongly monotone system with respect to $\leq$ in D. If*

(C.1) *has the form* (B.9) *where F and G are independent of t and* (B.12) *holds in D, then* π *is a monotone dynamical system with respect to* $\leq_K$. *If, in addition, the Jacobian matrix of f is irreducible at every point of D, then* π *is strongly monotone with respect to* $\leq_K$.

Hereafter, results will be stated only for the partial order $\leq_K$; it will be understood that they hold as well for the order $\leq$.

Although the most important and well-known results of the theory hold only for strongly monotone dynamical systems, there are some significant results that hold when π is merely a monotone dynamical system. For example, Theorem 4.2 of Chapter 4 implies that if the dimension of the state space is 2 ($x \in \mathbb{R}^2$) then every bounded forward (or backward) orbit of a monotone dynamical system converges to an equilibrium. This result no longer holds for higher-dimensional monotone dynamical systems. The next result gives two different sufficient conditions for a bounded solution to converge to a rest point.

THEOREM C.2. *Let* $\gamma^+(x)$ *be an orbit of the monotone dynamical system* (C.1) *which has compact closure in D. Then either of the following conditions is sufficient for* $\omega(x)$ *to be a rest point:*

(a) [S] $0 \leq_K f(x)$ $(f(x) \leq_K 0)$;
(b) [Hi3] $x <_K \pi(x, T)$ $(\pi(x, T) <_K x)$ *for some* $T > 0$.

Proof. If $0 \leq_K f(x)$ then, by Corollary B.8, $x \leq_K \pi(x, t)$ for all $t \geq 0$. By monotonicity, $\pi(x, s) \leq_K \pi(x, t+s)$ for $t, s \geq 0$. Therefore, $\pi(x, t)$ is monotone nondecreasing in t and $\lim_{t \to \infty} \pi(x, t) = e$ exists. The hypotheses ensure that e is a rest point.

If $x <_K \pi(x, T)$ then monotonicity implies that

$$\pi(x, nT) \leq_K \pi(x, (n+1)T) \quad \text{for} \quad n = 1, 2, \ldots,$$

and consequently there exists a point p such that $\pi(x, nT) \to p$ as $n \to \infty$. Furthermore, continuity of π implies that $\pi(p, T) = p$. It follows that $\omega(x)$ is the T-periodic orbit $\{\pi(p, t): t \in \mathbb{R}\}$. However, T may not be the minimal period of $\pi(p, t)$. Let $P = \{\tau: \pi(p, t+\tau) = \pi(p, t)\}$ be the set of all periods of the solution $\pi(p, t)$. It is easy to see that P is a closed set which is closed under addition and subtraction (it's a subgroup of $(\mathbb{R}, +)$) and which contains nT for every integer n. Since the strict inequality $x <_K \pi(x, T)$ holds, the continuity of π implies that for some $\epsilon > 0$, $x <_K \pi(x, T+s)$ holds for all s satisfying $|s| < \epsilon$. Arguing as before, this implies that $\omega(x)$ is a periodic orbit generated by a solution having period

$T + s$. But $\omega(x)$ is the orbit of the point p, so P must contain the interval $(T - \epsilon, T + \epsilon)$ and consequently it must contain the interval $(-\epsilon, \epsilon)$. Since P is closed under addition, it must contain an open interval of length 2ϵ centered on each of its points. Therefore, P is open. Since it is also closed, $P = \mathbb{R}$. This implies that p is a rest point and that $\pi(x, t) \to p$ as $t \to \infty$. $\square$

Theorem C.2(b) immediately leads to the following result, first proved by Hadeler and Glas [HG]. See also [Hi2].

THEOREM C.3. *A monotone dynamical system cannot have a nontrivial attracting periodic orbit.*

By a "nontrivial" periodic orbit we mean a periodic orbit that is not a rest point. Such an orbit is attracting if the omega limit set of each point of some neighborhood of the periodic orbit is the periodic orbit.

Proof of Theorem C.3. If there were an attracting periodic orbit, then one could find a point x in its domain of attraction such that $x <_K p$ for some point p on the periodic orbit. As p is a limit point of the positive orbit through x, there exists $T > 0$ such that $x <_K \pi(x, T)$. Consequently, $\pi(x, t)$ converges to a rest point by Theorem C.2(b), contradicting our assumption that it converges to a nontrivial periodic orbit. $\square$

Monotonicity of a dynamical system places restrictions on the basin of attraction of a rest point. Suppose that x_0 is a rest point of the monotone dynamical system generated by (C.1). Let B denote the basin of attraction of x_0:

$$B = \{x : \pi(x, t) \to x_0, \, t \to \infty\}.$$

If x_1 and x_2 are two points of B satisfying $x_1 \leq_K x_2$ and if $x_1 \leq_K z \leq_K x_2$, then $z \in B$. This follows from monotonicity, since $\pi(x_1, t) \leq_K \pi(z, t) \leq_K \pi(x_2, t)$ holds for all $t \geq 0$ and $\pi(x_i, t) \to x_0$ as $t \to \infty$. The next result exploits this property in the hyperbolic case.

THEOREM C.4. *The stable manifold of an unstable, hyperbolic rest point of a monotone dynamical system cannot contain two points that are related by the strict inequality $<_K$. If the system is strongly monotone, then the stable manifold cannot contain two distinct points that are related by $\leq_K$. In other words, the stable manifold is unordered.*

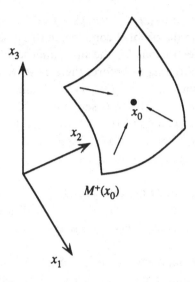

Figure C.1. Two-dimensional unordered stable manifold in $\mathbb{R}^3$.

Proof. If x_0 is a hyperbolic rest point then $B = M^+(x_0)$, the stable manifold of x_0. Since x_0 is hyperbolic and unstable, $M^+(x_0)$ has empty interior (see the proof of Theorem F.1). It follows that $M^+(x_0)$ cannot contain two points x_1 and x_2 satisfying $x_1 <_K x_2$, since then $M^+(x_0)$ would contain all points of the open set $\{z: x_1 <_K z <_K x_2\}$. The second assertion of the theorem follows from strong monotonicity and the positive invariance of the stable manifold. $\square$

Figure C.1 depicts a two-dimensional stable manifold in $\mathbb{R}^3$. The reader should sketch a one-dimensional stable manifold in $\mathbb{R}^2$.

A similar assertion to that of Theorem C.4 holds for a compact limit set of a monotone dynamical system.

THEOREM C.5. *A compact limit set of a monotone dynamical system cannot contain two points related by $<_K$. If the system is strongly monotone then the limit set is unordered.*

Proof. Consider first the case that the limit set L is the omega limit set of $\gamma^+(x_0)$. Suppose that L contains distinct points x_1 and x_2 satisfying $x_1 <_K x_2$. As the x_i are omega limit points of $\gamma^+(x_0)$, and since $\{x: x <_K x_2\}$ is a neighborhood of x_1, there exists $t_1 > 0$ such that $\pi(x_0, t_1) <_K x_2$. Similarly, $\{x: \pi(x_0, t_1) <_K x\}$ is a neighborhood of x_2, so there exists $t_2 > t_1$

such that $\pi(x_0, t_1) <_K \pi(x_0, t_2) = \pi(\pi(x_0, t_1), t_2 - t_1)$. By Theorem C.2(b), L is a rest point and the proof is complete in this case.

Consider now the case that L is the alpha limit set of $\gamma^-(x_0)$. Let $x(t) = \pi(x_0, t)$ for $t \leq 0$. Arguing as before, there exists $t_1 < 0$ such that $x_1 <_K x(t_1)$ and $t_2 < t_1$ such that $x(t_2) <_K x(t_1)$. Continue by choosing $t_3 < t_2$ such that $x(t_3) <_K x_2$ and $t_4 < t_3$ such that $x(t_3) <_K x(t_4)$. Therefore, the interval $I = [t_4, t_1]$ contains the interval $[t_4, t_3]$ on which $x(t)$ "falls" and the interval $[t_2, t_1]$ on which $x(t)$ "rises," and these intervals are disjoint. This contradicts the following lemma, proving the theorem in this case.

If the system is strongly monotone and x_1, x_2 are distinct points of L satisfying $x_1 \leq_K x_2$, then $y_i \equiv \pi(x_i, 1)$ for $i = 1, 2$ also belong to L and satisfy $y_1 <_K y_2$. This contradicts the first assertion of the theorem. □

Before stating the lemma required to complete the proof of Theorem C.5, the following definitions are needed. Let $x(t)$ be a solution of the monotone dynamical system (C.1) on an interval I. A subinterval $[a, b]$ of I is called a *rising* interval if $x(a) \leq_K x(b)$ and if equality does not hold (i.e. $x(a) \neq x(b)$); it is called a *falling* interval if $x(b) \leq_K x(a)$ and if equality does not hold. The definitions are due to Hirsch [Hil], who attributes the proof of the next lemma to L. Ito. Recall that in the proof of Theorem C.5 the interval $[t_4, t_3]$ is falling and the interval $[t_2, t_1]$ is rising.

LEMMA. *A solution $x(t)$ cannot have a rising interval and a falling interval that are disjoint.*

Proof. The most important observation is that if $[a, b]$ is a rising (falling) interval contained in I and if $s > 0$ is such that $[a+s, b+s]$ is contained in I, then the latter is also a rising (falling) interval in I. In fact, if $x(a) \leq_K x(b)$ and equality does not hold, then by monotonicity $x(a+s) = \pi(x(a), s) \leq_K \pi(x(b), s) = x(b+s)$ and equality does not hold. Consequently, rising and falling intervals remain such under right translation.

Suppose that I contains the falling interval $[a, r]$ and the rising interval $[s, b]$, and suppose that $a < r < s < b$. The other cases can be treated similarly. Let $A = \{t \in [s, b]: x(t) \leq_K x(s)\}$ and $s' = \sup A$. Then $s \leq s' < b$ and $[s', b]$ is a rising interval that contains no falling interval $[s', r]$ for any $r \in (s', b]$. Redefine $s = s'$ so that the rising interval $[s, b]$ has the same property as just described for $[s', b]$. A contradiction will be reached in each of the two cases $r - a \leq b - s$ and $r - a > b - s$.

If $r - a \le b - s$ then $[s, s + r - a]$ is a translate to the right of $[a, r]$, which is contained in I, and therefore is a falling interval. As $s + r - a \le b$, this contradicts that $[s, b]$ contains no such falling interval.

If $r - a > b - s$ then $a < a + b - r < s < b$, so $[a + b - r, b]$ is a translate to the right of $[a, r]$, contained in I, so it too is a falling interval. It follows that $x(s) \le_K x(b) \le_K x(a + b - r)$, where equality does not hold in either inequality. Let $c = \sup\{t \in [a + b - r, s] : x(b) \le_K x(t)\}$. Then $c < s < b$ and $x(b) \le_K x(c)$, so $[c, s]$ is a falling interval adjacent to the rising interval $[s, b]$. If $s - c \le b - s$ then $[s, 2s - c]$ is a translate to the right of the falling interval $[c, s]$, contained in $[s, b]$, and therefore is a falling interval. But this contradicts the earlier argument that $[s, b]$ contains no such interval. If $s - c > b - s$ then $c < c + b - s < s$ and $[c + b - s, b]$ is a right translate of the falling interval $[c, s]$, so it too is a falling interval. Therefore $x(b) \le_K x(c + b - s)$, with equality not holding, and this contradicts the definition of c. $\square$

Theorem C.5 places strong restrictions on how a limit set is imbedded in $\mathbb{R}^n$. Note in particular that a periodic orbit may always be considered as a limit set of one of its points and hence Theorem C.5 applies to a periodic orbit. This should convince the reader that periodic orbits are ruled out for two-dimensional monotone systems.

The following result of Hirsch [Hi1] exploits the strong restrictions on how a limit set is imbedded in space, described in the previous theorem.

THEOREM C.6. *A compact limit set of a monotone dynamical system in* $\mathbb{R}^n$ *can be deformed by a Lipschitz homeomorphism (with a Lipschitz inverse) to a compact invariant set of a Lipschitz system in* $\mathbb{R}^{n-1}$ *in such a way that trajectories are mapped to trajectories and such that the parameterization of solutions is respected.*

Proof. Consider a compact, unordered subset L of $\mathbb{R}^n$. Let v be a unit vector satisfying $0 <_K v$ and let H_v be the hyperplane orthogonal to v. This hyperplane H_v consists of vectors x such that $x \cdot v = 0$, where $\cdot$ is the standard dot (or scalar) product in $\mathbb{R}^n$. Let Q be the orthogonal projection onto H_v; that is, $Qx = x - (x \cdot v)v$. Since L is unordered, Q is one-to-one on L (this could fail only if L contains two ordered points). Therefore, Q_L, the restriction of Q to L, is a Lipschitz homeomorphism of L onto a compact subset of H_v. A straightforward argument by contradiction establishes the existence of $m > 0$ such that $|Q_L x_1 - Q_L x_2| \ge m |x_1 - x_2|$

whenever $x_1 \neq x_2$ are points of L. Therefore, Q_L^{-1} is Lipschitz on $Q(L)$. In any case, L is at most an $(n-1)$-dimensional set. Since L is a limit set, it is an invariant set for (C.1). It follows that the dynamical system restricted to L can be modeled on a dynamical system in H_v. In fact, if $y \in Q(L)$ then $y = Q_L(x)$ for a unique $x \in L$, and $\Pi(y, t) \equiv Q_L(\pi(x, t))$ is a dynamical system on $Q(L)$ generated by the vector field

$$F(y) = Q_L(f(Q_L^{-1}(y)))$$

on $Q(L)$. Since a Lipschitz vector field on an arbitrary subset of $\mathbb{R}^{n-1}$ can be extended to a Lipschitz vector field on all of $\mathbb{R}^{n-1}$ while preserving the Lipschitz constant [McS], it follows that F can be extended to all of H_v as a Lipschitz vector field. Therefore, $Q(L)$ is a compact invariant set for the $(n-1)$-dimensional dynamical system on H_v generated by the extended vector field. Since the restriction of this dynamical system to $Q(L)$ is equivalent to that of π on L, we have proved the theorem.   □

Theorem C.6 establishes that the dynamical system generated by the vector field (C.1) on $\mathbb{R}^n$, restricted to the limit set L, is topologically equivalent to the dynamical system generated by a Lipschitz vector field on $H_v \sim \mathbb{R}^{n-1}$ restricted to $Q(L)$. As a consequence, these two dynamical systems share common dynamical properties. Since L is a compact invariant set, so too is $Q(L)$ a compact invariant set. It is also known that limit sets are chain-recurrent (see [C] for the definition) and so $Q(L)$ has this property as well. Therefore, the dynamics on L are that of a compact, invariant, chain-recurrent set in one less dimension.

   Theorem C.6 makes plausible the following result of Hirsch [Hi1; Hi4] (see also [S3]).

THEOREM C.7. *A compact limit set of a monotone dynamical system in* $\mathbb{R}^3$ *which contains no rest points is a periodic orbit.*

According to Theorem C.6, the limit set can be deformed to a compact invariant set A, without rest points, of a planar vector field. By the Poincaré–Bendixson theorem, A must contain at least one periodic orbit and possibly entire orbits which have as their alpha and omega limits sets distinct periodic orbits belonging to A. Using the fact that A is chain-recurrent, Hirsch [Hi1] shows that these latter orbits cannot exist. Since A is connected it must consist entirely of periodic orbits; that is, it must be an annulus foliated by closed orbits. Monotonicity is used to show that the limit set is, in fact, a single periodic orbit (see the references listed just prior to Theorem C.7).

Theorem C.7 bears a strong resemblance to the Poincaré–Bendixson theorem stated in Chapter 1. It will be used in Chapter 4 for the case where (C.1) is a competitive system, that is, for a system (C.1) where $-f$ is cooperative. Note that the omega (alpha) limit set of a competitive system is the alpha (omega) limit set of the "time-reversed" cooperative system, so Theorems C.5, C.6, and C.7 apply to competitive systems. Unlike cooperative systems, competitive systems can have attracting periodic orbits. For more on the Poincaré–Bendixson theory of competitive and cooperative systems in $\mathbb{R}^3$, see [S3], [SW1], and [ZS].

A particularly important construction of Smale [Sm] will put the previous results in better context. Smale begins by fixing an arbitrary infinitely differentiable tangent vector field h on the $(n-1)$-dimensional simplex $S = \{x \in \mathbb{R}^n_+ : x_1 + x_2 + \cdots + x_n = 1\}$. By a *tangent vector field*, we mean that $\sum h_i(x) = 0$ so that h generates a (local) dynamical system on S by the differential equation $y' = h(y)$. Putting $P(x) = \prod x_i$, the differential equation $z' = P(z)h(z)$ generates a dynamical system on S which is equivalent to that generated by h in the interior of S (namely, in $\{x \in S : x_i > 0\}$) but which vanishes on the boundary of S. Smale shows that this latter vector field can be extended to a smooth vector field on $\mathbb{R}^n_+$ having the form $x_i' = x_i M_i(x)$ where $\partial M_i / \partial x_j < 0$; that is, the system is competitive. Furthermore, all solutions corresponding to positive initial data approach the invariant set S as $t \to \infty$. Therefore, the competitive system can have essentially arbitrary dynamics on S, consistent of course with the fact that S is an $(n-1)$-dimensional manifold! For example, if $n = 3$ then one could choose h so that in the interior of S there is a single equilibrium surrounded by a single periodic orbit. Several periodic orbits could easily be accommodated. If $n = 4$ then the celebrated Lorentz attractor could be imbedded in the interior of S as an attractor for the competitive system in $\mathbb{R}^4$.

By reversing time – that is, by replacing the vector field f constructed in the previous paragraph by $-f$ – we see that a monotone dynamical system can have essentially arbitrarily complex, $(n-1)$-dimensional dynamics. Of course, upon reversing time the invariant set S now becomes a repelling set and the equilibrium points 0 and ∞ become the attractors.

Now we state without proof some results that require π to be strongly monotone. They are all due to Hirsch. Let E be the set of rest points of π: $E = \{x : f(x) = 0\}$.

THEOREM C.8. *Let π be a strongly monotone dynamical system such that E has no points of accumulation in D. Suppose that $\gamma^+(x)$ has compact closure in D for every $x \in D$. Then the set of points x for which $\pi(x, t)$ does not converge to an equilibrium has Lebesgue measure zero.*

Recall that $x \in D$ is an accumulation point of E if every neighborhood of x contains a point of E different from x. The hypotheses of Theorem C.8 exclude such points. In most applications, E is a finite set and so this hypothesis holds.

Hirsch proves (a more general result than) Theorem C.8 by first obtaining the following result, which is useful in its own right, and then appealing to Fubini's theorem of integration theory.

THEOREM C.9. *Let the hypotheses of Theorem C.8 hold, except that E is assumed to be a finite set, and let J be a compact, totally ordered arc contained in D. Then $\omega(x)$ is an equilibrium point for all except possibly a finite subset of points of J.*

This result will typically be used where J is a line segment parallel to a positive vector: $J = \{a + tv : 0 \le t \le 1\}$ where $a \in D$, $a + v \in D$, and $0 <_K v$.

A more complete exposition of the theory of monotone dynamical systems can be found in the works of Hirsch [Hi1; Hi2; Hi3; Hi4], the review article [S6], and [ST1; ST2].

D

Persistence

As we have seen in the text, the equations governing interacting populations in a chemostat-like environment eventually take the form

$$x_i' = x_i f_i(x_1, x_2, \ldots, x_n),$$

$$x_i(0) = x_{i0} \geq 0, \tag{D.1}$$

$$i = 1, 2, \ldots, n.$$

To avoid technical conditions, assume that f is such that solutions of initial value problems are unique and extend to $[0, \infty)$. Thus (D.1) generates a semidynamical system. Of course, one could assume (as has been done before) that solutions extend to all of $\mathbb{R}$. However, checking the backward continuation of solutions presents a problem in one of the applications, so the results are stated for semidynamical systems. The form of the equations causes the positive cone to be invariant (Proposition B.7) and the coordinate axes and the bounding faces to be invariant (and represent lower-order dynamical systems).

The notion of persistence attempts to capture the idea that if the equations (D.1) represent a model ecosystem then all components of the ecosystem survive. The system (D.1) is said to be *persistent* if

$$\liminf_{t \to \infty} x_i(t) > 0, \quad i = 1, 2, \ldots, n$$

for every trajectory with positive initial conditions. The system (D.1) is said to be *uniformly persistent* if there exists a positive number ϵ such that

$$\liminf_{t \to \infty} x_i(t) \geq \epsilon, \quad i = 1, 2, \ldots, n$$

for every trajectory with positive initial conditions. The term "persistent" was first (?) used in this context by Freedman and Waltman, [FW1], with lim sup instead of lim inf. Other definitions are relevant; see Freedman and Moson [FM] for a discussion. A similar notion appears in Hofbauer [Ho] and in Schuster, Sigmund, and Wolf [SSW], where the term "cooperative" was used. (This later became "permanence.") There are two primary approaches to determining persistence; one can analyze the flow on the boundary, or one can develop a Liapunov-like technique. The former will be explained here in a general setting. Survey articles on the subject include [HS] and [W3]; see also [T2].

The general setting is that of topological dynamics in a metric space. Although the results will not be needed in this generality, this is the most elegant presentation of the results. The reader who does not wish to be concerned with this level of abstraction should just read "$\mathbb{R}^n$" for "locally compact metric space."

We review the basic definitions and set up the semidynamical system appropriate for systems of the form (D.1). Let X be a locally compact metric space with metric d, and let E be a closed subset of X with boundary ∂E and interior $\dot{E}$. The boundary, ∂E, corresponds to extinction in the ecological problems. Let π be a semidynamical system defined on E which leaves ∂E invariant. (A set B in X is said to be *invariant* if $\pi(B, t) = B$.) Dynamical systems and semidynamical systems were discussed in Chapter 1. The principal difficulty for our purposes is that for semidynamical systems, the backward orbit through a point need not exist and, if it does exist, it need not be unique. Hence, in general, the alpha limit set needs to be defined with care (see [H3]) and, for a point x, it may not exist. Those familiar with delay differential equations are aware of the problem. Fortunately, for points in an omega limit set (in general, for a compact invariant set), a backward orbit always exists. The definition of the alpha limit set for a specified backward orbit needs no modification. We will use the notation $\alpha_\gamma(x)$ to denote the alpha limit set for a given orbit γ through the point x.

Denote the flow on the boundary (the restriction of π to $\partial E \times \mathbb{R}^+$) by π_∂. The flow is said to be *dissipative* if for each $x \in E$, $\omega(x)$ is not empty and there exists a compact set G in E such that the invariant set $\Omega = \bigcup_{x \in E} \omega(x)$ lies in G. A nonempty invariant subset M of X is called an *isolated invariant set* if it is the maximal invariant set in some neighborhood of itself. Such a neighborhood is called an *isolating neighborhood*.

The *stable* (or *attracting*) set of a compact invariant set A is denoted by W^+ and is defined as

$$W^+(A) = \{x \mid x \in X, \omega(x) \neq \emptyset, \omega(x) \subset A\}.$$

The *unstable* set W^- is defined by

$W^-(A) = \{x \mid x \in X,$ *there exists a backward orbit* $\gamma^-(x)$ *such that*

$$\alpha_\gamma(x) \neq \emptyset \text{ and } \alpha_\gamma(x) \subset A\},$$

where $\alpha_\gamma(x)$ is the alpha limit set of the orbit γ through x. The *weakly* stable and unstable sets are defined as

$$W_w^+(A) = \{x \mid x \in X, \omega(x) \neq \emptyset, \omega(x) \cap A \neq \emptyset\}$$

and

$$W_w^-(A) = \{x \mid x \in X, \alpha_\gamma(x) \neq \emptyset, \alpha_\gamma(x) \cap A \neq \emptyset\}.$$

The stable and unstable sets correspond to the stable and unstable manifolds introduced for rest points and periodic orbits in Chapter 1. Unfortunately, if the attractors are more complex than rest points or periodic orbits, the question of the existence of stable and unstable manifolds becomes a difficult topological problem. In the applications that follow, these more complicated attractors do not appear, so one can simply deal with the stable manifold theorem. The Butler–McGehee lemma (used in Chapter 1) played a critical role in the first uses of persistence. The following lemma is a generalization of this work. It can be found (with slightly different hypotheses) in [BW], [DRS], and [HaW]. (In particular, the local compactness is not needed if a stronger condition – asymptotic smoothness – is placed on the semidynamical system.)

LEMMA D.1. *Let M be a compact isolated invariant set for the dynamical system* π, *defined on a locally compact metric space. Then for any* $x \in W_w^+(M) \backslash W^+(M)$ *it follows that*

$$\omega(x) \cap (W^+(M) \backslash M) \neq \emptyset \quad \text{and} \quad \omega(x) \cap (W^-(M) \backslash M) \neq \emptyset.$$

A similar statement holds for $\alpha_\gamma(x)$.

The following definitions are motivated by the technique used in the proofs in [FW1; FW2; FW3]. Let M, N be isolated invariant sets (not necessarily distinct). The set M is said to be *chained* to N, written $M \rightarrow N$, if there exists an element x, $x \notin M \cup N$, such that $x \in W^-(M) \cap W^+(N)$. A finite sequence $M_1, M_2, ..., M_k$ of isolated invariant sets will be called a *chain* if $M_1 \rightarrow M_2 \rightarrow \cdots \rightarrow M_k$ ($M_1 \rightarrow M_1$ if $k = 1$). The chain will be called a *cycle* if $M_k = M_1$. (These definitions were used for rest points instead of invariant sets in Chapter 1.)

The system π will be said to be *persistent* if for all $x \in \mathring{E}$,

$$\liminf_{t \rightarrow \infty} d(\pi(x, t), \partial E) > 0;$$

π will be said to be *uniformly persistent* if there exists an ϵ such that for all $x \in \dot{E}$,

$$\liminf_{t \to \infty} d(\pi(x, t), \partial E) \geq \epsilon > 0.$$

The boundary flow π_∂ is said to be *isolated* if there exists a covering $M = \bigcup_{i=1}^{k} M_i$ of $\Omega(\pi_\partial) = \bigcup_{x \in \partial E} \omega(x)$ by pairwise disjoint, compact, isolated invariant sets $M_1, M_2, \ldots, M_k$ for π_∂ such that each M_i is also an isolated invariant set for π. This is a sort of "hyperbolicity" assumption; for example, it prevents interior rest points (or other invariant sets) from accumulating on the boundary. In this case M is called an *isolated covering*. The boundary flow π_∂ is called *acyclic* if there exists some isolated covering $M = \bigcup_{i=1}^{k} M_i$ of π_∂ such that no subset of the M_is forms a cycle. An isolated covering satisfying this condition is also said to be *acyclic*.

The following theorem provides a criterion for uniform persistence in terms of the flow on the boundary (see [BFW; BW; HaW]).

THEOREM D.2. *Let π be a semidynamical system defined on a subset E, the closure of an open set, in a locally compact metric space X. Suppose that ∂E, the boundary of E, is invariant under π. Assume that π is dissipative and that the boundary flow π_∂ is isolated and acyclic with acyclic covering M. Then π is uniformly persistent if and only if*

$$W^+(M_i) \cap \dot{E} = \emptyset \quad \text{for each } M_i \in M. \tag{H}$$

The assumption of the invariance of the boundary is stronger than needed, and the assumption that the space is locally compact can be removed at the expense of further assumptions on the dynamical system. The combination of dissipative and uniform persistence allows the use of fixed-point theorems. The following is sufficient for our applications in $\mathbb{R}^n$.

THEOREM D.3. *Let the hypotheses of Theorem D.2 hold for a dynamical system π, with $X = \mathbb{R}^n$ and $E = \mathbb{R}^n_+$ (thus the invariant boundary ∂E is composed of the coordinate faces), and let (H) hold. Then there is a rest point in the interior of E.*

The proof can be made to follow from [HaW, thm. 3.2] and [BS, thm. 2.8.6]. A rigorous discussion would involve introducing new (and otherwise unneeded) concepts. However, the intuition is easy. The orbits of points inside the positive cone must eventually be inside a ball (the dissipative hypothesis) and "outside" a strip along the boundary (uniform persistence). This region is homeomorphic to a ball in $\mathbb{R}^n$. Thus it has the fixed-point property.

E

Some Techniques in Nonlinear Analysis

In this section, we compile some results from nonlinear analysis that are used in the text. The implicit function theorem and Sard's theorem are stated. A brief overview of degree theory is given and applied to prove some results stated in Chapters 5 and 6. The section ends with an outline of the construction of a Poincaré map for a periodic solution of an autonomous system of ordinary differential equations and the calculation of its Jacobian (Lemma 6.2 of Chapter 3 is proved).

In working with nonlinear differential equations, one often faces the problem of solving nonlinear equations of the form

$$f(x) = 0, \tag{E.1}$$

where f is a map from one Euclidean space $\mathbb{R}^p$ to another one $\mathbb{R}^m$. When $p > m$ and a particular solution can be found, the implicit function theorem then gives a method of finding all nearby solutions. Because this result is used so frequently, we state it here (following [H2]).

IMPLICIT FUNCTION THEOREM. *Suppose that $F: \mathbb{R}^m \times \mathbb{R}^n \to \mathbb{R}^m$ has continuous first partial derivatives and satisfies $F(0, 0) = 0$. If the Jacobian matrix of $F(x, y)$ with respect to x satisfies*

$$\det \frac{\partial F}{\partial x}(0, 0) \neq 0,$$

then there exist neighborhoods U of $0 \in \mathbb{R}^m$ and V of $0 \in \mathbb{R}^n$ such that for each fixed $y \in V$ the equation $F(x, y) = 0$ has a unique solution $x \in U$. Furthermore, this solution can be given as a function $x = g(y)$, where $g(0) = 0$ and g has continuous first partial derivatives.

More generally [Smo], g is as smooth as F. For example, if F has continuous second partial derivatives then so does g.

Sard's theorem is referred to on several occasions in the text. It states that certain sets are "small" in the sense of Lebesgue measure. For completeness, we state the theorem here, following [CH, chap. 2, thm. 10.1].

SARD'S THEOREM. *Let $f: U \to \mathbb{R}^p$ be r-times continuously differentiable on $U \subset \mathbb{R}^n$ with $r > \max\{0, n - p\}$. Then*

$$f(\{x \in U : \operatorname{rank} Df(x) < p\})$$

has Lebesgue measure zero.

Recall that the rank of the Jacobian matrix $Df(x)$ at x is the dimension of its column space (or range). When $p = n$, r must be at least 1 and the rank condition becomes $\det Df(x) = 0$.

We now return to the problem of solving (E.1). When the Euclidean spaces have the same dimension, a more topological approach to solving equation (E.1) is often appropriate. In what follows we describe such an approach, called (topological) degree theory. Degree theory is useful because it gives an algebraic count of the solutions of the equation (E.1) that is stable under small perturbations of the function f. We will often refer to a solution of (E.1) as a *zero of f*. For example, if f is the right-hand side of a differential equation then we will be concerned with the set of rest points of that equation. A full account of degree theory is beyond the scope of this brief appendix. Instead, we will simply state some of its important definitions and properties and then go ahead and use it for our purposes. The reader may consult [Smo] for a more thorough account of the theory.

Let f be continuously differentiable on an open bounded subset O of $\mathbb{R}^n$ and continuous on the closure, $\bar{O}$, and suppose that f has no zeros on the boundary of O. If it is assumed that every solution of (E.1) is non-degenerate – in the sense that the Jacobian $J(x)$ of f at the zero $x \in O$ is nonsingular – then the degree of f relative to O, denoted $\deg(f, O)$, is defined by

$$\deg(f, O) = \sum_{f(x) = 0} \operatorname{sgn} \det J(x) \tag{E.2}$$

where sgn denotes the sign ($+1$ or -1), det denotes determinant, and the sum is taken over all zeros of f in O. This sum is finite by the inverse function theorem [Smo] and because $\bar{O}$ is compact and f does not vanish on the boundary of O. As an example, let $f(x) = x^2 - \epsilon$ on $O = (-1, +1)$ for small positive ϵ. In this case, $\deg(f, O) = 0$ is a warning that while f has zeros, they could be perturbed away.

Sard's theorem – and the fact that continuous functions can be uniformly approximated by continuously differentiable ones on $\bar{O}$ – allows one to extend this definition to continuous functions that do not vanish on the boundary of O. So defined, the degree of a mapping has many useful properties, among which the following will be particularly important for our purposes.

HOMOTOPY INVARIANCE. If $H(x, t) = 0$ has no solutions (x, t) for which x belongs to the boundary of O and $0 \le t \le 1$, then $\deg(H(\cdot, t), O)$ is defined and independent of $t \in [0, 1]$ provided $H(x, t)$ is continuous.

DOMAIN DECOMPOSITION. If $\{O_i\}$ is a finite collection of disjoint open subsets of O and if (E.1) has no solutions $x \in (\bar{O} - \bigcup_i O_i)$, then

$$\deg(f, O) = \sum_i \deg(f, O_i).$$

SOLUTION PROPERTY. If $\deg(f, O) \ne 0$, then (E.1) has at least one solution in O.

Clearly, the last property accounts for the usefulness of the theory. The example following (E.2) demonstrates that the converse is false, in general.

One purpose of this section is to obtain two results concerning the rest points of certain vector fields that have the monotonicity properties described in Appendix C. Consequently, we will use the notation from that appendix. As the applications of our results come from Chapters 5 and 6 and it will frequently be necessary to refer to numbered equations from one of these chapters, we use the following convention: equation number (5.2.4) will refer to equation (2.4) in Chapter 5.

In order for the degree of a mapping to be stable to small perturbations of the mapping, the restriction that f not vanish on the boundary of O is obviously necessary. However, for the problems we have in mind, it turns out that f does vanish on the boundary of the appropriate open set O; our goal is to show that there must exist solutions in O. This dilemma occurs frequently in the applications. One often has quite a bit of knowledge about the zeros of f on the boundary of O and very little information about those inside O. For example, in Chapters 5 and 6, we may know that rest points E_0, E_1, E_2 lie on the boundary of the region Γ and we may also know the stability properties of these rest points. We would like to know if there is a positive rest point in the interior of Γ. Of course, one is free to choose the open set O in a different way so that no zeros of f belong to its boundary, but it is often more convenient to do the following instead. Perturb the function f by adding a small term depending on a single parameter. Then use the implicit function theorem to determine the fate of the rest points of f that lie on the boundary of O under the perturbation. Certain of these rest points will enter into O while others will leave O. If the perturbation is carefully chosen, the perturbed function will have no zeros on the boundary of O, so its degree relative to O can be computed and compared with an algebraic count of those zeros of f on the boundary of O which entered O under perturbation. If there is a discrepancy then there must exist other zeros, belonging to O, for the perturbed function. Information about the zeros of f in O can then be obtained by a limiting argument as the perturbation parameter tends to zero. We will use such an argument here; similar ones appear in [HoS] and [JST].

One first application of the theory will be to show that, roughly speaking, between two asymptotically stable rest points of a cooperative system there must exist another rest point which is, generically, unstable. This improves a result in [S2, prop. 3.7] which requires that the system be cooperative in an open region containing the order interval. In Theorem E.1, it is assumed only that (E.1) be cooperative on the order interval. Recall that f is "cooperative" if the off-diagonal entries of the Jacobian of f are nonnegative.

THEOREM E.1. *Let f be continuously differentiable and cooperative on the order interval $[x_1, x_2] \equiv \{x \in \mathbb{R}^n : x_1 \leq x \leq x_2\}$, where $x_1 < x_2$ and $f(x_i) = 0$, $i = 1, 2$. If $s(Df(x_i)) < 0$ for $i = 1, 2$, then there exists a rest point $x_0 \in [x_1, x_2]$ distinct from x_1 and x_2 such that $s(Df(x_0)) \geq 0$.*

REMARK 1. A similar result holds if f has the form of the right side of (B.9) and (B.12) holds in $[x_1, x_2]_K$ where $x_1 <_K x_2$.

REMARK 2. If, in addition, $Df(x)$ is also irreducible in $[x_1, x_2]$, then x_0 cannot belong to the boundary of $[x_1, x_2]$. In fact, if $x_1 \leq x_0 \leq x_2$ and equality does not hold, then strong monotonicity of the dynamical system induced by f implies that $x_1 < x_0 < x_2$.

REMARK 3. Generically, one expects x_0 to be nondegenerate, which implies that $s(Df(x_0)) > 0$ or that x_0 is unstable.

Proof of Theorem E.1. Let O denote the interior of $[x_1, x_2]$, and let y be a point of the line segment joining x_1 and x_2 so that $y - x_1 > 0$ and

$x_2 - y > 0$. Define $F(x, s) = f(x) - s(x - y)$ and note that $F(x, s)$ is cooperative in $[x_1, x_2]$ for $s \geq 0$. Since $F(x_1, s) = s(y - x_1) > 0$ and $F(x_2, s) = -s(x_2 - y) < 0$ for $s > 0$, Corollary B.8 implies that $[x_1, x_2]$ is positively invariant for the autonomous differential equation $x' = F(x, s)$ with fixed positive parameter s. Let $\pi(x, t, s)$ denote the solution operator corresponding to this one-parameter system. Theorem C.2 and the inequalities just listed imply that $\pi(x_1, t, s)$ monotonically increases to a rest point in O as t increases. Similarly, $\pi(x_2, t, s)$ monotonically decreases to a rest point in O as t increases. We use the implicit function theorem to study these two rest points.

Since $D_x F(x_1, 0) = Df(x_1)$ is nonsingular by hypothesis, the implicit function theorem implies that there exists a smooth branch of rest points $x = X_1(s)$, $X_1(0) = x_1$, $F(X_1(s), s) \equiv 0$. Implicit differentiation of the last identity at $s = 0$ gives

$$\frac{dX_1}{ds}(0) = -Df(x_1)^{-1}(y - x_1).$$

As $-Df(x_1)^{-1} \geq 0$ (see proof of Theorem A.12) and is nonsingular, and since $y - x_1 > 0$, it follows that $(dX_1/ds)(0) > 0$. Consequently, for small $s > 0$, $x_1 < X_1(s) < x_2$. A similar analysis at x_2 shows that there exists a smooth branch of rest points $x = X_2(s)$, $X_2(0) = x_2$, $F(X_2(s), s) \equiv 0$, and $(dX_2/ds)(0) < 0$. By continuity, for small $s > 0$, $x_1 < X_1(s) < X_2(s) < x_2$.

It follows by a comparison argument, using the monotonicity properties of the differential equation $x' = F(x, s)$, that $F(\cdot, s)$ has no zeros on the boundary of O and so $\deg(F(\cdot, s), O)$ is defined for $s > 0$. Define $H(x, s, t) = tf(x) - s(x - y)$ for $0 \leq t \leq 1$ and $x \in O$. The same considerations establishing that F does not vanish on the boundary of O also show that $H(x, s, t) = 0$ has no solutions x belonging to the boundary of O for $s > 0$ and $t \in [0, 1]$. By the homotopy property of the degree,

$$\deg(H(\cdot, s, 0), O) = \deg(H(\cdot, s, 1), O) = \deg(F(\cdot, s), O).$$

A simple calculation yields

$$\deg(H(\cdot, s, 0), O) = \operatorname{sgn} \det(-sI) = (-1)^n.$$

Thus, $\deg(F(\cdot, s), O) = (-1)^n$ for small positive s.

Let $O_i(s) \subset O$ be a sufficiently small neighborhood of $X_i(s)$ such that there are no other zeros of $F(\cdot, s)$ in $\bar{O}_i(s)$ except $X_i(s)$. Such a neighborhood exists because $X_i(s)$ is a nondegenerate zero of F for small positive s. Then

$$\deg(F(\cdot, s), O_i) = \operatorname{sgn} \det D_x F(X_i(s), s) = \operatorname{sgn} \det Df(x_i) = (-1)^n$$

for sufficiently small $s > 0$, by continuity of $D_x F$ and the fact that

$$s(Df(x_i)) < 0.$$

Let $O_3(s) = O\backslash(\bar{O}_1(s) \cup \bar{O}_2(s))$. By the domain decomposition property of the degree,

$$\deg(F(\cdot, s), O) = \deg(F(\cdot, s), O_1) + \deg(F(\cdot, s), O_2) + \deg(F(\cdot, s), O_3)$$
$$= 2(-1)^n + \deg(F(\cdot, s), O_3).$$

Consequently,

$$\deg(F(\cdot, s), O_3) = (-1)^{n+1}$$

and therefore $F(x, s) = 0$ has a solution $x = X_0(s) \in O_3(s)$ for small positive s, by the solution property of the degree. Clearly, $X_0(s) \neq X_i(s)$ for $i = 1, 2$, and the comparison argument (using the differential equation) shows that $X_0 \in [X_1, X_2]$.

If every zero x of $F(\cdot, s)$ in $O_3(s)$ satisfies $s(D_x F(x, s)) < 0$ then, using the domain decomposition property as before, we would have

$$\deg(F(\cdot, s), O_3) = p(-1)^n,$$

where $p \geq 1$ is the number of such zeros. Since this contradicts the previous displayed formula, $X_0(s)$ may be chosen to satisfy $s(D_x F(X_0(s), s)) \geq 0$.

Since $[x_1, x_2]$ is compact, we can select a sequence s_n such that $s_n \to 0$ and $X_0(s_n) \to x_0$ as $n \to \infty$. Continuity of F implies that $f(x_0) = 0$. Furthermore, $x_0 \neq x_i$ for $i = 1, 2$, since if (for example) $x_0 = x_1$ then $(X_0(s_n), s_n)$ would be a branch of solutions of $F = 0$ distinct from the branch $(X_1(s), s)$, both converging to $(x_1, 0)$. This violates the uniqueness of the latter branch, as provided by the implicit function theorem. This establishes the existence of a zero x_0 for f distinct from the x_i. Finally, $s(Df(x_0)) \geq 0$ by continuity of $D_x F(x, s)$ and the fact that $s(D_x F(X_0(s), s)) \geq 0$. This completes the proof. □

We now turn our attention to the zeros of the map $F: \Gamma \to \mathbb{R}^{2n}$, $F(x) = (F_1(u, v), F_2(u, v))$ where $x = (u, v)$, $\Gamma = \{(u, v) \in \mathbb{R}_+^{2n} : u + v \leq z\}$, $z > 0$, and

$$F_1(u, v) = [A + F_u(z - u - v)]u,$$

$$F_2(u, v) = [A + F_v(z - u - v)]v.$$

See Chapter 6, and in particular (6.2.1), for further details. In the case of the simple gradostat model (5.2.4), we have

$$A = \begin{pmatrix} -2 & 1 \\ 1 & -2 \end{pmatrix}, \qquad z = (\tfrac{2}{3}, \tfrac{1}{3}),$$

$$F_u(z - u - v) = \begin{pmatrix} f_u(\tfrac{2}{3} - u_1 - v_1) & 0 \\ 0 & f_u(\tfrac{1}{3} - u_2 - v_2) \end{pmatrix},$$

with an analogous formula for $F_v(z - u - v)$.

Our goal is to show that for both the simple gradostat model (5.2.4) and the general gradostat model (6.2.1), a positive rest point representing the coexistence of the two populations must exist if both single-species rest points, E_1 and E_2, exist and are asymptotically stable in the linear approximation. Furthermore, this positive rest point is generically unstable. This result is used in Chapter 5 to exclude the possibility that both E_1 and E_2 are asymptotically stable, since it is shown by Lemma 5.5.2 that any positive rest point must be asymptotically stable. It is also used in Chapter 6, where the case that both E_1 and E_2 are asymptotically stable cannot be excluded.

The existence of an unstable positive rest point when both E_1 and E_2 are stable does not follow immediately from Theorem E.1 and Remark 1 for two reasons. First, F is defined on Γ, which may not include the order interval $[E_2, E_1]_K$. This is only a technical point, however, as we shall see. Second, even if Γ contained the order interval, the rest point E_0 also belongs to the order interval and it cannot be excluded that E_0 is the zero of F (which Theorem E.1 guarantees) since it is unstable. However, the proof of Proposition E.2 uses the same ideas as the proof of Theorem E.1.

Since the general gradostat model includes the simple gradostat model as a special case, we consider the latter. Notation and results from Chapters 5 and 6 will be used as required. Suppose that F vanishes at E_0, E_1, and E_2; that is, suppose these are the rest points for (5.2.4) or (6.2.1), where

$$E_0 = (0, 0), \quad E_1 = (\hat{u}, 0), \quad E_2 = (0, \bar{v}).$$

Let J_i denote the Jacobian matrix of F at E_i, $i = 0, 1, 2$.

PROPOSITION E.2. *Suppose that $s(J_i) < 0$, $i = 1, 2$; that is, suppose E_1 and E_2 are asymptotically stable rest points of (6.2.1) in the linear approximation. Then there exists a positive rest point $E_* \in \Gamma$ satisfying $E_2 <_K E_* <_K E_1$ and $s(J_*) \geq 0$, where J_* is the Jacobian of F at E_*.*

Proof. Add an "immigration" term ϵe to the second and third equations of (6.1.6), where $\epsilon > 0$ is small and $e = (1, 1, \ldots, 1)$. The arguments of

Lemma 6.2.2 lead to the following vector field $F(x, \epsilon)$, the analog of (6.2.1):

$$F_1(u, v, \epsilon) = [A + F_u(S(\epsilon))]u + \epsilon e,$$

$$F_2(u, v, \epsilon) = [A + F_v(S(\epsilon))]v + \epsilon e,$$

where $S(\epsilon) = z(\epsilon) - u - v$ and $z(\epsilon) > 0$ satisfies

$$Az + e_0 + 2\epsilon e = 0.$$

Note that $z(0) = z$. The domain for $F(\cdot, \epsilon)$ is $\Gamma(\epsilon) \equiv \{x = (u, v) \in \mathbb{R}_+^{2n} : u + v \le z(\epsilon)\}$. The vector field $F(x, \epsilon)$ possesses all the monotonicity properties of $F(x) = F(x, 0)$.

The Jacobian J_1 of F (with respect to x) at E_1, with $\epsilon = 0$, is given in the proof of Lemma 6.4.2. It is nonsingular by hypothesis (see Lemma 6.4.2). Therefore, by the implicit function theorem, there exists a zero $E_1(\epsilon)$ of $F(\cdot, \epsilon)$ which is smooth in ϵ for all small ϵ and $E_1(0) = E_1$. The important question is whether or not $E_1(\epsilon)$ belongs to $\Gamma(\epsilon)$ for small $\epsilon > 0$. By implicit differentiation,

$$J_1 \frac{dE_1}{d\epsilon}(0) + \frac{\partial F}{\partial \epsilon}(E_1, 0) = 0.$$

Setting $E_1(\epsilon) = (u(\epsilon), v(\epsilon))$, we find that $v(\epsilon)$ satisfies

$$(A + F_v(z - \hat{u})) \frac{dv}{d\epsilon}(0) + e = 0.$$

Therefore

$$\frac{dv}{d\epsilon}(0) = -(A + F_v(z - \hat{u}))^{-1}e > 0,$$

where the inequality follows from Theorem A.12 and the fact that $s(J_1) = s(A + F_v(z - \hat{u})) < 0$. It follows immediately that $E_1(\epsilon)$ belongs to the interior of $\Gamma(\epsilon)$ for small positive ϵ. By continuity, $s(J_1(\epsilon)) < 0$ where $J_1(\epsilon)$ is the Jacobian of $F(\cdot, \epsilon)$ at $E_1(\epsilon)$.

A similar analysis applies to E_2 and we obtain $E_2(\epsilon)$, a smooth branch of zeros of $F(\cdot, \epsilon)$ belonging to the interior of $\Gamma(\epsilon)$ with $s(J_2(\epsilon)) < 0$ for small positive ϵ, where $J_2(\epsilon)$ is the Jacobian of $F(\cdot, \epsilon)$ at $E_2(\epsilon)$. Clearly, $E_2(\epsilon) <_K E_1(\epsilon)$ for small positive ϵ.

The Jacobian J_0 of $F(\cdot, \epsilon)$ at E_0, with $\epsilon = 0$, is described in the text following (6.4.1). It is nonsingular, so the implicit function theorem implies that there exists a zero $E_0(\epsilon)$ of $F(\cdot, \epsilon)$ which is smooth in ϵ and satisfies $E_0(0) = E_0$. We must determine whether $E_0(\epsilon)$ belongs to $\Gamma(\epsilon)$ for small positive ϵ. Letting $E_0(\epsilon) = (u(\epsilon), v(\epsilon))$, implicit differentiation yields

$$(A + F_u(z)) \frac{du}{d\epsilon}(0) + e = 0,$$

$$(A + F_v(z)) \frac{dv}{d\epsilon}(0) + e = 0.$$

It is easy to see that both derivatives have at least one negative component. For example, if the derivative of u were nonnegative then one easily sees, by writing the equations in component form, that it is necessarily positive. But this violates Theorem A.12 since $s(A + F_u(z)) > 0$. Consequently, $E_0(\epsilon)$ does not belong to $\Gamma(\epsilon)$ for small positive ϵ.

We will now use degree theory to show that there exist zeros of F in $\Gamma(\epsilon)$ distinct from the $E_i(\epsilon)$. Let $O(\epsilon)$ denote the interior of $\Gamma(\epsilon)$. Define

$$H(x, \epsilon, t) = (Au + \epsilon e, Av + \epsilon e) + t(F_u(S(\epsilon))u, F_v(S(\epsilon))v)$$

for $x \in \Gamma(\epsilon)$ with $\epsilon > 0$ and $0 \leq t \leq 1$. Observe that if $u_i = 0$ then

$$H(x, \epsilon, t)_i \geq \epsilon > 0,$$

so H has no zeros x with $u_i = 0$. Similarly, it has no zeros with $v_i = 0$. Suppose that H vanishes at a point $x \in \Gamma(\epsilon)$ such that $u_i + v_i = z_i(\epsilon)$ for some i. Let I be the set of all such indices i and let J be the complementary set of indices. If $j \in J$, then $u_j + v_j < z_j(\epsilon)$. Taking components of the identity $H = 0$ we find that, for $i \in I$,

$$\sum_j A_{ij} u_j + \epsilon = 0 \quad \text{and} \quad \sum_j A_{ij} v_j + \epsilon = 0.$$

Adding these equations gives

$$0 = \sum_{j \in I} A_{ij} z_j + \sum_{j \in J} A_{ij}(u_j + v_j) + 2\epsilon$$

$$\leq \sum_j A_{ij} z_j + 2\epsilon = -(e_0)_i \leq 0.$$

Since $e_0 \neq 0$, it follows that J is non-empty. Furthermore, the inequality implies that $A_{ij} = 0$ for every $i \in I$ and $j \in J$ since otherwise a strict inequality would occur, a contradiction. But this contradicts the irreducibility of A. We conclude that no zero of H can belong to the boundary of $\Gamma(\epsilon)$. The homotopy property of degree implies that

$$\deg(F(\cdot, \epsilon), O(\epsilon)) = \deg(H(\cdot, \epsilon, 1), O(\epsilon))$$

$$= \deg(H(\cdot, \epsilon, 0), O(\epsilon))$$

$$= \text{sgn}(\det A)^2$$

$$= +1,$$

since $(Au + \epsilon e, Av + \epsilon e)$ vanishes only at $-\epsilon(A^{-1}e, A^{-1}e) \in O(\epsilon)$.

Let $O_i(\epsilon)$ be an open subset of $O(\epsilon)$ such that $E_i(\epsilon) \in O_i(\epsilon)$ and $F(\cdot, \epsilon)$ has no zeros in $O_i(\epsilon)$ other than $E_i(\epsilon)$. Then

$$\deg(F(\cdot, \epsilon), O_i(\epsilon)) = (-1)^{2n} = +1,$$

since $s(J_i(\epsilon)) < 0$ by continuity. By the domain decomposition property,

$$\deg(F(\cdot, \epsilon), O_3(\epsilon)) = -1,$$

where $O_3(\epsilon) = O(\epsilon) \backslash (\bar{O}_1(\epsilon) \cup \bar{O}_2(\epsilon))$. Therefore, by the solution property of degree, there is a zero of F, denoted by $x(\epsilon)$, in $O_3(\epsilon)$. An argument similar to that in Theorem E.1 shows that we may assume $s(J(\epsilon)) \geq 0$, where $J(\epsilon)$ denotes the Jacobian of $F(\cdot, \epsilon)$ at $x(\epsilon)$. Clearly, $x(\epsilon) \neq E_i(\epsilon)$ for $i = 1, 2$.

Let $x_n = x(1/n)$ for $n = 1, 2, \ldots$. We may assume that x_n converges to E_* as $n \to \infty$ by passing to a subsequence if necessary. By continuity, E_* is a zero of F. We claim that $E_* \neq E_0, E_1, E_2$. Indeed, this follows by the uniqueness part of the implicit function theorem and the fact that $x_n \neq E_i(1/n)$ for $i = 0, 1, 2$. Therefore, by Lemma 6.4.1(d), $E_* > 0$. Since every solution of (6.2.1) is attracted to $[E_2, E_1]_K$, by a comparison argument (see e.g. the last part of the proof of Theorem 6.4.4 or Lemma 5.6.1), the inequality $E_2 <_K E_* <_K E_1$ is easy to see.

Since $s(J(\epsilon)) \geq 0$ for $\epsilon > 0$ and $x_n \to E_*$ as $n \to \infty$, continuity of the Jacobian of F in (x, ϵ) implies that $s(J_*) \geq 0$. $\qquad\square$

We conclude this appendix by constructing the Poincaré map corresponding to a nonconstant periodic solution $x(t) = x(t + T)$ of the autonomous system of differential equations $x' = f(x)$ and by verifying Lemma 6.2 of Chapter 3. We assume that $T > 0$ is the smallest period of $x(t)$ and, by translating and rotating our coordinate system, we assume that $x(0) = 0$ and $x'(0) = f(0) = a e_1$, $a > 0$. Here, e_i ($1 \leq i \leq n$) denotes the standard basis vectors in $\mathbb{R}^n$. The notation $\pi(x, t)$, introduced in Chapter 1 for the solution map of our differential equation, will be used. With this notation, $x(t) = \pi(0, t) = \pi(0, t + T)$ and

$$\frac{\partial \pi}{\partial t}(0, T) = f(\pi(0, T)) = f(0) = a e_1.$$

The variational equation (see Section 4 of Chapter 3) about the periodic solution $\pi(0, t)$ is given by

$$u' = Df(\pi(0, t))u.$$

Denote by $\Phi(t)$ the fundamental matrix solution $\Phi(0) = I_n$, where I_n is the $n \times n$ identity matrix. Since $u = x'(t)$ is a periodic solution of the variational equation, it follows that $x'(t) = a\Phi(t)e_1$ and therefore

$$\Phi(T)e_1 = e_1.$$

Let the Floquet multipliers (eigenvalues of $\Phi(T)$) of the variational equation be $1, \rho_1, \rho_2, \ldots, \rho_{n-1}$, where the terms are listed according to multiplicity and the first one corresponds to the eigenvector e_1. Finally, recall from the fundamental theory of ordinary differential equations [H2, chap. 1, thm. 3.3] that

$$\Phi(t) = \frac{\partial \pi}{\partial x}(0, t).$$

The Poincaré map will be defined in a neighborhood of zero in the hyperplane

$$\Sigma \equiv \{x : x_1 = 0\} = \mathbb{R}^{n-1}$$

as follows. Since we know that solutions starting at points on Σ near $x = 0$ remain near $x(t) = \pi(0, t)$, at least on bounded t intervals, they will return to Σ after approximately time T. Consequently, we look for solutions of $\pi_1(x, t) = 0$ for x near 0 and t near T. For the time being, we do not restrict x to lie on Σ. Since

$$\frac{\partial \pi_1}{\partial t}(0, T) = f_1(0) = a > 0,$$

the implicit function theorem implies that we can solve for $t = \tau(x)$ as a function of x for x in some neighborhood N of zero: $\pi_1(x, \tau(x)) \equiv 0$ satisfying $\tau(0) = T$. Restricted to Σ, τ is called the *first-return-time map*. Finally, we define Q to be the orthogonal projection onto Σ along e_1 – that is,

$$Qx = (x_2, x_3, \ldots, x_n) \in \mathbb{R}^{n-1}$$

for $x = (x_1, \ldots, x_n)$ – and R to be the injection of $\mathbb{R}^{n-1}$ into $\mathbb{R}^n$ defined by

$$R(x_2, x_3, \ldots, x_n) = (0, x_2, \ldots, x_n).$$

Then the Poincaré map P is defined by

$$P = Q \circ H \circ R,$$

where $H(x) = \pi(x, \tau(x))$; or, in simpler terms,

$$P(x_2, \ldots, x_n) = Q\pi(0, x_2, x_3, \ldots, x_n, \tau(0, x_2, \ldots, x_n))$$

for $(x_2, \ldots, x_n) \in N \cap \Sigma$. By the chain rule and the fact that Q and R are linear, the Jacobian of P at $0 \in \mathbb{R}^{n-1}$ is given by

$$DP(0) = QDH(0)R.$$

The ith column of the $n \times n$ matrix $DH(0)$ is

$$\frac{\partial \pi}{\partial x_i}(0, T) + \frac{\partial \pi}{\partial t}(0, T)\frac{\partial \tau}{\partial x_i}(0).$$

As $(\partial \pi/\partial t)(0, T) = ae_1$ and $Qe_1 = 0$, we have

$$DP(0) = Q\Phi(T)R.$$

Note that the first column of $\Phi(T)$ is e_1. Consequently, the eigenvalues $\rho_1, \rho_2, \ldots, \rho_{n-1}$ of $\Phi(T)$ are the eigenvalues of the $(n-1) \times (n-1)$ lower right block B of $\Phi(T)$. Using $QR = I_{n-1}$, we find that

$$DP(0) = B.$$

It follows that $\rho_1, \ldots, \rho_{n-1}$ are the eigenvalues of $DP(0)$, as asserted in Lemma 6.2 of Chapter 3.

F

A Convergence Theorem

In many of the arguments involving chemostats it was shown that the omega limit set had to lie in a restricted set, and the equations were analyzed on that set: one simply could choose initial conditions in the restricted set at time zero. The equation defining the restricted set – in effect, a conservation principle – allowed one variable to be eliminated from the system. We want to abstract this idea and make it rigorous. The omega limit set lies in a lower-dimensional set, and the trajectories in that set satisfy a smaller system of differential equations. However, it is not clear (and, indeed, not true [T3]) that the asymptotic behavior of the two systems is necessarily the same. (A very nice paper of Thieme [T1] gives examples and helpful theorems for asymptotically autonomous systems. A classical result in this direction is a paper of Markus [M].) In this appendix, a theorem is presented which justifies the procedure on the basis of stability.

Consider two systems of ordinary differential equations of the form

$$z' = Az, \qquad y' = f(y, z), \tag{F.1}$$

and

$$x' = f(x, 0), \tag{F.2}$$

where

$$z \in \mathbb{R}^m, \qquad (y, z) \in D \subset \mathbb{R}^n \times \mathbb{R}^m,$$

$$x \in \Omega = \{x \mid (x, 0) \in D\} \subset \mathbb{R}^n.$$

It will be assumed that f is continuously differentiable, D is positively invariant for (F.1), and (F.1) is dissipative in the sense that there is a compact subset of D into which every solution eventually enters and remains. The following additional hypotheses will be used.

(H1) All of the eigenvalues of A have negative real parts.

(H2) Equation (F.2) has a finite number of rest points in Ω, each of which is hyperbolic for (F.2). Denote these rest points by $x_1, x_2, \ldots, x_p$.

(H3) The dimension of the stable manifold of x_i is n for $1 \leq i \leq r$, and the dimension of the stable manifolds of x_j is less than n for $j = r+1, \ldots, p$. In symbols, $\dim(M^+(x_i)) = n$, $i = 1, \ldots, r$; $\dim(M^+(x_j)) < n$ for $j = r+1, \ldots, p$.

(H4) $\Omega = \bigcup_{i=1}^p M^+(x_i)$.

(H5) Equation (F.2) does not possess a cycle of rest points.

Note first that the only rest points of (F.1) are of the form $(x_i, 0)$, and that each is hyperbolic for (F.1). To avoid confusion between (F.1) and (F.2), denote the stable and unstable manifolds for (F.1) by Λ^+ and Λ^-, respectively. Then

(a) $\dim \Lambda^+(x_i, 0) = m + \dim M^+(x_i)$ and

(b) $M^+(x_i) \times \{0\} = \Lambda^+(x_i, 0) \cap \{(y, z) \in D \,|\, z = 0\}$.

Note also that x_i and $(x_i, 0)$ are locally asymptotically stable for (F.1) and (F.2), respectively, $i = 1, \ldots, r$. By (H4), every point of Ω is attracted to one of the rest points x_i, $i = 1, \ldots, p$.

The following theorem is a very special case of a general result of Thieme [T1].

THEOREM F.1. *Let* (H1)–(H5) *hold and let* $(y(t), z(t))$ *be a solution of* (F.1). *Then, for some* i,

$$\lim_{t \to \infty}(y(t), z(t)) = (x_i, 0).$$

In other words, $D \subset \bigcup_{i=1}^p \Lambda^+(x_i, 0)$. *Furthermore,* $\bigcup_{i=r+1}^p \Lambda^+(x_i, 0)$ *has Lebesgue measure zero.*

Proof. Let $(y(t), z(t))$ be a solution of (F.1), and assume that the first assertion of the theorem is false. If O denotes the omega limit set of this solution, then $O \neq \{(x_i, 0\}$ for $i = 1, 2, \ldots, p$. Let $(x, 0) \in O$. By (H4), either $x = x_i$ for some i or the solution of (F.2) through x converges to some x_i. By the invariance of O we conclude that $(x_i, 0)$ belongs to O for some i. Clearly, $i \geq r+1$ because points of the form $(x_i, 0)$, $1 \leq i \leq r$, are asymptotically stable, and a limit set containing an asymptotically stable rest point *is* that rest point, a contradiction to our hypothesis. Since $O \neq \{(x_i, 0)\}$, the Butler–McGehee theorem says that O must contain a point $(x, 0)$ with $x \in M^-(x_i)$ and $x \neq \{(x_i, 0)\}$. By (H4), $x \in M^+(x_j)$ for some j, and consequently x_i is chained to x_j, $x_i \to x_j$, in O. Continuing the argument a finite number of steps leads to a cycle, contradicting (H5). This proves the first assertion of the theorem, $D \subset \bigcup_{i=1}^{p} \Lambda^+(x_i, 0)$.

It is well known that the stable manifold $\Lambda^+(x_i, 0)$ of a hyperbolic, unstable rest point $(x_i, 0)$ has Lebesgue measure zero. This follows from Sard's theorem (see Appendix E) and the fact that the stable manifold is the image of a smooth one-to-one map of $\mathbb{R}^{l_i}$ into $\mathbb{R}^n \times \mathbb{R}^m$, where l_i is the dimension of the stable subspace of the linearization of (F.1) about $(x_i, 0)$ and consequently $l_i < n + m$ [Pi, p. 43]. It follows that $\bigcup_{i=r+1}^{p} \Lambda^+(x_i, 0)$ has measure zero. $\square$

The theorem states that almost all trajectories converge to one of the asymptotically stable rest points $(x_i, 0)$. In most of our applications, $r = 1$: there is exactly one asymptotically stable rest point. In this case, an observer would conclude that all trajectories converge to $(x_i, 0)$, since the probability of an initial condition being in the exceptional set is zero.

To illustrate the procedure, Theorem F.1 will be applied to the systems (5.1) and (5.2) of Chapter 1. The system corresponding to (F.1) is

$$\Sigma' = -\Sigma,$$

$$x_1' = x_1 \left(\frac{m_1(1 - \Sigma - x_1 - x_2)}{1 + a_1 - \Sigma - x_1 - x_2} - 1 \right),$$

$$x_2' = x_2 \left(\frac{m_2(1 - \Sigma - x_1 - x_2)}{1 + a_2 - \Sigma - x_1 - x_2} - 1 \right),$$ (F.3)

$$D = \{(\Sigma, x_1, x_2) \mid x_i > 0, \, x_1 + x_2 + \Sigma \leq 1\}.$$

The system corresponding to (F.2) is

$$x_1' = x_1 \left(\frac{m_1(1 - x_1 - x_2)}{1 + a_1 - x_1 - x_2} - 1 \right),$$

$$x_2' = x_2 \left(\frac{m_2(1 - x_1 - x_2)}{1 + a_2 - x_1 - x_2} - 1 \right),$$ (F.4)

$$\Omega = \{(x_1, x_2) \mid x_i > 0, \, x_1 + x_2 \leq 1\},$$

which is the same as (5.2) of Chapter 1.

In this case, $m = 1$ and $n = 2$. The matrix A is 1×1, or $A = [-1]$. Clearly (H1) is satisfied. System (F.4) has three rest points, which were labeled E_0, E_1, E_2 in Section 5 of Chapter 1. We restrict the analysis to the situation of Theorem 5.1 of Chapter 1. (The other cases can be done similarly.) The set Ω is just the portion of the nonnegative cone in $\mathbb{R}^2$ which lies on the origin side of the line $x_1 + x_2 = 1$ (Ω is a positively invariant set). All of the rest points are hyperbolic. Since E_1 is a local attractor, the dimension of its stable manifold is 2. The stability of E_1 prevents it from being a part of any chain of equilibria. Similarly, E_0 is a repeller, has no stable manifold, and cannot be part of any chain of equilibria. The rest point E_2 is not chained to itself, so (H4) is satisfied. The dimension of the stable manifold of E_2 is 1. This shows that (H3) is satisfied. The part of the stable manifold of E_2 which lies in Ω is just a portion of the x_2 axis; thus (H5) is satisfied. The proof given of Theorem 5.1 in Chapter 1 establishes (H4) by showing that every solution of (F.4) converges to a rest point. Theorem F.1 then allows one to conclude that every trajectory of (F.3) tends to a rest point of (F.4). Orbits through all initial conditions excluding those in the exceptional set converge to E_1.

The exceptional set is just the Σ-x_2 face of the positive cone in $\mathbb{R}^3$. No interesting initial conditions lie here.

References

[ALGM] A. A. Andronov, E. A. Leontovich, I. I. Gordon, and A. G. Maier (1973), *Qualitative Theory of Second-Order Dynamical Systems.* New York: Wiley.

[AH] R. Aris and A. E. Humphrey (1977), "Dynamics of a chemostat in which two organisms compete for a common substrate," *Biotechnology and Bioengineering* 19: 1375–86.

[AM] R. A. Armstrong and R. McGehee (1980), "Competitive exclusion," *American Naturalist* 115: 151–70.

[BaWo] M. M. Ballyk and G. S. K. Wolkowicz (1993), "Exploitative competition in the chemostat for two perfectly substitutable resources," *Mathematical Biosciences* 118: 127–80.

[BFr] B. C. Baltzis and A. G. Fredrickson (1983), "Competition of two microbial populations for a single resource in a chemostat when one of them exhibits wall attachment," *Biotechnology and Bioengineering* 25: 2419–39.

[BWu] B. C. Baltzis and M. Wu (1994), "Coexistence of three pure and simple competitors in a four member reactor network," *Mathematical Biosciences* (to appear).

[BT] J. V. Baxley and H. B. Thompson (1994), "Nonlinear boundary value problems and competition in the chemostat," *Nonlinear Analysis* 22: 1329–44.

[BP] A. Berman and R. J. Plemmons (1979), *Nonnegative Matrices in the Mathematical Sciences.* New York: Academic Press.

[BS] N. P. Bhatia and G. P. Szego (1967), "Dynamical systems: Stability theory and applications," in *Lecture Notes in Mathematics,* vol. 35. Berlin: Springer.

[BD] J. L. Brooks and S. I. Dodson (1965), "Predation, body size and composition of plankton," *Science* 150: 28–35.

[BB] H. R. Bungay and M. L. Bungay (1968), "Microbial interactions in continuous culture," *Advances in Applied Microbiology* 10: 269–90.

[BWh] L. P. Burton and W. M. Whyburn (1952), "Minimax solutions of ordinary differential systems," *Proceedings of the American Mathematical Society* 3: 794–803.

[BC] A. W. Bush and A. E. Cook (1975), "The effect of time delay and
      growth inhibition in the bacterial treatment of wastewater," *Journal
      of Theoretical Biology* 63: 385–95.

[BFW] G. Butler, H. I. Freedman, and P. Waltman (1986), "Uniformly
      persistent systems," *Proceedings of the American Mathematical
      Society* 96: 425–30.

[BHW1] G. J. Butler, S. B. Hsu, and P. Waltman (1983), "Coexistence of
      competing predators in a chemostat," *Journal of Mathematical
      Biology* 17: 133–51.

[BHW2] G. J. Butler, S. B. Hsu, and P. Waltman (1985), "A mathematical
      model of the chemostat with periodic washout rate," *SIAM Journal
      on Applied Mathematics* 45: 435–49.

[BW] G. Butler and P. Waltman (1986), "Persistence in dynamical systems,"
      Journal of Differential Equations 63: 255–63.

[BWo1] G. J. Butler and G. S. K. Wolkowicz (1985), "A mathematical model
      of the chemostat with a general class of functions describing nutrient
      uptake," *SIAM Journal on Applied Mathematics* 45: 138–51.

[BWo2] G. J. Butler and G. S. K. Wolkowicz (1987), "Exploitative competi-
      tion in a chemostat for two complementary, and possibly inhibitory,
      resources," *Mathematical Bioscience* 83: 1–48.

[Ca1] R. P. Canale (1969), "Predator–prey relationships in a model for the
      activated process," *Biotechnology and Bioengineering* 11: 887–907.

[Ca2] R. P. Canale (1970), "An analysis of models describing predator–prey
      interaction," *Biotechnology and Bioengineering* 12: 353–78.

[Cap] J. Caperon (1969), "Time lag in population growth response of
      Isochrysis galbana to a variable nitrate environment," *Ecology* 50:
      188–92.

[CL] E. A. Coddington and N. Levinson (1955), *Theory of Ordinary
      Differential Equations.* New York: McGraw-Hill.

[C]   C. Conley (1978), "Isolated invariant sets and the Morse index," in
      Conference Board of Mathematical Sciences, vol. 38. Providence,
      RI: American Mathematical Society.

[Co] W. A. Coppel (1965), *Stability and Asymptotic Behavior of Differ-
      ential Equations.* Boston: D. C. Heath.

[CB] S.-W. Chang and B. C. Baltzis (1989), "Impossibility of coexistence
      of three pure and simple competitors in configurations of three
      interconnected chemostats," *Biotechnology and Bioengineering*
      33: 460–70.

[CH] S. N. Chow and J. K. Hale (1982), *Methods of Bifurcation Theory.*
      New York: Springer.

[CM] A. Cunningham and P. Mass (1978), "Time lag and nutrient storage
      effects in the transient growth response of *Chlamydomonas reinhardii*
      in nitrogen limited batch and continuous culture," *Journal of General
      Microbiology* 104: 227–31.

[CN1] A. Cunningham and R. M. Nisbet (1980), "Time lag and co-operativity
      in the transient growth dynamics of microalgae," *Journal of Theoret-
      ical Biology* 84: 189–203.

[CN2] A. Cunningham and R. M. Nisbet (1983), "Transients and oscilla-
      tions in continuous cultures," in M. J. Bazin (ed.), *Mathematics in
      Microbiology.* New York: Academic Press, pp. 77–103.

[Cu1] J. M. Cushing (1977), "Periodic time-dependent predator prey sys-
      tems," *SIAM Journal on Applied Mathematics* 23: 972–9.

[Cu2] J. M. Cushing (1989), "A competition model for size-structured species," *SIAM Journal on Applied Mathematics* 49: 838–58.

[DKG] G. D'Ans, P. V. Kokotovic, and D. Gottlieb (1971), "A nonlinear regulator problem for a model of biological waste treatment," *IEEE Transactions on Automatic Control* AC-16: 341–7.

[DSu] I. W. Dawes and I. W. Sutherland (1976), *Micro-organism Physiology*. New York: Springer.

[De] A. M. Dean (1985), "The dynamics of microbial commensalisms and mutualisms," in D. H. Boucher (ed.), *The Biology of Mutualism, Ecology, and Evolution*. Oxford: Oxford University Press, pp. 270–304.

[DS] P. deMottoni and A. Schiaffino (1982), "Competition systems with periodic coefficients: A geometric approach," *Journal of Mathematical Biology* 11: 319–35.

[DMKH] O. Diekmann, J. A. J. Metz, A. L. M. Kooijman, and J. J. A. M. Heijmans (1984), "Continuum population dynamics with an application to *Daphnia magna*," *Nieuw Archief voor Wiskunde* (4) 2: 82–109.

[D1] M. R. Droop (1973), "Some thoughts on nutrient limitation in algae," *Journal of Phycology* 9: 264–72.

[D2] M. R. Droop (1974), "The nutrient status of algae cells in continuous culture," *Journal of the Marine Biology Association U.K.* 54: 825–55.

[DRS] S. R. Dunbar, K. P. Rybakowski, and K. Schmitt (1986), "Persistence in models of predator–prey populations with diffusion," *Journal of Differential Equations* 65: 117–38.

[EMP] D. Erle, H. Mayer, and T. Plesser (1979), "The existence of stable limit cycles for enzyme catalyzed reactions with positive feedback," *Mathematical Biosciences* 44: 191–208.

[E] S. F. Ellermeyer (1994), "Competition in the chemostat: Global asymptotic behavior of a model with delayed response in growth," *SIAM Journal on Applied Mathematics* 54: 456–65.

[Er] B. Ermentrout (1990), "Phase Plane: The Dynamical Systems Tool," ver. 3.0. Pacific Grove, CA: Brooks Cole.

[FSt] A. G. Frederickson and G. Stephanopoulos (1981), "Microbial competition," *Science* 213: 972–9.

[FM] H. I. Freedman and P. Moson (1990), "Persistence definitions and their connections," *Proceedings of the American Mathematical Society* 109: 1025–33.

[FSW1] H. I. Freedman, J. W.-H. So, and P. Waltman (1989a), "Coexistence in a model of competition in the chemostat incorporating discrete delays," *SIAM Journal on Mathematical Analysis* 49: 859–70.

[FSW2] H. I. Freedman, J. W.-H. So, and P. Waltman (1989b), "Chemostat competition with time delays," in J. Eisenfeld and D. S. Levine (eds.), *Biomedical Modelling and Simulation*. J. C. Baltzer AG Scientific Publishing, pp. 171–3.

[FW1] H. I. Freedman and P. Waltman (1977), "Mathematical analysis of some three-species food chain models," *Mathematical Bioscience* 33: 257–76.

[FW2] H. I. Freedman and P. Waltman (1984), "Persistence in a model of three interacting predator–prey populations," *Mathematical Bioscience* 68: 213–31.

[FW3] H. I. Freedman and P. Waltman (1985), "Persistence in a model of three competitive populations," *Mathematical Bioscience* 73: 89-101.

[Ga] G. F. Gause (1934), *The Struggle for Existence*. Baltimore: Williams and Wilkens.

[G1] J. P. Grover (1991), "Resource competition in a variable environment: Phytoplankton growing according to the variable-internal-stores model," *American Naturalist* 138: 811-35.

[G2] J. P. Grover (1992), "Constant- and variable-yield models of population growth: Responses to environmental variability and implications for competition," *Journal of Theoretical Biology* 158: 409-28.

[HG] K. P. Hadeler and D. Glas (1983), "Quasimonotone systems and convergence to equilibrium in a population genetic model," *Journal of Mathematical Analysis and Applications* 95: 297-303.

[H1] J. K. Hale (1977), *Theory of Functional Differential Equations*. New York: Springer.

[H2] J. K. Hale (1980), *Ordinary Differential Equations*. Malabar, FL: Krieger.

[H3] J. K. Hale (1988), *Asymptotic Behavior of Dissipative Systems*. Providence, RI: American Mathematical Society.

[HaS] J. K. Hale and A. S. Somolinas (1983), "Competition for fluctuating nutrient," *Journal of Mathematical Biology* 18: 255-80.

[HaW] J. K. Hale and P. Waltman (1989), "Persistence in infinite dimensional systems," *SIAM Journal on Mathematical Analysis* 20: 388-95.

[HH] S. R. Hansen and S. P. Hubbell (1980), "Single nutrient microbial competition: Agreement between experimental and theoretical forecast outcomes," *Science* 207: 1491-3.

[Har] P. Hartman (1964), *Ordinary Differential Equations*. New York: Wiley.

[Ha] J. D. Harvey (1983), "Mathematics of microbial age and size distribution," in M. Bazin (ed.), *Mathematics and Microbiology*. London: Academic Press, pp. 1-35.

[HKW] B. D. Hassard, N. D. Kazarinoff, and Y.-H. Wan (1981), *Theory and Application of Hopf Bifurcation*. London: Cambridge University Press.

[HP] A. Hastings and T. Powell (1991), "Chaos in a three species food chain," *Ecology* 72: 896-903.

[HET] D. Herbert, R. Elsworth, and R. C. Telling (1956), "The continuous culture of bacteria; a theoretical and experimental study," *Journal of Canadian Microbiology* 14: 601-22.

[Hi1] M. Hirsch (1982), "Systems of differential equations which are competitive or cooperative, I: Limit sets," *SIAM Journal on Applied Mathematics* 13: 167-79.

[Hi2] M. Hirsch (1984), "The dynamical systems approach to differential equations," *Bulletin of the American Mathematical Society* 11: 1-64.

[Hi3] M. Hirsch (1985), "Systems of differential equations which are competitive or cooperative, II: Convergence almost everywhere," *SIAM Journal on Mathematical Analysis* 16: 423-39.

[Hi4] M. Hirsch (1990), "Systems of differential equations that are competitive or cooperative, IV: Structural stability in three dimensional systems," *SIAM Journal on Mathematical Analysis* 21: 1225-34.

[Ho] J. A. Hofbauer (1980), "A general cooperation theorem for hypercycles," *Monatshefte für Mathematik* 91: 233-40.

[HoS] J. Hofbauer and K. Sigmund (1988), *Dynamical Systems and the Theory of Evolution.* Cambridge: Cambridge University Press.

[HSo] J. Hofbauer and J. W.-H. So (1994), "Competition in the gradostat: The global stability problem," *Nonlinear Analysis* 22: 1017-33.

[Hol] C. S. Holling (1959), "Some characteristics of simple types of predation and parasitism," *Canadian Entomologist* 91: 385-95.

[Hsu1] S. B. Hsu (1978), "Limiting behavior for competing species," *SIAM Journal on Applied Mathematics* 34: 760-3.

[Hsu2] S. B. Hsu (1980), "A competition model for a seasonally fluctuating nutrient," *Journal of Mathematical Biology* 9: 115-32.

[HCH] S. B. Hsu, K. S. Cheng, and S. P. Hubbell (1981), "Exploitative competition of micro-organisms for two complementary nutrients in continuous culture," *SIAM Journal on Applied Mathematics* 41: 422-44.

[HsH] S. B. Hsu and S. Hubbell (1979), "Two competitors competing for two prey species: An analysis of McArthur's model," *Mathematical Biosciences* 47: 143-72.

[HHW] S. B. Hsu, S. P. Hubbell, and P. Waltman (1977), "A mathematical theory for single nutrient competition in continuous cultures of micro-organisms," *SIAM Journal on Applied Mathematics* 32: 366-83.

[HSW] S. B. Hsu, H. L. Smith, and P. Waltman (1994), "Dynamics of competition in the unstirred chemostat," preprint, Department of Mathematics and Computer Science, Emory University, Atlanta.

[HW1] S. B. Hsu and P. Waltman (1992), "Analysis of a model of two competitors in a chemostat with an external inhibitor," *SIAM Journal on Applied Mathematics* 52: 528-40.

[HW2] S. B. Hsu and P. Waltman (1993), "On a system of reaction–diffusion equations arising from competition in an unstirred chemostat," *SIAM Journal on Applied Mathematics* 53: 1026-44.

[HWE] S. B. Hsu, P. Waltman, and S. F. Ellermeyer (1994), "A remark on the global asymptotic stability of a dynamical system modeling two species competition," *Hiroshima Mathematical Journal* 24: 435-46.

[HWW] S. B. Hsu, P. Waltman, and G. S. K. Wolkowicz (1994), "Global analysis of a model of plasmid-bearing, plasmid-free competition in the chemostat," *Journal of Mathematical Biology* (to appear).

[HS] V. Hutson and K. Schmitt (1992), "Permanence in dynamical systems," *Mathematical Biosciences* 111: 1-71.

[JST] W. Jäger, H. Smith, and B. Tang (1991), "Some aspects of competitive coexistence and persistence," in S. Busenberg and M. Martelli (eds.), *Delay Differential Equations and Dynamical Systems.* Berlin: Springer, pp. 200-9.

[JSTW] W. Jäger, J. W.-H. So, B. Tang, and P. Waltman (1987), "Competition in the gradostat," *Journal of Mathematical Biology* 25: 23-42.

[JW] W. Jäger and P. Waltman (1990), "Coexistent steady states for competitive populations with a spatial gradient and convection," preprint, Department of Mathematics and Computer Science, Emory University, Atlanta.

[JM] H. W. Jannash and R. T. Mateles (1974), "Experimental bacterial ecology studies in continuous culture," *Advances in Microbial Physiology* 11: 165-212.

[JDFT] J. L. Jost, J. F. Drake, A. G. Fredrickson, and H. M. Tsuchiya (1973), "Interaction of *Tetrahymena pyriformis, Escherichia coli, Azobacter vinelandii* and glucose in a minimal medium," *Journal of Bacteriology* 113: 834–40.

[Ka] E. Kamke (1932), "Zür Theorie der Systems Gewöhnlicher Differential-gleichungen II," *Acta Mathematica* 58: 57–85.

[Ke1] J. P. Keener (1983), "Oscillatory coexistence in the chemostat: A codimension-two unfolding," *SIAM Journal on Applied Mathematics* 43: 1005–9.

[Ke2] J. P. Keener (1985), "Oscillatory coexistence in a food chain model with competing predators," *Journal of Mathematical Biology* 22: 123–35.

[Ko] A. L. Koch (1985), "The macroeconomics of bacterial growth," in M. M. Fletcher and G. D. Floodgate (eds.), *Bacteria in Their Natural Environments*. London: Academic Press, pp. 1–41.

[K1] Y. Kuang (1989), "Limit cycles in a chemostat-related model," *SIAM Journal on Applied Mathematics* 49: 1759–67.

[K2] Y. Kuang (1993), "Delay differential equations with applications in population dynamics," in *Mathematics in Science and Engineering*, vol. 191. Boston: Academic Press.

[LT] P. Lancaster and M. Tismenetsky (1985), *The Theory of Matrices*. Orlando, FL: Academic Press.

[La] J. W. M. La Riviere (1977), "Microbial ecology of liquid waste," *Advances in Microbial Ecology* 1: 215–59.

[LC] D. Lauffenberger and B. Calcagno (1983), "Competition between two microbial populations in a non-mixed environment: Effect of cell random motility," *Biotechnology and Bioengineering* 25: 2103–5.

[LH] R. E. Lenski and S. Hattingh (1986), "Coexistence of two competitors on one resource and one inhibitor: A chemostat model based on bacteria and antibiotics," *Journal of Theoretical Biology* 122: 83–93.

[LeT] J. A. Leon and D. B. Tumpson (1975), "Competition between two species for two complementary or substitutable resources," *Journal of Theoretical Biology* 50: 185–201.

[LW1] R. W. Lovitt and J. W. T. Wimpenny (1979), "The gradostat: A tool for investigating microbial growth and interactions in solute gradients," *Society of General Microbiology Quarterly* 6: 80.

[LW2] R. W. Lovitt and J. W. T. Wimpenny (1981), "The gradostat: A bidirectional compound chemostat and its applications in micro-biological research," *Journal of General Microbiology* 127: 261–8.

[McD] N. MacDonald (1989), *Biological Delay Systems: Linear Stability Theory*. Cambridge: Cambridge University Press.

[MLW] C. A. Macken, S. A. Levin, and R. Waldstätter (1994), "The dynamics of bacteria-plasmid systems," *Journal of Mathematical Biology* 32: 123–45.

[McS] E. J. McShane (1934), "Extension of the range of functions," *Bulletin of the American Mathematical Society* 40: 837–42.

[M] L. Markus (1953), "Asymptotically autonomous differential systems," in *Contributions to the Theory of Nonlinear Oscillation*, vol. 3. Princeton, NJ: Princeton University Press, pp. 17–29.

[MM] J. E. Marsden and M. McCracken (1976), *The Hopf Bifurcation Theorem and Its Applications*. New York: Springer.

[MD] J. A. J. Metz and O. Diekmann (1986), "The dynamics of physiologically structured populations," in *Lecture Notes in Biomathematics,* vol. 68. Berlin: Springer.

[Mo1] J. Monod (1942), *Recherches sur la croissance des cultures bacteriennes.* Paris: Herman.

[Mo2] J. Monod (1950), "La technique de culture continue; theorie et applications," *Annales de l'Institute Pasteur* 79: 390–401.

[Mü] M. Müller (1926), "Über das Fundamentaltheorem in der Theorie der gewöhnlicher Differentialgleichungen," *Mathematische Zeitschrift* 26: 619–45.

[MH] A. Murray and T. Hunt (1993), *The Cell Cycle: An Introduction.* New York: Freeman.

[NG] R. M. Nisbet and W. S. C. Gurney (1982), *Modelling Fluctuating Populations.* New York: Wiley.

[NS] A. Novick and L. Szilard (1950), "Experiments with the chemostat on spontaneous mutations of bacteria," *Proceedings of the National Academy of Science* 36: 708–19.

[PK] S. Pavlou and J. G. Kevrekidis (1992), "Microbial predation in a periodically operated chemostat: A global study of the interactions between natural and externally imposed frequencies," *Mathematical Biosciences* 108: 1–55.

[Pi] S. Y. Pilyugin (1992), *Introduction to Structurally Stable Systems of Differential Equations.* Basel: Birkhäuser.

[Pos] J. Postgate (1992), *Microbes and Man.* Cambridge: Cambridge University Press.

[P] E. O. Powell (1958), "Criteria for the growth of contaminants and mutants in continuous culture," *Journal of General Microbiology* 18: 259–68.

[Po] G. E. Powell (1988), "Structural instability of the theory of simple competition," *Journal of Theoretical Biology* 132: 421–35.

[SSW] P. Schuster, K. Sigmund, and R. Wolf (1979), "Dynamical systems under constant organization, III: Cooperative and competitive behavior of hypercycles," *Journal of Differential Equations* 32: 357–86.

[S] J. Selgrade (1980), "Asymptotic behavior of solutions to single loop positive feedback systems," *Journal of Differential Equations* 38: 80–103.

[Se] G. Sell (1977), "What is a dynamical system?" in J. Hale (ed.), *Studies in Ordinary Differential Equations,* Studies in Mathematics, vol. 14. Washington, DC: Mathematical Association of America.

[Si] L. Simonsen (1991), "The existence conditions for bacterial plasmids: Theory and reality," *Microbial Ecology* 22: 187–205.

[Sm] S. Smale (1975), "On the differential equations of species in competition," *Journal of Mathematical Biology* 3: 5–7.

[S1] H. L. Smith (1981), "Competitive coexistence in an oscillating chemostat," *SIAM Journal on Applied Mathematics* 40: 498–522.

[S2] H. L. Smith (1986a), "Competing sub-communities of mutualists and a generalized Kamke theorem," *SIAM Journal on Mathematical Analysis* 4: 856–74.

[S3] H. L. Smith (1986b), "Periodic orbits of competitive and cooperative systems," *Journal of Differential Equations* 65: 361–73.

[S4] H. L. Smith (1986c), "Periodic solutions of periodic competitive and cooperative systems," *SIAM Journal on Mathematical Analysis* 17: 1289–1318.

[S5] H. L. Smith (1986d), "Periodic competitive differential equations and the discrete dynamics of competitive maps," *Journal of Differential Equations* 64: 165–94.

[S6] H. L. Smith (1988), "Systems of ordinary differential equations which generate an order preserving flow: A survey of results," *SIAM Review* 30: 87–113.

[S7] H. L. Smith (1990), "Microbial growth in periodic gradostats," *Rocky Mountain Journal of Mathematics* 20: 1173–94.

[S8] H. L. Smith (1991a), "Competition in a modified gradostat," in O. Arino, S. E. Axelrod, and M. Kimmel (eds.), *Mathematical Population Dynamics*. New York: Marcel Dekker, pp. 233–43.

[S9] H. L. Smith (1991b), "Equilibrium distribution of species among vessels of a gradostat," *Journal of Mathematical Biology* 30: 31–48.

[STa] H. L. Smith and B. Tang (1989), "Competition in the gradostat: The role of the communication rate," *Journal of Mathematical Biology* 27: 139–65.

[STW] H. L. Smith, B. Tang, and P. Waltman (1991), "Competition in an n-vessel gradostat," *SIAM Journal on Applied Mathematics* 5: 1451–71.

[ST1] H. L. Smith and H. Thieme (1990), "Quasi convergence for strongly order preserving semiflows," *SIAM Journal on Mathematical Analysis* 21: 673–92.

[ST2] H. L. Smith and H. Thieme (1991), "Convergence for strongly order preserving semiflows," *SIAM Journal on Mathematical Analysis* 22: 1081–1101.

[SW1] H. Smith and P. Waltman (1987), "A classification theorem for three dimensional competitive systems," *Journal of Differential Equations* 70: 325–32.

[SW2] H. Smith and P. Waltman (1991), "The gradostat: A model of competition along a nutrient gradient," *Microbial Ecology* 22: 207–26.

[SW3] H. Smith and P. Waltman (1994), "Competition for a single limiting resource in continuous culture: The variable-yield model," *SIAM Journal on Applied Mathematics* 54: 1113–31.

[Smo] J. Smoller (1983), *Shock Waves and Reaction Diffusion Equations*. New York: Springer.

[SoW] J. W.-H. So and P. Waltman (1989), "A nonlinear boundary value problem arising from competition in the chemostat," *Applied Mathematics and Computation* 32: 169–83.

[SFA] G. Stephanopoulos, A. G. Fredrickson, and R. Aris (1979), "The growth of competing microbial populations in a CSTR with periodically varying inputs," *American Institute of Chemical Engineering Journal* 25: 863–72.

[SLa] G. Stephanopoulos and G. Lapidus (1988), "Chemostat dynamics of plasmid-bearing plasmid-free mixed recombinant cultures," *Chemical Engineering Science* 43: 49–57.

[SL1] F. M. Stewart and B. R. Levin (1973), "Partitioning of resources and the outcome of interspecific competition: A model and some general considerations," *American Naturalist* 107: 171–98.

[SL2] F. M. Stewart and B. R. Levin (1977), "The population biology of bacterial plasmids: A priori conditions for the existence of conjugationally transmitted factors," *Genetics* 87: 209–28.

[Ta] B. Tang (1986), "Mathematical investigations of growth of micro-organisms in the gradostat," *Journal of Mathematical Biology* 23: 319–39.

[TaW] B. Tang and G. S. K. Wolkowicz (1992), "Mathematical models of microbial growth and competition in the chemostat regulated by cell-bound extra-cellular enzymes," *Journal of Mathematical Biology* 31: 1–23.

[TW] P. A. Taylor and J. L. Williams (1975), "Theoretical studies on the coexistence of competing species under continuous-flow conditions," *Canadian Journal of Microbiology* 21: 90–8.

[T1] H. R. Thieme (1992), "Convergence results and a Poincaré–Bendixson trichotomy for asymptotically autonomous differential equations," *Journal of Mathematical Biology* 30: 755–63.

[T2] H. R. Thieme (1993), "Persistence under relaxed point-dissipativity (with an application to an epidemic model)," *SIAM Journal on Mathematical Analysis* 24: 407–35.

[T3] H. R. Thieme (1994), "Asymptotically autonomous differential equations in the plane," *Rocky Mountain Journal of Mathematics* 24: 351–80.

[Ti] D. Tilman (1982), *Resource Competition and Community Structure.* Princeton, NJ: Princeton University Press.

[TDJF] H. M. Tsuchiya, J. F. Drake, J. L. Jost, and A. G. Fredrickson (1972), "Predator–prey interactions of *Dictyostelium discoideum* and *Escherichia coli* in continuous culture," *Journal of Bacteriology* 110: 1147–53.

[V] H. Veldcamp (1977), "Ecological studies with the chemostat," *Advances in Microbial Ecology* 1: 59–95.

[W1] P. Waltman (1983), *Competition Models in Population Biology.* Philadelphia: Society for Industrial and Applied Mathematics.

[W2] P. Waltman (1990), "Coexistence in chemostat-like models," *Rocky Mountain Journal of Mathematics* 20: 777–807.

[W3] P. Waltman (1991), "A brief survey of persistence," in S. Busenberg and M. Martelli (eds.), *Delay Differential Equations and Dynamical Systems.* Berlin: Springer, pp. 31–41.

[WHH] P. Waltman, S. P. Hubbel, and S. B. Hsu (1980), "Theoretical and experimental investigations of microbial competition in continuous culture," in T. Burton (ed.), *Modeling and Differential Equations in Biology.* New York: Marcel Dekker, pp. 107–52.

[Wi] F. M. Williams (1971), "Dynamics of microbial populations," in B. C. Patten (ed.), *Systems Analysis and Simulation in Ecology.* New York: Academic Press, pp. 197–265.

[WL] J. W. T. Wimpenny and R. W. Lovitt (1984), "The investigation and analysis of heterogeneous environments using the gradostat," in J. M. Grainger and J. M. Lynch (eds.), *Microbiological Methods for Environmental Biotechnology.* Orlando, FL: Academic Press, pp. 295–312.

[WLu] G. S. Wolkowicz and Z. Lu (1992), "Global dynamics of a mathematical model of competition in the chemostat: General response functions and differential death rates," *SIAM Journal on Applied Mathematics* 52: 222–33.

[ZS] H.-R. Zhu and H. Smith (1994), "Stable periodic orbits for a class of three dimensional competitive systems," *Journal of Differential Equations* 110: 143–56.

Author index

Andronov, A. A., 9
Aris, R., 16, 159, 161
Armstrong, R. A., 32, 250

Ballyk, M. M., 27
Baltzis, B. C., 25, 157, 250
Baxley, J. V., 232, 237
Berman, A., 133, 260
Bhatia, N. P., 280
Brooks, J. L., 229
Bungay, H. R., 2
Bungay, M.L., 2
Burton, L. P., 261, 265
Bush, A.W., 238
Butler, G. J., 12, 27, 37, 38, 40, 41, 42, 43, 159, 161, 181, 248, 279

Calcagno, B., 238
Canale, R. P., 48
Caperon, J., 238
Chang, S.-W., 157, 250
Cheng, K. S., 27
Chow, S. N., 283
Coddington, E. A., 7, 51
Conley, C., 274
Cook, A. E., 238
Coppel, W. A., 255, 261
Cunningham, A., 2, 48, 182, 185, 207
Cushing, J. M., 5, 181, 208, 214, 215, 219, 229, 230, 251

D'Ans, G., 2
Dawes, I. W., 5
Dean, A. M., 2
DeMottoni, P., 169, 180
Diekmann, O., 44, 208, 209, 213, 214, 229, 251
Dodson, S. I., 229
Drake, J. F., 48, 73, 76

Droop, M. R., 182
Dunbar, S. R., 12, 279

Ellermeyer, S. F., 240, 243
Elsworth, R., 4
Erle, D., 51
Ermentrout, B., 156

Fredrickson, A. G., 2, 25, 48, 73, 76, 159, 181
Freedman, H. I., 12, 238, 241, 242, 277, 279

Glas, D., 270
Gordon, I. I., 9
Gottlieb, D., 2
Grover, J. P., 181, 206
Gurney, W. S. C., 207, 238, 250

Hadeler, K. P., 270
Hale, J. K., 7, 12, 29, 159, 169, 181, 235, 239, 267, 278, 279, 280, 282
Hansen, S. R., 19, 20, 21, 22, 78, 99
Hartman, P., 62
Harvey, J. D., 213
Hastings, A., 71
Hattingh, S., 79, 80, 81, 97
Heijmans, J. J. A. M., 208
Herbert, D., 4, 269
Hirsch, M., 1, 95, 143, 169, 269, 270, 272, 273, 274, 276
Hofbauer, J., 140, 150, 157, 285
Holling, C. S., 44
Hsu, S. B., 2, 16, 27, 34, 43, 79, 99, 100, 159, 161, 181, 232, 237, 243, 244, 248, 251
Hubbell, S. P., 2, 16, 19, 20, 21, 22, 27, 37, 78, 99
Humphrey, A. E., 16

309

Hunt, T., 6
Hutson, V., 278

Jäger, W., 103, 126, 250, 252, 285
Jannash, H. W., 2, 73
Jost, J. L., 43, 73, 76

Kamke, E., 261
Keener, J. P., 77
Kevrekidis, J. G., 77
Koch, A. L., 5
Kokotović, P. V., 2
Kooijman, A. L. M., 208
Kuang, Y., 239, 248

Lancaster, P., 256, 257, 258
Lapidus, G., 2, 244
La Riviere, J. W. M., 2
Lauffenberger, D., 238
Lenski, R. E., 79, 80, 81, 97
Leon, J. A., 27
Leontovich, E. A., 9
Levin, B. R., 2, 16, 244
Levin, S. A., 244
Levinson, N., 7,51
Lovitt, R. W., 101, 102, 133
Lu, Z., 29, 37, 42, 248

McCracken, M., 60
MacDonald, N., 238
McGehee, R., 32, 250
Macken, C. A., 244
McShane, E. J., 274
Maier, A. G. ,9
Marsden, J. E., 60
Markus, L., 294
Mass, P., 207
Mateles, R. T., 2
Mayer, H., 51
Metz, J. A. J., 44, 208, 209, 213, 214, 229, 251
Monod, J., 4
Moson, P., 278
Müller, M., 261
Murray, A., 6

Nisbet, R. M., 2, 48, 182, 185, 207, 238, 250
Novick, A., 2

Pavlou, S., 77
Pilyugin, S. Y., 296
Plemmons, R. J., 133, 260
Plesser, T., 51
Postgate, J., 244

Powell, E. O., 16
Powell, G. E., 25, 27
Powell, T., 71

Rybakowski, K. P., 12, 279

Schiaffino, A., 169, 180
Schmitt, K., 12, 278, 279
Schuster, P., 278
Selgrade, J., 269
Sell, G., 48
Sigmund, K., 278
Simonsen, L., 244
Smale, S., 275
Smith, H. L., 1, 51, 120, 126, 127, 128, 129, 131, 133, 147, 151, 153, 156, 157, 159, 169, 181, 183, 232, 237, 250, 251, 259, 261, 265, 274, 275, 276, 285
Smoller, J., 60, 63, 282
So, J. W.-H., 103, 126, 140, 150, 157, 232, 237, 238, 241, 242, 282, 285
Somolinas, A. S., 159, 169, 181
Stephanopoulos, G., 2, 159, 181, 244
Steward, F. M., 2, 16, 244
Sutherland, I. W., 5
Szego, G. P., 280
Szilard, L., 2

Tang, B., 103, 126, 127, 128, 129, 138, 147, 157, 207, 250, 285
Taylor, P. A., 2
Telling, R. C., 4
Thieme, H. R., 12, 276, 278, 294
Thompson, H. B., 232, 237
Tilman, D., 27
Tismenetsky, M., 256, 257, 258
Tsuchiya, H. M., 48, 73, 76
Tumpson, D. B., 27

Veldcamp, H., 2

Waltman, P., 2, 3, 12, 16, 37, 43, 79, 99, 100, 103, 126, 127, 129, 147, 157, 161, 181, 183, 232, 237, 238, 241, 244, 251, 252, 275, 277, 279, 280
Waldstätter, R., 244
Whyburn, W. M., 261, 265
Williams, F. M., 2, 208, 227
Williams, J. L., 2
Wimpenny, J. W. T., 101, 102, 133
Wolf, R., 278
Wolkowicz, G. S. K., 27, 29, 37, 38, 40, 41, 42, 207, 244, 248
Wu, M., 57, 250

Zhu, H.-R., 51, 275

Subject index

acyclic flow, covering, 280
alpha limit point, 8
alpha limit set, 8
antibiotic, 78
asymptotically autonomous system, 294
asymptotically orbitally stable solution, 53, 55, 59, 240
asymptotically stable solution, 10
attractor, 9; chaotic, 71
Azotobacter vinelandii, 73

basin of attraction, 40, 270
batch culture, 4
bifurcation point, 60
bifurcation theory, 60, 147
body size, 208
break-even concentration, 14, 218
Butler–McGehee theorem, 12, 13, 49, 96, 97, 113, 295

Caperon–Droop model, 182
cell cycle, 4
cell quota, 182
chain: rest points, 11; invariant sets, 279
chain-recurrent set, 274
chaotic attractor, 71
characteristic polynomial, 63, 216, 255
chemotaxis, 238
Chlamydomonas reinhardii, 207
ciliates, 48
collection vessel, 3
communication rate, 127
comparison theorem, 120
competitive exclusion, 20,,40, 81, 159, 203, 221, 228
competitive system, 17, 93, 120, 169, 263
conservation principle, 25, 35, 81, 105, 135, 164, 183, 192, 225, 243
constant-yield model, 182, 206
consumption term, 4, 5

contaminant, 41
continuous culture, 2, 4
continuous dynamical system, 7
cooperative system, 95, 120, 169, 263
culture vessel, 3
cycle: of equilibria, 11; of invariant sets, 279

degree theory, 283
delay differential equation, 239
differential difference equation, 241
diffusion coefficient, 232
dilution rate, 4, 127, 130
discrete gradient, 102
dissipative system, 8, 105, 235, 278
Droop, Caperon–Droop model, 182
Dulac criterion, 9, 188, 223
dynamical system, 7; discrete, 172

E. coli, 20, 21, 78
enzyme kinetics, 4
equilibrium point, 8
exploitative competition, 1, 26
exponential kinetics, 31

falling interval, 272
feed bottle, 3
first-return-time map, 292
Floquet exponent, 52, 162, 166, 168
Floquet multiplier, 52, 67, 162, 292
Floquet theory, 53, 60, 162
flow rate, 3, 131
food chain, 47, 53, 73, 248; long, 69; three-level, 43
functional differential equations, 231, 252
functional response, 44
fundamental matrix, 52, 163, 263

generalized type-K function, 264
genetically altered organism, 2, 78, 243

Gerschgorin circle theorem, 256
Gerschgorin disk, 256
global attractor, 17, 38, 247
global bifurcation, 249
global stable manifold, 164
glucose, 73
gradostat, 101, 102
Green's theorem, 9, 55, 56
growth efficiency, 215

half-saturation constant, 4, 80, 104
handling time, 44
heteroclinic orbit, 200
Hopf bifurcation, 248
hyperbolic equilibrium, 11, 43, 85, 295;
    periodic solution, 53; fixed point, 163
hyperbolic kinetics, 31

implicit function theorem, 60, 61, 62, 217,
    282
inadequate competitor, 16, 32, 46
inadequate prey, 46
inhibitor, 20, 78
integro-differential equations, 214
internal-stores model, 182, 238
invariance principle, 28
invariant set, 8
irreducible dynamical system, 95; matrix,
    121, 132, 256
isolated covering, 280
isolated invariant set, 278
isolating neighborhood, 278

Jacobian matrix, 10
Jordan canonical form, 217

knife-edge effect, 18, 24

LaSalle corollary, 28, 29, 32, 33, 37
Leibniz rule, 211
Liapunov function, 29
limit cycles, 112, 249
limiting nutrient, 3
linearization, 10, 63

mean value, 161
Michaelis-Menten: dynamics, 5, 37;
    constant, 4, 80
moment functions, 214
Monod function, 185
monotone dynamical system, 94, 103, 106,
    121, 268
monotone dynamics, 140
monotone response function, 30

Nalidixic acid, 20, 21, 78
negative definite matrix, 258

negative orbit, 8
negative trajectory, 8
non-autonomous equations, 160
nondegenerate rest point, 283, 285
nutrient gradient, 101, 102, 129, 231

omega limit point, 8
omega limit set, 8
open system, 2
orbit, 8
oscillations, 85, 239, 243
oscillatory washout, 160
overflow vessel, 3, 130

p-convex set, 262
partial differential equation, 209, 214, 231,
    232
partial order relation, 120, 261
period-doubling bifurcation, 71
periodic differential equation, 162; planar,
    169
periodic orbit, 8
Perron-Frobenius theory, 95, 215, 257
persistent dynamical system, 96, 277, 279
physiological efficiency coefficient, 217,
    221, 222
piecewise linear kinetics, 31
plasmid, 243, 244
Poincaré map, 62, 63, 66, 67, 163, 169, 170,
    172, 250, 292
Poincaré-Bendixson theorem, 9, 17, 51, 59,
    95, 111, 112, 188, 223
Poincaré-Bendixson trichotomy, 9
pollutants, 78
positive definite matrix, 258
positive orbit, 7
positive trajectory, 7
positively invariant set, 8, 32
predator-prey equations, 43
principal minors, 117, 259
Pseudomonas aeruginosa, 21

quasipositive matrix, 134

repeller, 11
reproductive efficiency, 215
reservoir, 130, 131
resistance (to an antibiotic), 78
rest point, 8
rising interval, 272
Routh-Hurwitz criterion, 91, 97, 255

Sard's theorem, 283
semidynamical system, 7, 235, 239
simple eigenvalue, 63
singularly perturbed boundary value
    problem, 152

size distribution, 226
size-structured competitors, 228
size-efficiency hypothesis, 229
solution map, 62, 63
spectral radius, 257
spectrum, 257
stability modulus, 134, 257
stable limit cycle, 43
stable manifold, 11, 117, 163, 205
stable set, 278
stable solution, 10
standard n-vessel gradostat, 130, 131
steady state, 8
stored nutrient, 182, 241
strongly connected graph, 256
strongly monotone dynamical system, 137, 142, 197, 268
structure variables, 208
structured population model, 208
Sturm–Liouville problem, 236, 237
substrate, 4
symmetric matrix, 258

Tetrahymena pyriformis, 73, 76
trajectory, 8
tryptophan, 21
tubular reactor, 232
type-K function, 261

uniform persistence, 121, 143, 175, 237, 277, 280
uniformly asymptotically stable periodic solution, 162, 163
unstable manifold, 11, 117, 163, 164
unstable set, 278

variable-yield model, 182, 206
variational matrix, 11

washout rate, 4 , 159
wastewater treatment, 2, 159
weakly stable set, 279
weakly unstable set, 279
"well mixed" hypothesis, 101, 231

yield constant, 5, 182